DNA AND RNA: PROPERTIES AND MODIFICATIONS, FUNCTIONS AND INTERACTIONS, RECOMBINATION AND APPLICATIONS

ENCYCLOPEDIA OF DNA RESEARCH

VOLUME 3

DNA and RNA: Properties and Modifications, Functions and Interactions, Recombination and Applications

Additional books in this series can be found on Nova's website under the Series tab.

Additional E-books in this series can be found on Nova's website under the E-books tab.

DNA AND RNA: PROPERTIES AND MODIFICATIONS, FUNCTIONS AND INTERACTIONS, RECOMBINATION AND APPLICATIONS

ENCYCLOPEDIA OF DNA RESEARCH

VOLUME 3

SAMUEL J. DUNCAN
AND
PATRICIA H. WILEY
EDITORS

Nova Science Publishers, Inc.
New York

For permission to use material from this book please contact us:
Telephone 631-231-7269; Fax 631-231-8175
Web Site: http://www.novapublishers.com

Additional color graphics may be available in the e-book version of this book.

Library of Congress Cataloging-in-Publication Data

Encyclopedia of DNA research / editors, Samuel J. Duncan and Patricia H. Wiley.
p. cm.
ISBN 978-1-61324-305-3 (hardcover)
1. DNA. I. Duncan, Samuel J. II. Wiley, Patricia H.
QP624.E53 2011
572.8'603--dc22

2011010126

Published by Nova Science Publishers, Inc. † New York

CONTENTS

In: Encyclopedia of DNA Research (3 Volume Set)
Editors: S. J. Duncan and P. H. Wiley

ISBN 978-1-61324-305-3

Chapter 22

Nucleotide-Excision Repair Deficiency: A Causative Factor for Accelerated Aging Diseases in Humans?

Ming Yin and Qingyi Wei[1]

Departments of Epidemiology, The University of Texas, M. D. Anderson Cancer Center, 1515 Holcombe Blvd, Houston, TX77030, U. S.

Abstract

Nucleotide excision repair (NER) recognizes and removes bulky DNA damage that leads to DNA double-helix distortion. Inherited defects in the NER pathway have been associated with aging symptoms and premature death. In this review, we discussed the mechanisms of how NER maintains genomic integrity and affects clinical outcomes, if NER is compromised, and the role of NER as a potential causative factor for mammalian aging. We summarized current evidence of the association between NER and aging process through a discussion of most updated research findings from genetic models of mice and molecular epidemiological studies from humans. Several possible mechanisms by which NER influences aging are suggested, including the GH/IGF1 axis, p53-induced apoptosis/senescence in response to DNA damage, reactive oxygen species (ROS), and maintenance of telomerase activities. A comparison between normal aging and NER deficiency-related aging is also made to illustrate their similarities and discrepancies. We conclude with a number of issues needed to be resolved to further clarify the relationship between NER and aging process in humans.

Introduction

It has long been asked why we are aging? Aging is the time-dependent deterioration of physical conditions of an individual, involving both genetic and environmental factors, and

[1] Correspondence to: Department of Epidemiology, Box 189, The University of Texas M. D. Anderson Cancer Center, 1515 Holcombe Boulevard, Houston, TX 77030. Phone: 713-792-3020; Fax: 713-563-0999; E-mail: qwei@mdanderson.org.

thus aging decreases fitness and ultimately causes death. The aging process is complex and characteristic of a variety of different mechanisms, which are interactive rather than mutually exclusive. The DNA-damage theory of aging proposes that aging is a consequence of accumulation of unrepaired DNA damage as supported by the following observations: (1) In addition to the inherent instability, DNA is under constant attack by both exogenous and endogenous agents, such as ultraviolet light, chemicals and oxidative stress. It is estimated that approximately 800 DNA lesions occur per hour in each cell, or about 19,200 per cell per day [1]. Such a huge number of DNA lesions, if left unrepaired, will inevitably lead to biological and functional decline and finally disastrous consequence; (2) Both antioxidant defense mechanisms and DNA repair response decline with age, resulting in decreased genomic stability and reduced capacity of removing deleterious lesions. Robust evidence supports the importance of genomic integrity for proper functioning of vital biological activities in mammals. For example, accumulated DNA damage and subsequent mutations are often found in brain, muscle, liver, kidney, and other long-lived stem cells, acting as a major culprit in the causation of cancer and aging-related diseases [2-4]; (3) There is a direct evidence to associate increased DNA damage or decreased DNA repair capacity with reduced life expectancy. For example, it was found that 8-hydroxydeoxyguanosine (8OHdG), a biomarker of oxidative DNA damage, elevated with age, showing an inverse association with lifespan in the models of different species [5-7]; and (4) The most compelling evidence comes from diseases with inherited DNA repair defects that lead to phenotypes resembling the normal aging process [8].

To protect against harmful effects of DNA lesions, cells are equipped with a set of complementary repair systems, with both sufficient specificity and redundancy. There are many forms of DNA damage and DNA repair, but not all of them are necessarily significant in causing aging. Base-excision repair (BER) and nucleotide-excision repair (NER) are two major pathways responsible for the repair of DNA single strand lesions. Different from genes involved in BER, whose absence is either lethal or unharmful due to redundant genes that protect against obvious phenotypes [9], mutations in NER genes are associated with severe clinical consequences, including cancer predisposition and accelerated aging characterized by a subset of age-related phenotypes. In this chapter, we will discuss the potential role of NER as a causative factor for aging and possible underlying mechanisms.

1. Nucleotide Excision Repair: Functional and Clinical Association

The well-studied NER is an evolutionarily conserved pathway, consisting of at least 23 components/proteins that are responsible for removal of bulky monoadducts (such as those induced by chemical agents) and photoproducts (as caused by exposure to ultraviolet radiation) that distort DNA double helix [10, 11]. NER has two branches: global genomic repair and transcription-coupled repair, which differ only in the step of recognizing damaged nucleotides. Global genomic NER (GG-NER) employs several "damage sensing" proteins, such as the XPC-Rad23B complex, to scan for DNA helix distortion in the whole genome (both active and inactive genes) and repairs damage in both transcribed and untranscribed DNA strands, whereas transcription-coupled NER (TC-NER) directs the repair machinery preferentially to DNA lesions, which block transcription, that occur on the transcribed strand. In TC-NER, CSA, CSB and XAB2 proteins, instead of XPC-Rad23B, bind to the site of DNA damage and facilitate DNA repair by stabilizing RNA polymerase II at or near the site of damage [12]. Once a lesion is identified, both GG-NER and TC-NER utilize a common mechanism to repair it (Figure 1). Briefly, repair of the damaged DNA strand includes making an incision at the 5' and 3' of the lesion, removing a 30-nucleotide section containing the damage, and ligating the gap by pairing DNA synthesis [13]. The excision step involves at least 18 components, including proteins encoded by *XPA-G* genes (representing seven XP complementation groups), the ERCC1-hHR23B-RPA trimmers, and additional subunits of TFIIH [14]. XPC-hHR23B is an NER factor that detects a damaged site in DNA and then recruits TFIIH to the damaged site [15]. Two helicase subunits of TFIIH (i.e., XPB and XPD), together with XPA, RPA, and XPG, form a 30 base-pair bubble around the lesion [16]. XPA and RPA seem to play a role in verifying damage and are crucial for the correct positioning of two endonucleases (i.e., XPG and XPF-ERCC1) before incision [17]. The incision made at the 3' to the lesion by XPG precedes the 5' incision made by XPF-ERCC1; the latter requires the structural presence but not the catalytic activity of XPG [18]. The incisions made by XPG and XPF-ERCC1 are at the double- and single-stranded DNA border in the incision complex [19]. Observation of the mobility of various NER proteins in living cells suggests that NER proceeds by the sequential assembly of individual factors involved, rather than through the action of a preassembled repairosome [20]. Recently, there is growing evidence demonstrating the involvement of some NER components in double-strand break (DSB) repair [21], which still requires further investigation.

The importance of the NER pathway is evidenced by three rare, autosomal recessive human diseases that result from germline mutations of NER genes, including xeroderma pigmentosum (XP), Cockayne syndrome (CS) and trichothiodystrophy (TTD) syndrome. Patients with these syndromes have all shown reduced lifespan (Table 1). XP is a consequence of defects in the GG-NER, with or without deficiencies in transcription-coupled repair (TCR) of DNA-distorting lesions, characterized by increased photosensitivity, hyperpigmentation and ichthyosis in the sun-exposed skin and resulted in a 2000-time increased risk of skin cancer than the general population [8]. There are seven groups of XP, corresponding to mutations in the genes from XPA to XPG. Cockayne syndrome is caused by defects in TC-NER, resulting from mutations in CSA and CSB genes, which cause growth failure, impaired development of the nervous system, photosensitivity, and premature aging [22]. In contrast with XP, CS patients are not susceptible to cancer. The third disease trichothiodystrophy, is caused by mutations in XPB or XPD that encode subunits of TFIIH leading to destabilization or dysfunction of the transcription factor [23]. TTD shares overlapped symptoms with CS, such as progeroid features of cachectic dwarfism, sensorineural deafness, retinal degeneration, and CNS calcification, but displays additional cutaneous symptoms, such as brittle hair and ichthyotic skin [24]. It appears that NER defects lead to two distinct phenotypes: skin cancer-associated disorder (XP) and aging-associated disorder without cancer predisposition (CS and TTD). XP is considered a disease of skin photoaging, because nuclear changes in XP cells are similar to those found in normal or pathologic senescence [25].

Table 1. Comparisons of NER disorders: Xeroderma pigmensosum, Cockayne syndrome and Trichothiodystrophy

	Xeroderma pigmentosum	Cockayne syndrome	Trichothiodystrophy
Defective genes	XPA to XPG	CSA, CSB, XPD and XPG	XPB, XPD
Defective pathway	GG-NER	TC-NER	Transcription
Clinical phenotype			
Photosensitivity	Strong	Moderate	Weak
Cancer predisposition	Strong	Null	Null
Neurological degereration	A few	Common	Rare
Sensorineural deafness	A few	Common	A few
Mental retardation	A few	Common	Common
Ichthyosis	Rare	Null	A few
Brittle hair	Null	Null	Common
Retinopathy	Null	Common	Null
Life expectancy	Reduced by 30 years	< 20 year	< 10 year

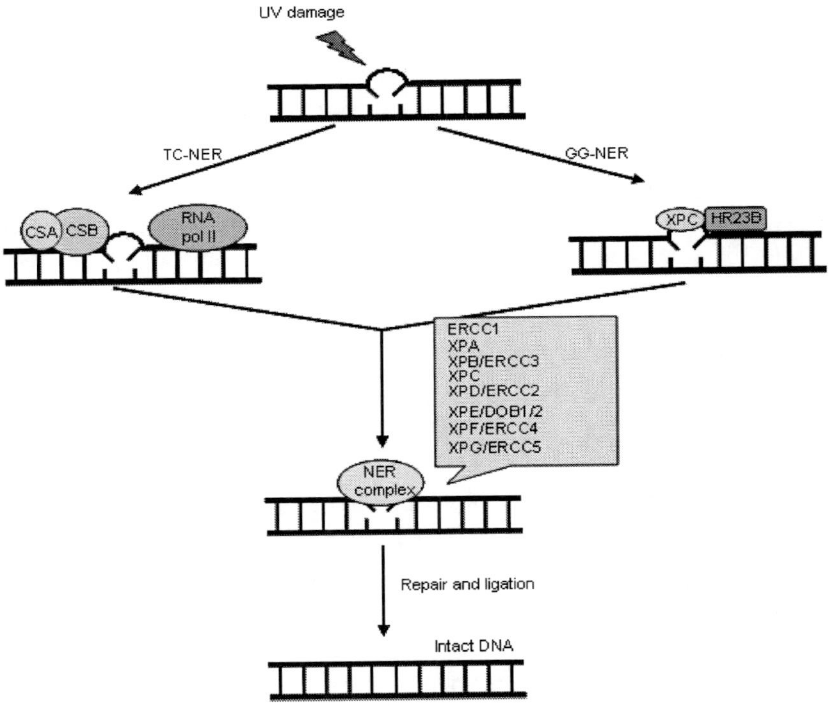

Figure 1. Repair of UV-induced DNA damage by transcription-coupled (TC)-NER and global genomic (GG)-NER.

2. NER and Premature Aging: Evidence from Mouse Models

At present, a variety of genetically engineered mouse mutants are available with defects of NER in an attempt to mimick either premature aging or cancer predisposition. In most cases, the mutation is recessive, indicating that only the homozygote mutants are affected. There are 8 so-called core NER factors (ERCC1, XPA, XPB/ERCC3, XPC, XPD/ERCC2, XPE/DDB1, XPF/ERCC4, and XPG/ERCC5) that actively participate in NER activities, each performing independent, yet integrated functions. We will discuss the biological significance of each of NER core components through respective mouse models.

2.1. ERCC1-Deficient Mice

The ERCC1 protein is involved in the 5' incision step of NER. Mutations in the *ERCC1* gene in humans have not been reported, possibly because it is not compatible with life. Two types of *ERCC1* mutant mice are available, including *ERCC1* knockout mice, in which ERCC1 is totally inactivated, and ERCC1 truncated mice, in which the protein is partly inactivated by deletion of seven amino acids within the carboxyl terminal [26]. Phenotypically, homozygous *ERCC1* knockout mice were runted at birth and died before weaning with liver failure. Organ biopsy revealed severe liver aneuploidy and elevated p53 levels in liver, brain and kidney, suggesting the presence of DNA damage [27]. Mice with truncated ERCC1 showed a similar phenotype but a longer lifespan, compared with *ERCC1* knockout mice (6 months vs. 4 weeks), reflecting a relatively milder defect in NER capacity. An unexpected finding in this mouse model was that there was no clear increase in cellular apoptosis, a common event seen in DNA-damaged cells. In contrast, the *ERCC1*-mutant mouse embryonic fibroblasts (MEFs) showed substantially increased doubling time and dividing failure at late passages (passages 6-7), suggesting an early onset of cellular replicative senescence of *ERCC1*-mutant primary cells.

2.2. XPF-Deficient Mice

XPF (ERCC4) protein is functionally associated with ERCC1 through constituting an endonuclease complex that makes an incision 5' to the DNA bulky damage. The XPF-deficient mice were generated by mutating G-445 in exon 8 of the mouse *XPF* gene into a stop codon, which mimicks the frameshift mutation in humans [28]. It is not surprising to see that XPF-deficient mice had a phenotype almost identical to that of *ERCC1* knockout mice, including a severe postnatal growth defect and premature death at approximately 3 weeks old. Histological examination revealed that the liver of mutant animals contained abnormal cells with enlarged nuclei. In addition, *XPF*-mutant MEFs were hypersensitive to UV irradiation and mitomycin C treatment. Although the investigators did not further explore the mechanisms underlying these phenotypes, it is reasonable to expect some similarities between XPF-deficient and ERCC1-deficient mice.

2.3. XPA-Deficient Mice

The XPA protein plays a crucial role within NER through multiple interactions with other NER factors, absence of which completely abolished the complex assembly [29]. XPA-deficient mice were generated by replacing exons 3 and 4 of the *XPA* gene with a neo marker cassette, thus abrograting its DNA-binding ability [30, 31]. The XPA-deficient mice developed normally, were healthy and fertile, and did not show histological abnormalities in various tissues, except for a low incidence of benign hepatocellular adenomas (± 15% of the mice, between the age of 15 and 20 months). This hepatocellular adenoma could be a consequence of oxidative DNA damage, because the liver is metabolically highly active and produces a substantial amount of free oxygen radicals. Compared with humans, the XPA-deficient mice displayed almost all the characteristics, such as increased skin sensitivity to UV and chemical carcinogens, but lacked neurological symptoms. In accordance, histopathological analysis of the brain tissue also revealed no signs of neurodegeneration. Overall, XPA is important for the normal function of NER, involving in the photoaging of skin tissue.

2.4. XPB-Deficient Mice

XPB and XPD are subunits of the protein complex TFIIH, which has a dual role in NER and basal transcription initiation. The XPB-deficient mice were generated by Andressoo, *et al.*, including a knock-in mouse model (Xpb^{fs}) and a *XPB* gene-frameshifted mouse model ($Xpb^{\Delta 43}$) that had 90% of the truncated XPB protein lacking the last 43 amino acids [32]. Homozygous Xpb^{fs} mutant mice died at the embryonic stage, whereas heterozygous mutant mice appeared to be phenotypically normal, consistent with a recessive mode of inheritance. The homozygous $Xpb^{\Delta 43}$ mice were viable and fertile with litter sizes comparable to those of wild-type littermates, probably because of the low expression of normal XPB protein. Surprisingly, investigation of $Xpb^{\Delta 43}$ mice did not reveal any detectable CS-like developmental or accelerated segmental aging features in both phenotypic and pathological examinations, except for an increased sensitivity to UV radiation. It may imply that small amount of normal XPB activity (10%) is sufficient to prevent the progeroid symptoms, or a longer time is required to translate the genetic defect into obvious phenotype, which may be beyond the mouse lifespan.

2.5. XPC-Deficient Mice

The XPC protein binds to HR23B and forms the XPC-HR23B complex, which is essential in the DNA damage recognition and DNA repair initiation in the NER pathway. XPC-deficient mice were generated by deletion of the coding portion corresponding to +59 to +546 of the *XPC* cDNA, resulting in a frameshift mutation and two stop codons 151 nucleotide downstream of the deletion [33]. XPC-deficient mice were viable and fertile and did not show an increased incidence of spontaneous tumors at one year of age. Compared with heterozygous mutants and wild-type controls, homozygous XPC-deficient mice demonstrated high tumor susceptibility to UV but not chemical carcinogens, such as 4-nitroquinoline 1-oxide. These results suggest that XPC is particularly involved in photoaging of skin tissue induced by UV light.

2.6. XPD-Deficient Mice

XPD is one of the two helicase subunits of transcription/repair complex TFIIH (the other is XPB). *XPD* mutations are associated with two types of different diseases, XP and TTD. Two models of XPD-deficient mice were generated, including *XPD* knockout mice and XPD^{R722W} (arginine 722 to tryptophan single amino acid substitution) mice, mimicking a point mutation found in TTD patients [34]. Similar to XPB-deficient mice, homozygous *XPD* knockout mice died early in embryonic two-cell stage, probably because of its essential function for basal transcription [35]. However, $XPD^{R722W/R722W}$ mice were totally viable, indicating that at least part of the XPD function still remains. Compared with the wild-type and heterozygous littermates, $XPD^{R722W/R722W}$ mice showed a slight but consistent growth retardation from birth, lower body weight, brittle hair and nails, a wizened and prematurely aged appearance, as well as significantly reduced lifespan (<1 year), but no spontaneous tumor formation. In addition, $XPD^{R722W/R722W}$ mice also exhibited other features indicative of premature aging, such as skeletal deformities and early hair graying. Regarding to tumor predisposition, $XPD^{R722W/R722W}$ mice appeared to be prone to UV- and DMBA-induced skin cancer with strikingly lower severity compared with lesions found in XPA- and CSB-deficient mice [34].

2.7. XPG-Deficient Mice

The XPG protein is reported to have a structure-specific DNA endonuclease activity and to function as a 3'-incision nuclease in NER [36, 37]. XPG-deficient mice were generated by an insertional mutation in exon 3 of the mouse *XPG* gene. This mutation interferes with the stability of *XPG* transcripts [38]. As a result, *XPG* transcripts were not detectable in the homozygous mutant mice that had the transcript level approximately half the amount detected in the heterozygous mutant mice, compared with the wild-type littermates. The *XPG* homozygous mutant mice were completely viable, indistinguishable from those of the wild-type or heterozygote pups, except for severely retarded growth and premature death before weaning (23 days postpartum). Fibroblasts of $XPG^{-/-}$ mice were sensitive to UV, but not X-rays or H_2O_2, indicating a defective NER activity. There is evidence supporting an association between loss of XPG function and premature replication senescence, because $XPG^{-/-}$ MEFs ceased growing by 4 to 5 weeks after the start of *in vitro* culture, while cells from the heterozygous or wild-type embryos kept their growth capability 3 to 4 weeks longer.

2.8. XPE-Deficient Mice

The *XPE* (*DDB1*) gene encodes a small subunit of a heterodimeric DNA binding protein with a high affinity to UV-damaged DNA, which appears to participate in the DNA repair pathway at the stage of binding and recognition [39]. In addition to its role in DNA repair, XPE has been identified as a transcriptional activator via an interaction with E2F1 and controls p53-dependent apoptosis in response to UV irradiation [40, 41]. Two types of XPE-deficient mice were available, generated by respective deletion of *XPE* exons 4-5 or 4-7, resulting in loss of XPE expression [41, 42]. $XPE^{-/-}$ mice were viable, fertile, but smaller than heterozygous and wild-type littermates. Like in human XPE disease, mice exhibited increased skin tumorigenesis in response to UV irradiation. In their late life stage (20-26 months old), *XPE*$^{-/-}$ mice showed high percentages of spontaneous tumors, suggesting a role of XPE in inhibiting age-dependent carcinogenesis.

To summarize, these genetically engineered mouse models provide valuable information to examine the impact of individual genes of the NER machinery on a physiological condition. NER factors are not equally important, as demonstrated by the fact that some mice with particular types of gene inactivation were not viable or died prematurely before weaning, whereas others showed a near normal phenotype and lifespan. When challenged by external mutagens, such as UV, X-ray radiation or chemical carcinogens, all mouse models showed increased sensitivity to at least one of the mutagens, reflecting a reduced DNA repair capacity. The NER progeroid mice appeared to recapitulate most of the aging-related symptoms seen in human NER deficient diseases. For example, mutations in *XPC*, *XPG*, and *XPE* genes lead to accelerated aging predominantly in the skin and eyes, driven by environmental factors, which is consistent with XP diseases. In contrast, mutations in *XPD* led to accelerated aging in multiple tissues without elevation of spontaneous tumor incidence, which is a good mimicking of CS and TTD. Given these results, NER progeroid mice have provided experimental evidence that mutations arising from NER defects could be one of the driving forces in the aging process.

3. NER and Longevity: From Animals to Humans

In humans, current investigations have primarily focused on NER-related aging syndromes (i.e. XP, CS and TTD) and the increased cancer predisposition as a result of reduced NER capacity. However, these are pathological conditions, which may not be representative of the natural aging process. Two questions remain to be answered: (1) Are similar changes intrinsic to natural aging as well as NER-related aging? and (2) Is individual longevity influenced by the NER capacity in the general population?

For the first question, it should be known that no progeroid syndrome exactly resembles "physiological" aging: (1) Compared with natural aging, NER-related progeria accelerates aging in several types, but not all, of tissues; (2) Although the NER progeria shares some common features with natural aging, the histopathologic changes are not the same. For example, the photoaged skin showed dramatic elastosis, increased glycosaminoglycans, disorganization and atypia of keratinocytes, inflammation of the dermis, hyperplasia of dermal fibroblasts and loss of vascularization with telangiectatic changes; whereas the skin in sun-unexposed areas showed loss and disorganization of extracellular elastin and collagen fiber, reduction in extracellular glycosaminoglycans, senescence of dermal fibroblasts and decreased vascularization [43, 44]; (3) In contrast to the NER-related aging, where only the NER pathway is compromised, pathophysiology of natural aging is more complex, involving genetic and environmental factors and affecting multiple physiological pathways. Despite these differences, comparative gene expression analysis revealed broad parallels between the transcriptomes of NER-deficient mice (*Csb*$^{m/m}$/*Xpa*$^{-/-}$ mice and *ERCC1*$^{-/-}$ mice) and naturally aged tissues on a genome-wide scale [45].

Regarding to the second question, it is a new field of research, which has been rarely touched. We will discuss it from an epidemiological perspective. In 1993, we first reported an age-related decline of NER capacity, amounting to around 0.61% per year in an apparently healthy population from 20 to 60 years of age [46]. A similar decline of NER capacity was found in other human studies [47, 48], as measured by the host-cell reactivation assay and western blotting analysis, as well as in different species, such as *C. elegans* [49]. In 2009, Kazimírová *et al.* reported micronuclei and chromosome aberrations, which were important biomarkers of aging, elevated in elderly people (65-70 years old) compared with the young group (20-25 years old), and was influenced by *XPC* and *XPD* polymorphisms [50]. In accordance, Polosak *et al.* showed that aging was associated with a progressive decrease of *XPD* mRNA expression and an increase of *XPD* promoter methylation [51]. Most importantly, they compared 149 centenarians (a mean age of 101.1 years) with 413 young subjects (a mean age of 27.1 years) and found that the *XPD* Lys751Gln polymorphism promoted extreme longevity, which provided a strong evidence that NER may influence longevity in the general population. Although an age-related decline in the general population was not supported by some studies [52], sufficient evidence has been accumulated to suggest that the functional decline of the NER capacity could pose a dramatic influence on natural aging process, whereas increased NER capacity counteracts the aging effect, thus contributing to longevity in the general population.

4. NER and Aging Modulation: The Underlying Mechanisms

In general, NER plays a critical role in the repair of bulky DNA damage that compromises genomic integrity and interferes with gene transcription. Recent investigations further have demonstrated that the influence of NER on aging is much more complicated than we originally thought. The complex cell-signaling network enables NER factors to impact a broad range of biological activities by interacting with multiple pathways. The mechanisms of how NER modulates the aging process are discussed as follows (Figure 2).

4.1. Maintenance of Genomic Integrity

The main source of genomic instability in mammalian cells is reactive oxygen species (ROS), which causes approximately 2×10^4 oxidative DNA lesions per human genome every day [53] and more than 100 different types of oxidative DNA lesions, ranging from base modifications to single- and double-strand DNA breaks [54]. ROS are produced as byproducts of mitochondrial electron transport, various oxygen-utilizing enzyme systems, peroxisomes, and other processes associated with normal aerobic metabolism. There is an impressive inverse correlation between metabolic rate and longevity, because a low metabolic rate leads to reduced ROS production. A series of experiments indicated that mice with a calories-restricted diet maintained longer youthful appearances and activity levels and showed delays in age-related diseases, compared with mice taking a normal diet [55, 56].

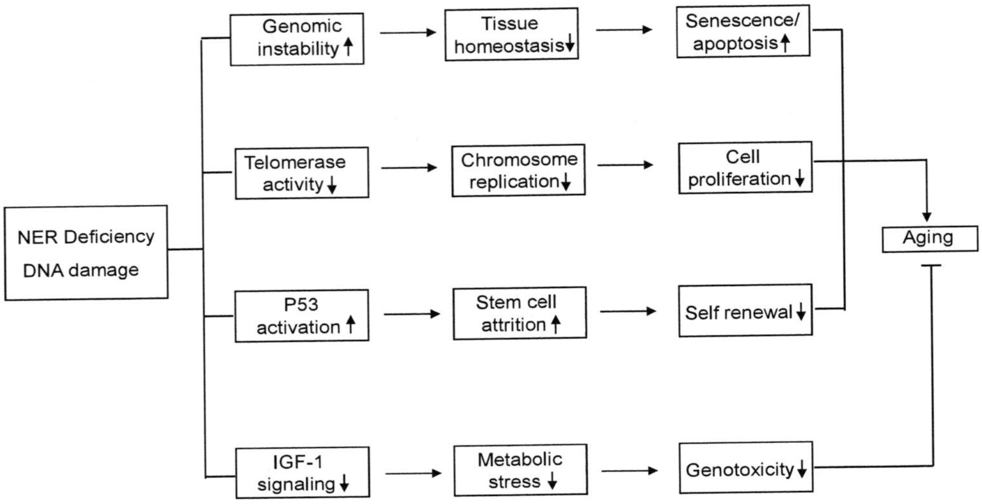

Figure 2. Mechanisms by which NER deficiency and DNA damage contribute to the aging process.

In humans, the harmful ROS effect is mainly counterbalanced by antioxidant defense systems (i.e. base-excision repair and ROS scavengers), while NER also seems to play a role. For example, Reardon *et al.* showed the removal of 8-oxodG, an oxidative lesion, by NER [57]. *XPA* was reported to be involved in the dual incision step of repairing ROS-induced 5′,8-purine cyclodeoxynucleosides [58]. More recently, *ERCC6* was suggested to participate in oxidative damage repair [59]. Taken together, efficient NER may help to reduce ROS-induced DNA damage and promote longevity.

4.2. Maintenance of Telomere

Telomeres are special functional complexes at the ends of eukaryotic chromosomes, which stabilize and protect the ends of chromosomes. It is maintained by the telomerase, an enzyme that helps to balance terminal DNA losses during replication by addition of tandemly repeated telomeric sequences, TTAGGG. In proliferating cells, the telomere progressively shortens because of the end replication problem [60], loss of telomere-binding proteins [61], or oxidative stress [62]. As a result, cells lose dividing ability and undergo senescence.

Current evidence suggests a dual role of NER in telomere maintenance. Patients with DNA repair deficiency are extremely vulnerable to genotoxic mutagens, which is associated with accelerated telomere shortening [63, 64]. Hence, intact function of NER may help maintain telomere length through repairing damaged telomeric DNA. Conversely, the NER factor, XPF-ERCC1, could also function as a negative mediator of telomere length maintenance. Telomere protection depends on TRF2, a small dimeric protein that binds along the duplex telomeric repeat tract, deletion of which leads to immediate degradation of the 3' telomeric overhangs, resulting in DNA damage and cell arrest [65]. Zhu *et al.* demonstrated that ERCC1/XPF was a component of the telomeric TRF2 complex and was responsible for the overhang removal, as a nucleotide excision repair endonuclease [66]. It was found that loss of XPF–ERCC1 in mouse cells led to an accumulation of telomeric DNA-containing doubleminute chromosomes, whereas ERCC1/XPF-deficient cells retained the telomeric overhang after TRF2 inhibition [66]. Although the exact role of NER in telomere maintenance has not been clarified, a phenotype of replicative senescence in ERCC1-deficient mice implies that accelerated telomere shortening may be dominant in situation of NER deficiency and may contribute to the aging process.

4.3. Influencing the GH/IGF-1 Signaling

The insulin-like growth factor-1 (IGF-1) signaling is one of the most important cell signaling pathways, which plays a critical role in the growth of essentially every normal organ both prenatally and postnatally [67]. Robust evidence indicated that reduced or defective IGF-1 signaling promoted longevity in different animal models, including nematodes, fruit flies and mice [68, 69]. In humans, the level of IGF-1 showed age-dependent changes, with an increasing level from birth to puberty and then declining afterwards [70]. A theory of metabolic stress is suggested to explain the aging-promoting effect of the GH/IGF-1 signaling. Briefly, the IGF-1 signaling has a powerful mitogenic potency that stimulates oxidative metabolisms that drive organismal growth [71, 72]. The intense metabolic activity inadvertently generates more free radicals, leading to increased genotoxic stress and accelerated cell senescence. If the IGF-1 signaling is reduced, energy use is shifted from growth and proliferation to protection and maintenance of tissues, thus decelerating aging process. A realistic example of this theory is the diseases of gigantism and acromegaly, which are caused by excessive secretion of GH from pituitary and are linked to a reduced life expectancy. However, a successful treatment towards normalization of IGF-1 levels may restore a normal life expectancy in these patients [73].

In NER-deficient mice, the down-regulation of IGF-1 signaling genes and the consequent decrease of serum IGF-1 are common findings, as observed in *ERCC1*$^{-/-}$ and *Csb*$^{m/m}$/*Xpa*$^{-/-}$ mouse models [45], and these seemed to be physiological responses to unrepaired DNA damage rather than a disorder of the GH/IGF-1 axis, because GH levels and pituitary function appeared to be normal. There are profound pathophysiological changes that could explain the IGF-1 signaling reduction. As mentioned previously, *ERCC1*$^{-/-}$ mice had severe aneuploidy and elevated p53 levels in the liver, the primary organ for the IGF-1 production. Hence, a function-compromised liver probably could only produce a lower amount of IGF-1 despite a normal GH level. Additionally, elevated stress responsive proteins, such as p53, p63, p73 and BRCA1, may also interact with Sp1, the major transcriptional factor of the *IGF-1R* gene and sequester it from binding to the promoter site, thus downregulating the *IGF-1R* gene expression [74-77]. It is believed that the reduced IGF-1 signaling plays a compensatory protective role against aging process, which can be overwhelmed by the massive unrepaired DNA damage.

4.4. Regulation of p53-Associated Senescence and Apoptosis

In cells with damaged DNA, p53 is frequently stabilized and activated to induce cell cycle arrest or apoptosis. If the damage is repaired, the cell cycle continues; if not, apoptosis is induced or cell cycle is permanently arrested, known as cellular senescence. A series of mouse models were found to have accelerated aging with an altered p53 expression or activity. For example, Tyner *et al.* and Maier *et al.* demonstrated that p53 hyperactivity resulted in cancer resistance and reduced lifespan [78, 79]. Nevertheless, p53 inactivation attenuated senescence phenotype in *Brca1*$^{-/-}$*p53*$^{+/-}$ and *Zmpste24*$^{-/-}$ mouse models [80, 81].

Taken together, p53 could be an important player in the aging process. Many of NER-deficient mouse models showed increased p53 levels as a result of accumulated DNA damage. The *Csb*$^{m/m}$/*Xpa*$^{-/-}$ and *ERCC1*$^{-/-}$ mice both had elevated p53, were growth-retarded, and had progeroid phenotypes, with a significantly shortened lifespan [27, 82]. It is hypothesized that p53-induced apoptosis and senescence in response to DNA damage produced deleterious effects to tissue homeostasis. Some experiments have associated DNA damage with tissue stem cell attrition [83, 84]. When stem cells are exhausted, tissues are no longer replenished, and aging phenotype emerges. Indeed, *ERCC1*$^{-/-}$ mice exhibited significantly reduced hematopoietic progenitor numbers and proliferative reserves, compared with wild-type littermates [85]. Hence, reduced NER capacity could contribute to accelerated aging mediated by the p53 pathway.

Conclusion

The aging process is a progressive, functional decline of a cell or an organism, which leads to systemic failure and culminates in death [86]. It is also a polygenic and multifactorial process, caused by accumulated defects resulting from both environmental and genetic factors. The DNA damage theory suggests that genomic instability is a major driving force of the aging process, which is supported by numerous clinical observations and experimental results. In this chapter, we summarized current research progress of the role of NER as a potential causative factor in a particular type of aging diseases associated with NER difficiency. Combining the evidence from animal models, human diseases and epidemiological studies, we propose that NER may heavily influence the aging process through not only its DNA repair capacity but also the interaction among multiple physiological pathways. In addition, it should be noted that the relationship between aging and DNA repair is an active field of research. With the rapidly expanded knowledge, we may be able to answer some fundamental questions: (1) To which extent does the NER deficiency-induced aging recapitulate the natural aging process? (2) Why does only part of the body show accelerated aging, considering the NER deficiency is a systematic disorder? and (3) How does NER modulate the aging process in the general population without apparent DNA repair defects?

References

[1] Vilenchik MM, Knudson AG, Jr. (2000). Inverse radiation dose-rate effects on somatic and germ-line mutations and DNA damage rates. Proceedings of the National Academy of Sciences of the United States of America 97, 5381-5386.

[2] Busuttil RA, Dolle M, Campisi J, Vijga J. (2004). Genomic instability, aging, and cellular senescence. *Ann. N Y Acad. Sci.* 1019, 245-255.

[3] Martin I, Grotewiel MS. (2006). Oxidative damage and age-related functional declines. *Mech. Ageing Dev.* 127, 411-423.

[4] Hosokawa M, Fujisawa H, Ax S, Zahn-Daimler G, Zahn RK. (2000). Age-associated DNA damage is accelerated in the senescence-accelerated mice. *Mech. Ageing Dev.* 118, 61-70.

[5] Svoboda P, Maekawa M, Kawai K, Tominaga T, Savela K, Kasai H. (2006). Urinary 8-hydroxyguanine may be a better marker of oxidative stress than 8-hydroxydeoxyguanosine in relation to the life spans of various species. *Antioxid Redox Signal* 8, 985-992.

[6] Agarwal S, Sohal RS. (1994). DNA oxidative damage and life expectancy in houseflies. Proceedings of the National Academy of Sciences of the United States of America 91, 12332-12335.

[7] Ksiazek K, Piatek K, Witowski J. (2008). Impaired response to oxidative stress in senescent cells may lead to accumulation of DNA damage in mesothelial cells from aged donors. Biochem Biophys Res Commun 373, 335-339.

[8] Hoeijmakers JH. (2009). DNA damage, aging, and cancer. *N. Engl. J. Med.* 361, 1475-1485.

[9] Schumacher B, Hoeijmakers JH, Garinis GA. (2009). Sealing the gap between nuclear DNA damage and longevity. *Mol. Cell Endocrinol* 299, 112-117.

[10] Friedberg EC. (2003). DNA damage and repair. Nature 421, 436-440.

[11] Reardon JT, Sancar A. (2005). Nucleotide excision repair. *Prog. Nucleic. Acid Res. Mol. Biol.* 79, 183-235.

[12] Laine JP, Egly JM. (2006). When transcription and repair meet: a complex system. Trends Genet 22, 430-436.

[13] Sancar A. (1995). DNA repair in humans. *Annu. Rev. Genet* 29, 69-105.

[14] de Boer J, Hoeijmakers JH. (1999). Cancer from the outside, aging from the inside: mouse models to study the consequences of defective nucleotide excision repair. *Biochimie* 81, 127-137.

[15] Yokoi M, Masutani C, Maekawa T, Sugasawa K, Ohkuma Y, Hanaoka F. (2000). The xeroderma pigmentosum group C protein complex XPC-HR23B plays an important role in the recruitment of transcription factor IIH to damaged DNA. *J. Biol. Chem.* 275, 9870-9875.

[16] Araujo SJ, Nigg EA, Wood RD. (2001). Strong functional interactions of TFIIH with XPC and XPG in human DNA nucleotide excision repair, without a preassembled repairosome. *Mol. Cell Biol.* 21, 2281-2291.

[17] Missura M, Buterin T, Hindges R, Hubscher U, Kasparkova J, Brabec V, Naegeli H. (2001). Double-check probing of DNA bending and unwinding by XPA-RPA: an architectural function in DNA repair. *Embo. J.* 20, 3554-3564.

[18] Wakasugi M, Shimizu M, Morioka H, Linn S, Nikaido O, Matsunaga T. (2001). Damaged DNA-binding protein DDB stimulates the excision of cyclobutane pyrimidine dimers in vitro in concert with XPA and replication protein A. *J. Biol. Chem.* 276, 15434-15440.

[19] Sijbers AM, de Laat WL, Ariza RR, Biggerstaff M, Wei YF, Moggs JG, Carter KC, Shell BK, Evans E, de Jong MC, Rademakers S, de Rooij J, Jaspers NG, Hoeijmakers JH, Wood RD. (1996). Xeroderma pigmentosum group F caused by a defect in a structure-specific DNA repair endonuclease. *Cell* 86, 811-822.

[20] Houtsmuller AB, Rademakers S, Nigg AL, Hoogstraten D, Hoeijmakers JH, Vermeulen W. (1999). Action of DNA repair endonuclease ERCC1/XPF in living cells. *Science* 284, 958-961.

[21] Ahmad A, Robinson AR, Duensing A, van Drunen E, Beverloo HB, Weisberg DB, Hasty P, Hoeijmakers JH, Niedernhofer LJ. (2008). ERCC1-XPF endonuclease facilitates DNA double-strand break repair. *Mol. Cell Biol.* 28, 5082-5092.

[22] Weidenheim KM, Dickson DW, Rapin I. (2009). Neuropathology of Cockayne syndrome: Evidence for impaired development, premature aging, and neurodegeneration. *Mech. Ageing Dev.* 130, 619-636.

[23] Stefanini M, Botta E, Lanzafame M, Orioli D. (2009). Trichothiodystrophy: From basic mechanisms to clinical implications. *DNA Repair* (Amst).

[24] Lambert WC, Gagna CE, Lambert MW. (2008). Xeroderma pigmentosum: its overlap with trichothiodystrophy, Cockayne syndrome and other progeroid syndromes. *Adv. Exp. Med. Biol.* 637, 128-137.

[25] Puvion-Dutilleul F, Sarasin A. (1989). Chromatin and nucleolar changes in Xeroderma pigmentosum cells resemble aging-related nuclear events. *Mutation research* 219, 57-70.

[26] Weeda G, Donker I, de Wit J, Morreau H, Janssens R, Vissers CJ, Nigg A, van Steeg H, Bootsma D, Hoeijmakers JH. (1997). Disruption of mouse ERCC1 results in a novel repair syndrome with growth failure, nuclear abnormalities and senescence. *Curr. Biol.* 7, 427-439.

[27] McWhir J, Selfridge J, Harrison DJ, Squires S, Melton DW. (1993). Mice with DNA repair gene (ERCC-1) deficiency have elevated levels of p53, liver nuclear abnormalities and die before weaning. *Nat. Genet* 5, 217-224.

[28] Tian M, Shinkura R, Shinkura N, Alt FW. (2004). Growth retardation, early death, and DNA repair defects in mice deficient for the nucleotide excision repair enzyme XPF. *Mol. Cell Biol.* 24, 1200-1205.

[29] Camenisch U, Nageli H. (2008). XPA gene, its product and biological roles. *Adv. Exp. Med. Biol.* 637, 28-38.

[30] Nakane H, Takeuchi S, Yuba S, Saijo M, Nakatsu Y, Murai H, Nakatsuru Y, Ishikawa T, Hirota S, Kitamura Y, et al. (1995). High incidence of ultraviolet-B-or chemical-carcinogen-induced skin tumours in mice lacking the xeroderma pigmentosum group A gene. *Nature* 377, 165-168.

[31] Miyamoto I, Miura N, Niwa H, Miyazaki J, Tanaka K. (1992). Mutational analysis of the structure and function of the xeroderma pigmentosum group A complementing protein. Identification of essential domains for nuclear localization and DNA excision repair. *J. Biol. Chem.* 267, 12182-12187.

[32] Andressoo JO, Weeda G, de Wit J, Mitchell JR, Beems RB, van Steeg H, van der Horst GT, Hoeijmakers JH. (2009). An Xpb mouse model for combined xeroderma pigmentosum and cockayne syndrome reveals progeroid features upon further attenuation of DNA repair. *Mol. Cell Biol.* 29, 1276-1290.

[33] Sands AT, Abuin A, Sanchez A, Conti CJ, Bradley A. (1995). High susceptibility to ultraviolet-induced carcinogenesis in mice lacking XPC. *Nature* 377, 162-165.

[34] Broughton BC, Thompson AF, Harcourt SA, Vermeulen W, Hoeijmakers JH, Botta E, Stefanini M, King MD, Weber CA, Cole J, et al. (1995). Molecular and cellular analysis of the DNA repair defect in a patient in xeroderma pigmentosum

complementation group D who has the clinical features of xeroderma pigmentosum and Cockayne syndrome. *Am. J. Hum. Genet.* 56, 167-174.

[35] de Boer J, Donker I, de Wit J, Hoeijmakers JH, Weeda G. (1998). Disruption of the mouse xeroderma pigmentosum group D DNA repair/basal transcription gene results in preimplantation lethality. *Cancer Res.* 58, 89-94.

[36] Cloud KG, Shen B, Strniste GF, Park MS. (1995). XPG protein has a structure-specific endonuclease activity. Mutation research 347, 55-60.

[37] Matsunaga T, Mu D, Park CH, Reardon JT, Sancar A. (1995). Human DNA repair excision nuclease. Analysis of the roles of the subunits involved in dual incisions by using anti-XPG and anti-ERCC1 antibodies. *J. Biol. Chem.* 270, 20862-20869.

[38] Harada YN, Shiomi N, Koike M, Ikawa M, Okabe M, Hirota S, Kitamura Y, Kitagawa M, Matsunaga T, Nikaido O, Shiomi T. (1999). Postnatal growth failure, short life span, and early onset of cellular senescence and subsequent immortalization in mice lacking the xeroderma pigmentosum group G gene. *Mol. Cell Biol.* 19, 2366-2372.

[39] Chu G, Chang E. (1988). Xeroderma pigmentosum group E cells lack a nuclear factor that binds to damaged DNA. *Science* 242, 564-567.

[40] Kulaksiz G, Reardon JT, Sancar A. (2005). Xeroderma pigmentosum complementation group E protein (XPE/DDB2): purification of various complexes of XPE and analyses of their damaged DNA binding and putative DNA repair properties. *Mol. Cell Biol.* 25, 9784-9792.

[41] Itoh T, Cado D, Kamide R, Linn S. (2004). DDB2 gene disruption leads to skin tumors and resistance to apoptosis after exposure to ultraviolet light but not a chemical carcinogen. *Proc. Natl. Acad. Sci. U S A* 101, 2052-2057.

[42] Yoon T, Chakrabortty A, Franks R, Valli T, Kiyokawa H, Raychaudhuri P. (2005). Tumor-prone phenotype of the DDB2-deficient mice. Oncogene 24, 469-478.

[43] Rabe JH, Mamelak AJ, McElgunn PJ, Morison WL, Sauder DN. (2006). Photoaging: mechanisms and repair. *J. Am. Acad. Dermatol.* 55, 1-19.

[44] Kligman LH, Kligman AM. (1986). The nature of photoaging: its prevention and repair. *Photodermatol.* 3, 215-227.

[45] Schumacher B, van der Pluijm I, Moorhouse MJ, Kosteas T, Robinson AR, Suh Y, Breit TM, van Steeg H, Niedernhofer LJ, van Ijcken W, Bartke A, Spindler SR, Hoeijmakers JH, van der Horst GT, Garinis GA. (2008). Delayed and accelerated aging share common longevity assurance mechanisms. *PLoS Genet* 4, e1000161.

[46] Wei Q, Matanoski GM, Farmer ER, Hedayati MA, Grossman L. (1993). DNA repair and aging in basal cell carcinoma: a molecular epidemiology study. Proceedings of the National Academy of Sciences of the United States of America 90, 1614-1618.

[47] Tyson J, Caple F, Spiers A, Burtle B, Daly AK, Williams EA, Hesketh JE, Mathers JC. (2009). Inter-individual variation in nucleotide excision repair in young adults: effects of age, adiposity, micronutrient supplementation and genotype. *Br. J. Nutr.* 101, 1316-1323.

[48] Goukassian DA, Bagheri S, el-Keeb L, Eller MS, Gilchrest BA. (2002). DNA oligonucleotide treatment corrects the age-associated decline in DNA repair capacity. *Faseb J.* 16, 754-756.

[49] Meyer JN, Boyd WA, Azzam GA, Haugen AC, Freedman JH, Van Houten B. (2007). Decline of nucleotide excision repair capacity in aging Caenorhabditis elegans. *Genome Biol.* 8, R70.

[50] Kazimirova A, Barancokova M, Dzupinkova Z, Wsolova L, Dusinska M. (2009). Micronuclei and chromosomal aberrations, important markers of ageing: possible association with XPC and XPD polymorphisms. *Mutation research* 661, 35-40.

[51] Polosak J, Roszkowska-Gancarz M, Kurylowicz A, Owczarz M, Dobosz P, Mossakowska M, Szybinska A, Puzianowska-Kuznicka M. (2009). Decreased expression and the Lys751Gln polymorphism of the XPD gene are associated with extreme longevity. *Biogerontology.*

[52] Shaddock JG, Feuers RJ, Chou MW, Casciano DA. (1993). Evidence that DNA repair may not be modified by age or chronic caloric restriction. *Mutation research* 301, 261-266.

[53] Ames BN, Shigenaga MK. (1992). Oxidants are a major contributor to aging. Ann N Y Acad Sci 663, 85-96.

[54] Hoeijmakers JH. (2007). Genome maintenance mechanisms are critical for preventing cancer as well as other aging-associated diseases. *Mech. Ageing Dev.* 128, 460-462.

[55] Weindruch R, Walford RL, Fligiel S, Guthrie D. (1986). The retardation of aging in mice by dietary restriction: longevity, cancer, immunity and lifetime energy intake. *J. Nutr.* 116, 641-654.

[56] Pugh TD, Oberley TD, Weindruch R. (1999). Dietary intervention at middle age: caloric restriction but not dehydroepiandrosterone sulfate increases lifespan and lifetime cancer incidence in mice. *Cancer Res*. 59, 1642-1648.

[57] Reardon JT, Bessho T, Kung HC, Bolton PH, Sancar A. (1997). In vitro repair of oxidative DNA damage by human nucleotide excision repair system: possible explanation for neurodegeneration in xeroderma pigmentosum patients. Proceedings of the National Academy of Sciences of the United States of America 94, 9463-9468.

[58] Kuraoka I, Bender C, Romieu A, Cadet J, Wood RD, Lindahl T. (2000). Removal of oxygen free-radical-induced 5',8-purine cyclodeoxynucleosides from DNA by the nucleotide excision-repair pathway in human cells. Proceedings of the National Academy of Sciences of the United States of America 97, 3832-3837.

[59] Gu J, Zhao H, Dinney CP, Zhu Y, Leibovici D, Bermejo CE, Grossman HB, Wu X. (2005). Nucleotide excision repair gene polymorphisms and recurrence after treatment for superficial bladder cancer. *Clin. Cancer Res.* 11, 1408-1415.

[60] Olovnikov AM. (1971). [Principle of marginotomy in template synthesis of polynucleotides]. Dokl Akad Nauk SSSR 201, 1496-1499.

[61] Smogorzewska A, van Steensel B, Bianchi A, Oelmann S, Schaefer MR, Schnapp G, de Lange T. (2000). Control of human telomere length by TRF1 and TRF2. *Mol. Cell Biol.* 20, 1659-1668.

[62] von Zglinicki T, Pilger R, Sitte N. (2000). Accumulation of single-strand breaks is the major cause of telomere shortening in human fibroblasts. *Free Radic. Biol. Med.* 28, 64-74.

[63] Ishibashi T, Lippard SJ. (1998). Telomere loss in cells treated with cisplatin. Proceedings of the National Academy of Sciences of the United States of America 95, 4219-4223.

[64] Tchirkov A, Lansdorp PM. (2003). Role of oxidative stress in telomere shortening in cultured fibroblasts from normal individuals and patients with ataxia-telangiectasia. *Hum. Mol. Genet* 12, 227-232.

[65] Wu Y, Mitchell TR, Zhu XD. (2008). Human XPF controls TRF2 and telomere length maintenance through distinctive mechanisms. *Mech. Ageing Dev.* 129, 602-610.

[66] Zhu XD, Niedernhofer L, Kuster B, Mann M, Hoeijmakers JH, de Lange T. (2003). ERCC1/XPF removes the 3' overhang from uncapped telomeres and represses formation of telomeric DNA-containing double minute chromosomes. *Mol. Cell* 12, 1489-1498.

[67] Jones JI, Clemmons DR. (1995). Insulin-like growth factors and their binding proteins: biological actions. *Endocr .Rev.* 16, 3-34.

[68] Giannakou ME, Partridge L. (2007). Role of insulin-like signalling in Drosophila lifespan. *Trends Biochem. Sci.* 32, 180-188.

[69] Kenyon C. (2001). A conserved regulatory system for aging. *Cell* 105, 165-168.

[70] Collett-Solberg PF, Cohen P. (2000). Genetics, chemistry, and function of the IGF/IGFBP system. *Endocrine* 12, 121-136.

[71] Carter CS, Ramsey MM, Sonntag WE. (2002). A critical analysis of the role of growth hormone and IGF-1 in aging and lifespan. *Trends Genet* 18, 295-301.

[72] Bartke A. (2003). Is growth hormone deficiency a beneficial adaptation to aging? Evidence from experimental animals. *Trends Endocrinol. Metab.* 14, 340-344.

[73] Melmed S. (2009). Acromegaly pathogenesis and treatment. *The Journal of clinical investigation* 119, 3189-3202.

[74] Ohlsson C, Kley N, Werner H, LeRoith D. (1998). p53 regulates insulin-like growth factor-I (IGF-I) receptor expression and IGF-I-induced tyrosine phosphorylation in an osteosarcoma cell line: interaction between p53 and Sp1. *Endocrinology* 139, 110[illegible]-1107.

[75] Idelman G, Glaser T, Roberts CT, Jr., Werner H. (2003). WT1-p53 interactions in insulin-like growth factor-I receptor gene regulation. *J. Biol. Chem.* 278, 3474-3482.

[76] Nahor I, Abramovitch S, Engeland K, Werner H. (2005). The p53-family members p63 and p73 inhibit insulin-like growth factor-I receptor gene expression in colon cancer cells. *Growth Horm IGF Res* 15, 388-396.

[77] Sarfstein R, Maor S, Reizner N, Abramovitch S, Werner H. (2006). Transcriptional regulation of the insulin-like growth factor-I receptor gene in breast cancer. *Mol. Cell Endocrinol.* 252, 241-246.

[78] Maier B, Gluba W, Bernier B, Turner T, Mohammad K, Guise T, Sutherland A, Thorner M, Scrable H. (2004). Modulation of mammalian life span by the short isoform of p53. *Genes Dev.* 18, 306-319.

[79] Tyner SD, Venkatachalam S, Choi J, Jones S, Ghebranious N, Igelmann H, Lu X, Soron G, Cooper B, Brayton C, Hee Park S, Thompson T, Karsenty G, Bradley A, Donehower LA. (2002). p53 mutant mice that display early ageing-associated phenotypes. *Nature* 415, 45-53.

[80] Cao L, Li W, Kim S, Brodie SG, Deng CX. (2003). Senescence, aging, and malignant transformation mediated by p53 in mice lacking the Brca1 full-length isoform. *Genes Dev.* 17, 201-213.

[81] Varela I, Cadinanos J, Pendas AM, Gutierrez-Fernandez A, Folgueras AR, Sanchez LM, Zhou Z, Rodriguez FJ, Stewart CL, Vega JA, Tryggvason K, Freije JM, Lopez-Otin C. (2005). Accelerated ageing in mice deficient in Zmpste24 protease is linked to p53 signalling activation. *Nature* 437, 564-568.

[82] van der Pluijm I, Garinis GA, Brandt RM, Gorgels TG, Wijnhoven SW, Diderich KE, de Wit J, Mitchell JR, van Oostrom C, Beems R, Niedernhofer LJ, Velasco S, Friedberg EC, Tanaka K, van Steeg H, Hoeijmakers JH, van der Horst GT. (2007). Impaired genome maintenance suppresses the growth hormone--insulin-like growth factor 1 axis in mice with Cockayne syndrome. *PLoS Biol.* 5, e2.

[83] Chambers SM, Goodell MA. (2007). Hematopoietic stem cell aging: wrinkles in stem cell potential. *Stem. Cell Rev.* 3, 201-211.

[84] Rando TA. (2006). Stem cells, ageing and the quest for immortality. *Nature* 441, 1080-1086.

[85] Prasher JM, Lalai AS, Heijmans-Antonissen C, Ploemacher RE, Hoeijmakers JH, Touw IP, Niedernhofer LJ. (2005). Reduced hematopoietic reserves in DNA interstrand crosslink repair-deficient Ercc1-/- mice. *Embo. J.* 24, 861-871.

[86] Kirkwood TB. (2005). Understanding the odd science of aging. *Cell* 120, 437-447.

In: Encyclopedia of DNA Research (3 Volume Set)
Editors: S. J. Duncan and P. H. Wiley
ISBN 978-1-61324-305-3

Chapter 23

Historical Development of Uncaria Preparations and Their Related Bioactive Components

Ronald W. Pero[1]
Institute of Clinical Medical Sciences, Section of Immunology
Lund University, SE-221 84 Lund, Sweden

1. Overview of Uncaria Sp. Bark and Its Commercial Preparations as a Medicinal Herb. The Oxindole Alkaloid /Water Extraction Reference Standard Story

Uncaria sp. is a well known herbal medicine used for generations by the Ashinka Indians native to the Amazon basin. There have been two set of bioactive ingredients for which Uncaria extracts have been developed and standardized. The first are oxindole alkaloids, initially studied and described in 1967; the second are a set of molecules know as Carboxy Alkyl Esters (CAEs ™) first identified and described in 1997 as the bioactive ingredients in AC-11® (formerly C-MED-100®). More recently (2005) it was shown that one of the acid moieties of CAEs is quinic acid. Chlorogenic acid also is present in Uncaria sp. water extracts, where it occurs as a natural ester of quinic acid. Oxindole alkaloids are much less soluble in water than quinic acid or chlorogenic acid; hence, they are absent (<0.05%) from AC-11®, which instead contains specifically CAEs. This overview concentrates on characterization of the AC-11® water extract because it is the only Uncaria extract standardized to a bioactive chemotype that enhances DNA repair, this chemotype, is not known to be present in Uncaria sp. extracts that are latent with oxindole alkaloids.

The overview herein includes mode of action studies, as well as clinical effects, some of which are common both to water insoluble products such as oxindole alkaloids and to water soluble products such as CAEs. Examples include: anti-inflammation, nuclear fraction kappa beta (NF-kB) inhibition, and immune enhancement. The salient characteristic that

[1] Tel: 802-375-8322, Fax: 802-375-6690.

differentiates between insoluble alkaloid based Uncaria products and water soluble CAE's, specific to AC-11 and DNA Repair enhancement, is in the nature of the bio-efficacy of each. Principally AC-11® , Uncaria tomentosa water extract, has been documented to: (i) increase serum protein thiols (antioxidant protection), (ii) increase urinary nicotinamide and tryptophan levels, (iii) increase the number of white blood cells(WBC), (iv) increase pneumococcal titer, (v) enhance DNA repair, (vi) increase lymphocyte growth response, (vii) reduce level of 8-OH quanine DNA adducts, (viii) associated to weight loss, (ix) reduce sunburn, (x) increase removal of UV-induced thymine T-T dimer DNA damage, (xi) improve life style-induced clinical responses, (xii) inhibit tumor cell growth and (xiii) inhibit inflammation; additionally other Uncaria water extracts exhibited the ability to (xiv) inhibit amyloid body formation related to Alzheimers, (xv) protect against ozone injury. (Castillo 2005,2006, Cisneros et al. 2005, Lamm et al. 2001, Mammone et al. 2006, Miller et al. 2005, Piscoya et al. 2001, Pero et al. 2002, 2005, 2009, Pero and Lund 2009, Sheng et al. 1998, 2000A,200B, 2001, Akesson et al. 2003A, 2003B, and also as cited elsewhere in this review).

2. Historical Background

Uncaria tomentosa species has had a long history of use as a folk medicine among the indigenous peoples of Central and South America. Traditional practice was to steep the bark and stems or leaves (1-3 gm) with boiling water for several hours, and then decant the resulting "tea" and drink 1-3 cups per day. According to anecdotal evidence, preparations of this type were believed to be especially useful from a historical perspective to treat a multitude of health problems associated with rheumatism, arthritis, gastric ulcers, gastrointestinal/inflammatory/autoimmune disorders, tumors, child birth trauma, psychotropic responses (mental state) and infections. (Blumenthal, M. 2003, Capasso, F. et al. 2003). Historically, most commercially produced water extracts were roughly equivalent to what is produced by the original "folk medicine" methods described above; such preparations were not further chemically altered except by what might be caused from spray drying , freeze drying or molecular sieving.

The literature documents how commercially available forms of alkaloid-based Uncaria sp. extracts are prepared. Over 50 bioactive ingredients have been identified and isolated from Uncaria sp., more than any other South American plant, since the early 1960's when Klaus Keplinger, Wagner and colleagues identified the oxindole alkaloids present in this plant (Keplinger 1982, Keplinger et al. 1999, Wagner et al. 1985, Stuppner et al. 1992). For example, quinovic acid glycosides (Aquino et al. 1989), polyhydroxylated triterpines (Aquino et al. 1991), steroids (Senatore et al. 1989), procyanidins (Wirth and Wagner 1992) and tannins (Jones 1995) have all been identified over the years since Keplinger, as bioactive components of Uncaria sp. extracts. Although biological properties associated with Uncaria sp. were found to exist, no one prior to 1998 was able to explain completely the apparent efficacy of these alkaloid based extracts when examined against the anecdotal record of the water extracts used by indigenous peoples in South America. Instead there was a general acceptance, with limited data, that oxindole alkaloids were responsible for the efficacious properties of Uncaria sp.

The folk medicine practice uses hot water extraction, because it was the only available method of preparation known to native Indians. This fact was ignored by Keplinger and others. Instead they used alcohol based extracts which became more common in recent decades. Therefore alcohol based solvents were used to extract the oxindole alkaloids, which were the first class of bioactives isolated from Uncaria. There are at least 17 indole alkaloids identified. Their content in plant extracts is greatly enhanced by organic solvent extraction such as with alcohol, ethyl acetate or other organic solvents, or by acidic extraction, because alkaloids are basic and they can easily be converted to salts. Nevertheless, these types of Uncaria sp. ingredients are difficult to reconcile as explanations for the historic medicinal use of Uncaria sp. based simply on their poor water solubility, thus severely limiting their presence in water extracts.

Although the resultant doses of crude Uncaria sp. extracts based on plant part concentration are more or less consistent with those produced by "folk medicine" practices, the commercially made forms vary considerable in their preparation processes. (Blumenthal 2003, Laus 2004, Montoro et al. 2004, Piscoya et al. 2001, Sheng et al. 2000A). Doses and applications described include: (i) Teas: One to 25 grams of root bark added to 250 milliliters of pH adjusted water, boiled for five to 10 minutes, then cooled and strained. One cup taken three times daily. (ii) Water Extracts: 100-700 mg/day; such as AC-11®. (iii) Alkaloid Extracts: An extract containing only pentacyclic oxindole alkaloids taken by mouth in a dose of 20 milligrams two or three times daily for the first 10 days and then 20 milligrams thereafter. Also, 1% to 7% oxindole alkaloids extracts have been used between 100 mgs and 1000 mgs daily. (iv) Tinctures: One to two milliliters taken orally two to three times daily, or 20 to 40 drops five times daily. (v) Decoctions: One tablespoon of pulverized root in one quart of hot water taken orally before breakfast.

Early research posited a theory that 2 different types of Uncaria sp. alkaloids (i.e.chemotypes) are contained in this plant: the "bad" ones, or *tetracyclic* oxindole alkaloids (TOA's) and the 'good' ones, or *pentacyclic oxindole* alkaloids (POA's) (Montoro et al. 2004; Laus et al. 1997). An inconclusive investigation was made into whether the "bad" alkaloids counteracted the immune benefits of the "good" alkaloids; the presence of TOA's together with POA's appear to be immune-suppressive. Either way, there was little doubt that POA's and TOA's differ considerably in their water solubility. This was thought to be a factor of significance of how the alkaloids interact with the body. POA's moreover were shown to be concentrated more in the roots and bark of the plant, whereas, TOAs were found to be located primarily in leaves and bark. As a result of the early research, it was theorized that either organic solvent extracts of Uncaria sp. should be abandoned as chemotypes for Uncaria sp. alkaloids, or else the TOA's should be separated from the POA's by water solubility criteria, or by plant part location or both.

3. The Water Extract Method

The oxindole alkaloid research was a pioneer scientific effort because it attempted to standardize the numerous bioactive indole alkaloids in Uncaria. However, in general, this class of alkaloids lacks water solubility. In addition, it is difficult to correlate the notion that these water insoluble components constitute the natural occurring active ingredients of

Uncaria sp. with field observations of how the aqueous form of the Uncaria extract was traditionally prepared in SA in use as a "folk medicine".

Accordingly, Pero and colleagues choose to focus their studies on the efficacy of water extracts or Uncaria prepared in a manner similar to the method used by the indigenous population of the Amazon Rainforest. Specifically, they did not rely on alcohol- based (i.e. organic solvent) extraction or concentration methods but sought in essence to duplicate, in a standardized setting, the "Amazon Recipe". The resulting aqueous extract first introduced by Pero was called C-MED-100®. The trade name was later changed to AC-11®. When C-Med-100®/AC-11® extract and other water-based extracts of Uncaria sp. were analyzed for oxindole alkaloids only trace amounts were found (Kuras et al. 2009) Sheng et al. 2000B, Sandoval et al. 2002). Furthermore, it was shown through a series of scientific studies from 1998 through 2009 that CAE's, the active ingredient in C-MED-100/AC-11, were responsible for the efficacy related to enhanced DNA Repair.

Currently, only two classes of active ingredients serve as the basis for standardizing the biological activity of Uncaria sp. extracts. The first ingredients, beginning in the 1960's, were the oxindole alkaloids already cited. These ingredients were standardized as tetracyclic oxindole alkaloids (TOAs), pentacyclic oxindole alkaloids (POAs) or combinations thereof at about 1.3% POAs - 2.96 % (i.e. for all oxindole alkaloids) of the acidic or organic solvent Uncaria extracts (Blumenthal 2003, Laus et al. 1999). The second class of ingredients, beginning in 1998 (Sheng et al. 1998), consists of Carboxy alkyl Esters (CAEs), which are found in abundance up to 8-10 % in the water soluble AC-11® extract. AC-11® has only traces of alkaloids present (< 0.05% POAs + TOAs) (Sheng et al. 2000B, 2005).

The importance of using water extracts of Uncaria products lies in the fact that the water soluble efficacious components of Uncaria are more readily available for uptake, distribution and cellular modulation. This view is confirmed by several laboratories that have used water soluble extracts of either Unacria tomentosa or Uncaria quianensis, a sub-species of Uncaria (Sandoval et at 2000, Kuras et al. 2009). Another commercially available water extract of Uncaria quianensis, called Vincaria has been used as an anti-inflammatory treatment, and also reported to contain reduced oxindole alkaloids (Miller et al. 2005, Piscoya et al. 2001).

Detailed chemical analyses of organic solvent extractions of Uncaria sp. using sophisticated HPLC-ES/MS analyses have determined POA's to be 26.895 mg/gm and TOAs to be 2.753 mg/gm. Added together the total alkaloid content on average would be 29.649 mg/gm (Montoro et al. 2004). POA's and POA content has normally been determined and identified by HPLC, whereas CAE content has been validated by a colorimetric procedure involving conversion of CAE to hydroxamic acids and reaction with ferric chloride (Bartos 1980, Sheng et al. 2005). When these alkaloid concentrations are compared with the alkaloid levels shown in AC-11® or Vincaria water extracts which were < 0.05 mg/gm, or about 593 times less alkaloid content than in solvent based extracts (i.e. 29.649/ 0.05 = 593). Therefore it follows the potent anti-inflammatory properties of Uncaria sp. water extracts cannot be reasonably explained by their oxindole alkaloid content, but rather must be due to other components perhaps not found in organic solvent extracted Uncaria products. It was determined by Pero and colleagues (Sheng et al. 2005) after identification and isolation of CAE's in 2000 that the CAE's were responsible for approximately 90 percent of the biological activity found in C-MED-100/AC-11. This conclusion was validated by the charcoal filtration (removal) of CAE's at 190-200nm and the subsequent decrease in biological activity.

The alkyl group and the carboxy group in ester linkage in CAEs together have now been identified as the active ingredients in water soluble Uncaria products. Further, they have been isolated as esters of quinic acid (i.e QAEs, Sheng et al. 2005). Due to the strong acidity of the stomach ($P^H = 1$), and combined with the fact that gastrointestinal microflora are efficient at metabolizing QAEs such as chlorogenic acid (Olthof et al. 2003; Gonthier et al. 2003), it is believed that these ingredients would be hydrolyzed to free quinic acid and the alcohol moieties for systemic uptake and absorption. It is clear from the data presented by Sheng and colleagues (Sheng et al. 2005) that quinic acid is one of the bioactive components of Uncaria tomentosa and can account for about 40% of the DNA repair enhancement observed from AC-11® in vivo (Table 4 in Sheng et al. 2005). This study focused on Doxorubicin-induced leucopenia in the rat model treated with 80 mg/kg of AC-11®, and measured by direct comparison to quinic acid or Quinmax™ treated with 200 mg/kg, found similar results in relationship to enhanced DNA repair capacity. Since the alcohol moiety of the QAEs in AC-11® also contributes to the biological activity of AC-11®, the alcohol moiety would yield additional biologic activity to that shown for quinic acid alone. Evidence supporting this interpretation is that when AC-11 or quinic acid analogs are base hydrolyzed with 2M NaOH, about 50% of growth inhibiton of cancer cells is lost, while base hydrolysis treatment of quinic acid or quinic acid lactone had no impact on in vitro cancer cell toxicity (Table 3 in Sheng et al. 2005). These recorded data (Sheng et al. 2005) are consistent with the fact that the QAE's in AC-11 are related to chlorogenic acid. Clearly both quinic acid and/or its alcohol moiety esterified to it; i.e. such as the phenolic, caffeic acid, are both naturally occurring in water extracts of Uncaria (i.e. 5-caffeoylquinic acid or chlorogenic acid), and are both hypothesized to be additional actives as DNA repair enhancers. AC-11® is a neutraceutical DNA repair enhancer because it contains at least two quite different sub classes under the CAE standardization : i.e. one having QAEs (e.g. such as chlorogenic acid) and the other a still yet unidentified DNA repair enhancer relating to the alcohol moiety of QAEs.

4. Mode of Action of Water Extracts of Uncaria Tomentosa Such as AC-11®

There are two products currently being sold in the marketplace based on the reference standard of quinic acid esters or quinic acid itself. They are AC-11®/C-Med-100® (CAEs and QAEs), and QuinMax™, the ammonium chelate of quinic acid (US patents 6,039,949; 6,238,675B1; 6,361,805 B2, 7,579,023; 7,595064; PCT/US2006/009394; PCT/US2007 /073261). AC-11® is commonly dosed at 350 mg per day (range = 250-700) (Lamm et al. 2001, Mammone et al. 2006, Pero et al. 2002, 2005, Sheng et al. 2000A, 2001), and quinic acid ammonium chelate (Quinmax™) at 1000 mg per day (range = 500-3000) (Pero et al. 2009, Pero and Lund 2009). Given that AC-11® has been quantified to contain 8-10% CAEs (Sheng et al. 2005), and assuming half the molecular weight of the QAEs is quinic acid (e.g. chorogenic acid), then the daily dose of quinic acid in AC-11® would be about 25-70 mg/day (i.e. 10% of AC-11® of daily doses of 250-700 mg/day divided by 2 = 12.5 to 35 mg /day), or considerably lower than 1000 mg/day recommended for Aqua Bimini (100% quinic acid ammonium chelate). Yet both products are well known to enhance DNA repair, anti-inflammation, and immunity as documented in his report. The logical explanation is that there are apparently other water soluble components found in AC-11® water extracts, that can either synergize the effects of quinic acid; e.g., the alcohol moiety portion of the QAEs such as the phenolic acid, caffeic acid, or yet an unidentified new bioactive that is additive or synergistic to the effects of quinic acid (postulated in this study). In other words, there are two different products. One, AC-11® that acts as a neutraceutical (chemically defined in part), and two, QuinMax™ that acts as a pharmaceutical (chemically defined), and yet both are based in large part on the efficacy of quinic acid.

a) DNA Repair Enhancement and Aging

AC-11® is essentially free of oxindole alkaloids have been shown to possess a broad spectrum of biological activity, both orally and topically, including DNA repair enhancement and anti-inflammatory benefits. These two biological mechanisms are key molecular targets to develop treatments that protect skin cells exposed to ultraviolet light from the sun (non-oxidative stress). For the reason that AC-11® is the only documented natural source of components which simultaneously up regulate DNA repair and stimulate a positive response to inflammation, studies were undertaken to show additional benefit to the skin from AC-11® treatment. The data clearly demonstrated that skin cultures co-incubation with AC-11® reduced skin cell death from UV exposure, and this protection was accounted for by a concomitant increase in DNA repair. (Mammone et al. 2006, Sheng et al. 2000A, Confidential Report #3). AC-11® has also demonstrated enhanced DNA repair in vivo, in vitro and in human studies after exposure to DNA- damaging chemicals (e.g. doxorubicin, hydrogen peroxide, ozone, environmental exposures). Presumably this phenomenon occurs by preventing or removing DNA damage that in turn blocks damaged cell replication (Cisneros et al. 2005, Pero et al. 2002, 2005, 2009A, 2009B Sheng et al. 2000B, 2001, 2005). Moreover, rat white blood cells (WBC) or mouse spleen WBC were increased in vivo after AC-11® and quinic acid chelate (QuinMax) oral administration either by gavage or in drinking water from 7.1 to > 8.1 x 10^6 cells/ml, or a 13-25% increase over untreated controls in the protection of DNA from becoming damaged and killing the cells (Sheng et al. 2005, Akesson et al. 2005). It has been scientifically established that there is a causal effect between DNA damage and accelerated aging. If the above data were extrapolated into a human lifespan, then living to an average age of 90 years would be increased to approximately 112 years by ingesting optimal amounts of AC-11® or quinic acid equivalents during one's lifetime.

The DNA repair studies cited above are based on comprehensive analysis of the type of DNA damage that is environmentally induced and the measurable benefits. It is not just a general enhancement for the excision DNA repair enzymatic process. Rather, the enzymes being used to repair DNA depend on the type of DNA lesions being inflicted, so quantifying DNA repair relevant to one type of DNA lesion does not necessarily indicate a general DNA repair enhancement of importance to a broader anti-aging effect. We reported in a number of studies a broad spectrum of DNA repair enhancement endpoints such as: (i) single strand breaks measured by alkaline elution, (ii) thymine T-T dimers, (iii) 8-OH guanine adducts, (iv) DNA replication synthesis/cell survival measured by thymidine incorporation into DNA or cell counting or number of UV induced sun burn cells, (v) recovery from ozone injury and

(vi) serum protein thiols a surrogate estimate of the DNA repair enzyme poly ADP ribose polymerase (PARP) (Pero et al. 1995) and lifespan (Pero et al. 2000).

(B) Immune Function Enhancement

It has been observed that both AC-11® and a Quinic Acid Chelate (QuinMax™) enhance immune cell function by increasing the number of fully functional lymphocytes without suppressing antigenic responses to growth stimuli or inducing cell lymphocyte death by apoptosis (Lamm et al. 2001, Sheng et al. 2000A, Åkesson et al. 2003A, 2003B, 2005). In addition, the immune modulating transcription factor, NF-kB, was inhibited by both AC-11® and a quinic acid chelate (QuinMax™). (Sandoval et al. 2002, Akesson et al. 2003B, 2005).

(C) Inhibition of Inflammation

As already pointed out earlier Uncaria extracts rich in alkaloids have consistently been reported for their effectiveness as anti-inflammatory treatments (Jones 1995, Reinhard 1999). There are now more than 50 dietary supplements standardized for the alkaloid content available commercially (Keplinger et al. 1999) as anti-inflammatory agents. However, none of these products were prepared as hot water extracts such as AC-11®, which follows the historical folk medicine practices. Whereas we have reported that alkaloid-containing Uncaria sp. products have greater toxicity than AC-11® (Sheng et al. 2000A), we have also determined that both AC-11® and quinic acid chelates are potent inhibitors of NF-kB and thus potent anti- inflammatory agents, along with reduced toxicity (Akesson et al. 2003B, 2005). This work has been confirmed by in vivo reports (Sandoval-Chacon et al. 1998, Sandoval et al. 2000) conducted on water extracts of Uncaria sp. not standardized to CAEs (Miller et al. 2005, Piscoya et al. 2001). These non-standardized water extracts also had greatly reduced alkaloid content providing further evidence that compounds other than alkaloids are primarily responsible for the anti-inflammatory activity observed.

(D) Anti-Oxidation

Endogenous generated oxidative stress in the form of hypochlorous acid strongly inhibits DNA repair when estimated by a variety of procedures for quantifying DNA repair; e.g unscheduled DNA synthesis (UDS), inhibition of poly ADP ribose polymerase (PARP, repair of strand breaks assayed by nucleoid sedimentation or alkaline elution (Pero et al. 1996). Moreover, oxidative stress evaluated by the level of thiols in serum correlates strongly to most human diseases including cancer, cardiovascular disease, diabetes, and inflammation (Banne et al. 2003) and prognosis of HIV infection (Marmor et al. 1997). Both AC-11® (Pero et al. 2002, 2005) and Quinmax™ (Pero et al. 2009) clinical treatments have demonstrated the ability to prevent oxidation via an increase in serum protein thiol analyses that estimates levels of in vivo oxidative stress. Ozone, an oxidative stress environmental toxin, causes cellular damage which in turn is reduced in mice by an aqueous extract of Uncaria tomentosa (Cisneros et al. 2005).

(E) Neurogenic Effects

Alzheimers, mood, and psychological stress (lifestyle factors) have been shown to be mediated by water extracts of Uncaria sp. Examples include amyloid body inhibition (Castello 1998, 2005, 2006) and increased serum tryptophan levels, the serotonin precursor (Pero et al. 2009). These psychotropic observed benefits were also independent of their indole alkaloid content.

4. Available Published Data in the Scientific Literature for Uncaria Sp

Water soluble extracts and/or water soluble components of Uncaria sp. have been shown to be responsible for medicinal benefits. There are 3 main sources of data available for evaluation of this efficacy. They are in vitro, rodent and human studies. Taken together the evidence is strong that aqueous components in Uncaria tomentosa, specifically AC-11®, provide effective clinical treatments for many ailments and disorders in humans.

First, there are peer reviewed in vitro data that HL-60 leukemic, K-563 leukemic, Jurkat, and Raji cells, MCF 7 beast cancer cells, murine macrophage RAW 264.7 cells, allium cells, Chinese hamster ovarian cells, and rat ascites hepatoma cells were all shown to be positively affected by Uncaria sp. water extracts (Kuras et al. 2009, Riva et al. 2001, Sandoval et al. 2002, Santa et al. 1997, Sheng et al. 1998, Yagasaki et al. 2000, Akesson et al. 200B).

Second, DNA repair, immune enhancement enhancement, anit-inflammation, recovery from doxorubicin chemotherapy, and improvement in overcoming Listeria monocytrogenes infection were all successfully treated in rodents with water extracts of Uncaria sp. (Aguilar et al. 2002, Cisneros et al. 2005, Eberlin et al. 2005, Sheng et al. 2000A, 2000B, 2005, Akesson et al. 2003A, 2003B, 2005)

Third, over 200 human volunteers in total were administered a daily dose of AC-11® or another water extract of Uncaria species between 250 to 7000 mg/day that stimulated immune response, DNA repair and/or anti-inflammation without presenting with any side effects (Lamm et al. 2001, Miller et al. 2005, Piscoya et al. 2001, Pero et al. 2002, 2005, 2009, Sheng et al. 2000A, 2001). It appears obvious that Uncaria sp. water extracts are safe and efficacious for further clinical evaluation.

Rodent total dose treatment ranges that have been tested for AC-11® and other water extracts of Uncaria ranged from 400 to12,000 mg/kg (Aguilar et al. 2002, Cisneros et al. 2005, Eberlin et al. 2005, Sheng et al. 2000A, 2000B, 2005, Akesson et al. 2003A, 2003B, 2005). Although weight loss has occurred at the total dose of AC-11® of 5,600 mg/kg in humans (Pero et al. 2005), rodent studies did not confirm weight loss by any mechanism (Pero et al. 2005), thus indicating a positive benefit from AC-11® treatment. In fact no significant toxicity was observed in rodents even when evaluated by a complete histopathological examinaton (Sheng et al. 2000A).

Furthermore, there were efficacious indications at all doses tested of Uncaria water extracts in rodents which were total doses ranging from 720-12,000 mg/kg (Aguilar et al. 2002, Cisneros et al. 2005, Eberlin et al. 2005, Sheng et al. 2000A, 2000B, 2005, Akesson et al. 2003A) . Compared to the dose range for human efficacy studies which was total doses ranging from 80-601 mg/kg (Lamm et al. 2001, Pero et al. 2002, 2005, Sheng et al. 2000A, 2001), the rodent model values were generally higher than was needed in humans presumably because of body surface area differences. It is well known that the greater the surface area of a mammal available for adsorption, the greater will be the efficiency of adsorption. The body volume in humans is much higher than in rodents, thus providing a higher absorptive potential being roughly 70 kg versus 0.250 kg, respectively. Therefore, it was concluded that the rodent studies generally supported the data obtained in humans for both toxicity and efficacy.

The human clinical studies performed so far after supplementation with either AC-11®, Quinmax™ or other water extracts of Uncaria sp., have demonstrated a broad range of efficacious responses such as: (i) increased serum protein thiols (antioxidant protection), (ii) increased urinary nicotinamide and tryptophane levels, (iii) increased number of WBC, (iv) increased pneumococal titer, (v) DNA repair enhancement, (vi) increased lymphocyte growth response, (vii) reduced level of 8-OH quanine DNA adducts, (viii) weight loss, (ix) reduced sunburn, (x) increased removal of UV-induced thymine T-T dimer DNA damage, (xi) improvement in life style-induced clinical responses and (x) osteoarthritis relief in the knee (Lamm et al. 2001, Mammone et al. 2006, Miller et al. 2005, Piscoya et al. 2001, Pero et al. 2002, 2005, 2009, Pero and Lund 2009, Sheng et al. 2000A, 2001). The data presented supports at least 4 different modes of action; namely DNA repair, immune enhancement, inhibition of inflammation and anti-oxidation after treatment with either AC-11® or QuinMax®. The broad spectrum of clinical responses documented by these peer-reviewed articles supports the health benefits associated with thee commercially available preparations.

5. Contraindications

There are over 30 reports dealing with the efficacy and toxicity of water soluble Uncaria sp. products which includes cells, tissues, rodents and humans. In all these studies there were no toxic side effects observed from the hundreds of observations in animals and humans. Apart from occasional weight loss documentation there have been no other symptoms recorded. On the other hand, indole alkaloid extracts often have been associated with toxic side effects (Aquilar et al. 2002; Pilarski et al. 2005).

6. Discussion

The scientific evidence behind water soluble Uncaria sp. extracts has now been presented independent from their oxindole alkaloid content. This was accomplished by standardizing water soluble AC-11® to the bioactive ingredients identified as CAEs (8-10 %) that was in turn depleted of containing any significant amounts of the bioactive oxindole alkaloids. (< 0.05%) As the water soluble Uncaria extract AC-11® became more clearly delineated chemical, first identified as CAEs (8-10%), then as QAEs (4-5 %) and then to QA (2-2.5 %) by assuming ½ the MW of QAEs was QA (Sheng et all 2005). Because "folk" medicine practice used water extracts comparable to AC-11®, and they have insignificant alkaloid content, there seems little doubt that QA and QAE-based Uncaria sp. extracts (e.g. AC-11®) are predominantly responsible for the observed Uncaria sp. efficacy over the years.

Although widely distributed in nature and plant foods, quinic acid has not previously been recognized to have any nutritional benefits until it was discovered to be a bioactive ingredient of C-MED-100®/AC-11®. This is an important point when considering quinic acid safety and efficacy. Although not previously recognized as a significant component to our food chain, Quinic Acid has nonetheless been present as a major health component in many foods affecting the basic nutritional health status and metabolism in man. (Pero et al. 2009).

QA is a well-known key metabolic intermediate in the synthesis of plant aromatic compounds including aromatic amino acids that is essential to animal life. It is called the shikimate pathway (Hermann et al), and relates directly to human metabolism because the GI tract microflora produces QA everyday or utilizes the QA content from food. Interestingly quinic acid equivalents are mainly found in brightly colored foods (i.e. reds, oranges, greens, yellows) vegetables and fruits which are well known to be exceptionally healthy food sources; for example, such as prune, kiwi, sea buckthorn, coffee, cranberry, lingonberry, blueberry, wortleberry, red/yellow tamarillo, sultana, quince, sunflower, nectarine, peach, pear, plum, honey, black currant, medlar, apricot, asparagus, mushroom and green olive (Van Gorsel et al. 1992, Beveridge et al. 1999, Englehardt et al. 1985, Graham et al. 1992, Jensen et al. 2002, Romero Rodrigues et al. 1992, Lewis et al. 1995, Silva et al. 2002, Mourgue et al. 1975).

The recent discovery that QAE's™/Quinmax™ can increase the synthesis of tryptophan and nicotinamide is a major development to the area of clinical nutrition. In fact, certain foods or AC-11® and QA chelates (QuinMax™) when administered to humans as a dietary supplement may be an effective way effective to enhance essential aromatic amino acid biosynthesis, via endogenous biosynthesis by gastrointestinal (GI) microflora (Hermann and Weaver 1993, Hermann 1995, Pero et al. 2009). Such a novel approach to clinical nutrition has never before been postulated or otherwise known. This concept would revolutionize how we treat most illnesses since their origins would have in common basic metabolic mechanisms that regulate energy, plus hormone and protein synthetic events. The data presented provides strong evidence that AC-11® contributes to human nutritional enhancement as a dietary supplement based upon its high content of QA and QAEs™. Due to the potential importance to human health, and since AC-11® has these basic nutritional compositional properties, studies are underway to confirm this nutritional utility of AC-11®. As a dietary supplement AC-11® supplies QA and QAE's™ to the diet. It would be more economical and practical, than supplying pure QA, which to date has not been synthesized on a commercial scale.

ACKNOWLEDGMENTS

The author is indebted to Dr Margaretha Lund-Pero a co-worker from the University of Lund and Tony Worth from Optigenex for critical review analysis of all aspects of this review.

REFERENCES

Adamson, RH, Bridges, JW, Evans, ME, Williams, RT. 1970. Species differences in the aromatization of quinic acid in vivo and the role of gut bacteria. *Biochemical Jour.* 116: 437-433.

Aguilar, JL, Rojas, P, Marcelo, A, Plaza, A, Bauer, R, Reininger, E, Klass, CA, Merfort, I. 2002. Anti-inflammatory activity of two different extracts of Unaria tomentosa. *J. Ethanopharmology* 81(2): 271-276

Åkesson, C, Pero, RW, Ivars, F. 2003A. C-Med-100, a hot water extract of Uncaria tomentosa, prolongs leukocyte survival in vivo. *Phytomedicine* 10: 25-33, 2003A

Åkesson, C, Lindgren, H, Pero, RW, Leanderson, T, Ivars, F. 2003B. An extract of Uncaria Tomentosa inhibits cell division and NF-kB activity without inducing cell death. *International Immunopharmacology* 3: 1889-1900.

Åkesson, C., Lindgren, H., Pero, R.W., Leanderson, T., Ivars, F. 2005. Quinic acid is a biologically active component of the Uncaria tomentosa extract C-Med 100®. International Immunopharmacology 5: 219-22.

Aquino, R., De Simone, F., Pizza, C., Conti, C. and Stein, M. L. 1989. Plant metabolites. Structure and in vitro antiviral activity of quinovic acid glycosides from Uncaria tomentosa and Guettarda platypoda. *J. Nat. Prod.* 52: 679-685.

Aquino, R., De Feo, V., De Simone, F., Pizza, C. and Cirino, G. 1991. Plant metabolites. New compounds and anti-inflammatory activity of Uncaria tomentosa. *J. Nat. Prod.* 54: 453-459.Jones, K. Cat's claw, healing vine of Peru. Sylvan Press, Seattle: 1995.

Banne, A, Amiri, A, Pero, RW. 2004. Reduced Level of Serum Thiols in Patients with a Diagnosis of Active Disease. *JAAM* 6(4): 325-32.

Beveridge, T, Li, TSC, Oomah, BD, Smith, A. 1999. Sea buckthorn products: manufacture and composition. *J. Agri. Food Chem.* 47: 3480-3488.

Blumenthal, M (ed). 2003. The ABC clinical guide to herbs. *Americian Botanical Council*, Austin, Texas. Pages 1-441.

Capasso, M, Gaginella, TS, Grandolini, G, Izzo, AA (eds). 2003. Phytotherapy. Springer-Verlag, Berlin Heideberg New York. ISBN 3540-00052-6. Pages 1-424

Castillo, G., et al. 1998. Pharmaceutical compositions containing *Uncaria tomentosa* extract for treating Alzheimer's disease and other amyloidoses. *Patent-Pct. Int. Paol.* 1998; 00 33,659: 67 pp.

Castillo, G. 2005. Methods of isolating amyloid-inhibiting compounds and use of compounds isolated from *Uncaria tomentosa* and related plants. US Patent No. 6,929,808, August 16, 2005.

Castillo, G. 2006. Methods of isolation of amyloid inhibitory ingredients within *Uncaria tomentosa*. US Patent No 7,029,710, April, 18, 2006.

Cisneros, FJ, Jayo, M, Niedziela, L. 2005. An Uncaria tomentosa (Cat's Claw) extract protects mice against ozone-induced lung inflammation. *Journal of Ethanopharmacology* 96: 355-364

Eberlin, S, dos Santos, LM, Queiroz, ML. 2005. Uncaria r\tomentos extract increases the number of myeloid progenitor cells in the bon marrow of mice infected with Listeria monocytogenes. *Int. Immunopharmacol.* 5(7): 1235-1246.

Emanuel, P, Noah Scheinfeld, N. 2007. A review of DNA repair and possible DNA-repair adjuvants and selected natural anti-oxidants. *Dermatology Online Journal* 13 (3): 10

Englehardt, UH, Maier, HG. 1985. Acids in coffee. The proportion of individual acids in the total titratable acid. Z Lebensm Unters Forsch 18(1): 20-23.

Gonthier, MP, Verny, MA, Besson, C, Remesy, C, Sealbert, A. 2003. Chlorogenic acid bioavailability largely depends on its metabolism by the gut microflora in rats. *J. Nutr.* 133(6): 1853-1859

Graham, HN. 1992. Green tea composition, consumption and polyphenol chemistry. *Pre Med.* 21: 334-350.

Herrmann KM. 1995. The shikimate pathway. Early steps in the biosynthesis of aromatic compounds. The Plant Cell 7: 907–919.

Herrmann KM, Weaver LM 1993. The shikimate pathway. *Ann. Rev. Plant Physiol. Plant Mol. Biol.* 50: 473–503.

Jensen, HD, Krogfelt, KA, Cornett, C, Hansen, SH, Cristensen, SB. 2002. Hydrophilic carboxylic acids and iridoid glycosides in the juice of American and European cranberries (Vaccinium macrocarpon and V. Oxycoccos), lingonberries (V. vitis-idaea) and blueberries (V. myrtillus). *J. Agri. Food Chem.* 50(23): 6871-6874

Kuras, M, Pilarski, R, Nowakowska, j, Zobel, A, Brzost, K, Antosiewicz, J, Guiewicz, K. 2009. Effect of alkaloid-free and alkaloid –rich preparations from Uncaria tomentosa bark on mitotic activity and chromosome morphology. *J. Ethanopharmacol.* 12(1): 140-147

Keplinger, K. 1982. Cytotoxic, contraceptive and anti-inflammatory agents from Uncaria tomentosa. *PCT Int. Appl.,* WO 8201130

Keplinger, K., Laus, G., Wurm, M., Dierich, MP, Teppner, H. 1999. Uncaria tomentosa (Willd.) DC.-Ethnomedicianl use and new pharmacological, toxicological and botanical results 1999. *Journal of Ethanopharmacology* 64:23-34.Jones, K. 1995. Cat's Claw-Healing vine of Peru. Sattle (WA). Sylvan Press

Lamm, S., Sheng, Y., Pero, R.W. 2001. Persistent response to pneumococcal vaccine in individuals supplemented with a novel water soluble extract of Uncaria tomentosa, C-Med-100. *Phytomedicine* 8(4): 267-274.

Laus, G. 2004. Advances in chemistry and bioactivity of the genus Unacaria. *Phytotherapy Res.* 18: 259-274

Laus, G, Brossner, D, Kelinger, K. 1997. Alkaloids of the Pervian Uncaria tomentosa. *Phytochemistry* 45(4): 855-860

Lewis, J, Milligan, G, Hunt, A. 1995. NUTTAB95 Nutrient Data Table for Use in Australia. Canberra: Food Standards Australia New Zealand (FRANZ). Organic acid components of foods (g/100g edible portion). ORGAFOOD.TXT, COFA index num.2.

Mammone, T, Akesson, C, Gan, D, Giampapa, V, Pero, RW. 2006. A water soluble extract from Uncaria tomentosa (Cat's claw) is a potent enhancer of DNA repair in primary organ cultures of human skin. *Phytotherapy Res.* 20: 178-183.

Marmor, M., Alcabes, P., Titus, S., Frenkel, K. Krasinski, K., Penn, A. and Pero. RW.1997. Low serum thiol levels predict shorter times-to-death among HIV-infected injecting drug users. *AIDS* 11:1389-1393

Miller, MJ, Mehta, K, Kunte, S, Raut, V, Gala, J, Dhumate, R, Shukla, A, Tupalli, Parikh, H, Bobrowski, P, Chaudhary, J. 2005. Early relief of osteoarthritis symptoms with a natural mineral supplement and a herbomineral combination: a randomized controlled trial (ISRCTN 38432711) *J. Inflamm. LOND* 2: 11

Montoro, P, Carbone, V de DiazZuniga Quioz, J, DeSimone, F, Pizza, C. 2004. Identification and quantification of components in extracts of Uncaria tomentosa by HPLC-ES/MS. *Phytochemical Analysis* 15:55 64

Mourgue, M, Lanet, J, Blanc A, Steinmetz, MD. 1975. Quinic acid and ioschlorogeniec acids in sunflower seeds (Helianthus annus Lin.). *C R Seances Soc. Biol. Fil.* 169(5): 1256-1259.

Olthof, MR, Holliman, PC, Zock, PL, Katan, MB. 2003. Chlorogenic acid, quercetin-3-rutinside and back tea phenols are extensively metabolized in human's bioavailability in humans. *J. Nutr.* 133(6): 1806-1814

Pero, R.W., Olsson, A., Sheng, Y., Hua, J., Möller, C., Kjellén, E., Killander, D. And armor, M. Progress in identifying clinical relevance of inhibition, stimulation and measurements of poly ADP ribosylation. *Biochimie* 77:385-393 (1995)

Pero, R.W., Sheng, Y., Olsson, A., Bryngelsson, C. and Lund-Pero, M. 1996. Hypochlorous acid/N chloramines are naturally produced DNA repair inhibitors.*Carcinogenesis* 17(1):13-18.

Pero, RW, Hoppe, C, Sheng.Y. 2000. Serum thiols as a surrogate estimate of DNA repair correlates to mammalian life span. *Anti-aging Med.* 3(3): 241-249

Pero, RW, Amiri, A, Welther, M, Rich, M. 2005. Formulation and clinical evaluation of combining DNA repair and immune enhancing nutritional supplements. *Phytomedicine* 12(4): 255-263.

.Pero, RW, Giampapa, V, Vojdani, A. 2002. Comparison of a broad spectrum anti-aging nutritional supplement with and without the addition of a DNA repair enhancing cat's claw extract. *J. Anti-aging Med.* 5(2): 345-353.

Pero, RW, Lund, H, Leanderson, T. 2009. Antioxidant metabolism induced by quinic acid. Increased urinary excretion of tryptophan and nicotinamide. *Phytotherapy Research* 23: 335-346

Pero, RW, Lund, H. 2009. DNA repair in serum correlates to clinical assessment of anti-aging lifestyle criteria in healthy volunteers treated with quinic acid ammonia chelate (Quinmax^{TM}, Aqua Bimini) in drinking water. *Submitted to Clincial Biochemistry.*

Piscoya J, Rodriguez Z, Bustamante SA, Okuhama NN et al. 2001. Efficacy and safety of freeze-dried cat's claw in osteoarthritis of the knee: mechanisms of action of the species Uncaria guianensis. *Inflamm. Res*. 2001; 50(9):442-8.

Riva L, Coradini D, Di Fronzo G et al. The antiproliferative effects of Uncaria tomentosa extracts and fractions on the growth of breast cancer cell line. *Anticancer. Res.* 2001; (4A):2457-61.

Romero Rodrigues, MA, Vazquez, ML, Lopez Hernandez, J, Lozano, S. 1992. Determination of Vitamin C and organic acids in various fruits by HPLC. *Jour Chromat. Sci.* 30: 433-437.

Reinhard, K.1999. Uncaria tomentose (Willd) D.C. Cat's Claw, Una de Gato, or Seventaro. *J.Alternative Complement Med.* 5(2): 143-151.

Sandoval-Chacon, M, Thompson, JH, Zhang, XJ, Liu, X, Mannick, EE. 1998. Anti-inflammatory actions of cat's claw: the role of NF-kappaB." *Aliment. Pharmacol. Ther.* 1998; 12(12): 1279–89.

Sandoval, M, Okuhama, NN, Zhang, XJ, Condezo, LA, Angeles, FM, Lao, J, Angeles, FM, Musah, RA, Bobrowski, P, Miller, MJ. 2002. Anti-inflammatory and antioxidant activities of cats claw (Uncaria tomentosa and Uncaria guinianensis are independent of their alkaloid content. .Phytomedicine 9: 325-337

Santa, M, Lopez, A, Diaz, MM, Alban, J, Galan de Mera, A, Vicante Orellana, JA, Pozuelo, JM. 1997. Evaluation of the toxicity of Uncaria tomentosaby bioasays in vitro. *J.* Ethanolopharmcology 57 (3): 183-187.

Senatore, A., Cataldo, A., Iaccarino, F. P. and Elberti, M. G. 1989. [Phytochemical and biological study of Uncaria tomentosa]. Boll. Soc. Ital .Biol. Sper. 65: 517-520.

Sheng, Y, Pero, RW, Amiri, A, Bryngelsson, C. 1998. Induction of apoptosis and clonogenic growth of human leukemic cell lines treated with aqueous extracts of Uncaria tomentosa. *Anticancer Res* .18: 3363-3368

Sheng, Y, Bryngelsson, C, Pero, RW._2000A. Enhanced DNA repair, immune function and reduced toxicity of C-Med-100^{TM}, a novel aqueous extract from Uncaria tomentosa. *Journal of Ethanopharmacology* 69: 115-126.

Sheng, Y, Pero RW, Wagner H. 2000B. Treatment of chemotherapy-induced leukopenia in the rat model with aqueous extract from Uncaria Tomentosa. *Phytomedicine* 7(2): 137-143.

Sheng, Y., Li, L., Holmgren, K., Pero, R.W. 2001. DNA repair enhancement of aqueous extracts of Uncaria Tomentosa in a human volunteer study. *Phytomedicine* 8(4): 275-282.

Sheng, Y, Åkesson, C, Holmgren, K, Bryngelsson, C, Giampapa, V, and Pero, RW. 2005 An active ingredient of Cats Claw water extracts. Identification and efficacy of quinic acid. *Submitted to J Ethanpharmacology* 96(3): 577-584.

Silva, BM, Andrade, PB, Mendes, GC, Seabra, RM, Ferreira, MA. Study of the organic acids composition of quince (Cydonia oblonga Miller) fruit and jam. 2002. *J. Agric. Food Chem.* 50(8): 2313-2317

Stuppner, H., Sturm, S., Konwalinka, G. 1992. HPLC analysis of the main oxindole alkaloids from Uncaria tomentosa. *Chromatographia.* 34: 597-600.

Van Gorsel, H, Li, C, Kerbel, EL, Smits, M, Kadar, AA. 1992. Composition characterization of prune juice. J Agric Food Chem 40: 784-789.

Wagner, H., Kreutzkamp, B. and Jurcic, K. 1985. [The alkaloids of Uncaria tomentosa and their phagocytosis-stimulating action]. *Planta Med.*419-423

Wirth, C., Wagner, H. 1997. Pharmacologically active procyanidins from the bark of Uncaria tomentosa. *Phytomedicine.* 4: 265-266.

Yagasaki, K, Furuse, T, Miura, Y, Oakauchi, R. 2000. Inhibitory effects of chlorogenic acid and its related compounds on the invasion of hepatoma cells in culture. *Cytotechnology* 33: (1-3): 229-235

In: Encyclopedia of DNA Research (3 Volume Set)
Editors: S. J. Duncan and P. H. Wiley
ISBN 978-1-61324-305-3

Chapter 24

DNA DAMAGE RESPONSE AND REPAIR: MECHANISMS IN AGING AND CANCER

Panagiotis J. Vlachostergios*[1] *and Ioannis A. Voutsadakis
Department of Medical Oncology, University Hospital of Larissa, University of Thessa y
School of Medicine, Larissa, Greece

ABSTRACT

The preservation of expected life span and longevity as well as the assurance of life succession among all species requires the integrity and faithful transcription of DNA, which is in constant danger from both environmental and intracellular stressors, such as radiation or products of cellular metabolism, respectively. Thus, a dedicated machinery of DNA repair factors is responsible for reversing DNA damage across the genome. Efficient response to various effectors of DNA damage is also dependent on a complex network of sensors and mediators of stress signals which are upstream of DNA repair activation and together constitute components of the DNA damage response (DDR). The final outcome for the cell may vary according to the severity and persistence of the induced lesions, resulting in either precancerous phenotypes from mutagenic lesions, or in premature age-related pathologies in the context of enhanced apoptosis or senescence. In this article we review the contribution of functional deterioration of the DDR in the evolution of aging and carcinogenesis. We describe how defects in the tumor suppressor-, hormonal- and DNA repair-related processes result in aging acceleration and emergence of malignancy with inferences from the genotypes and phenotypes of clinically relevant progeroid and cancer syndromes and also based on data from mouse p53 knock-out and human gene polymorphism studies. From a reverse point of view, we discuss observations of age-related alterations in DDR. We also discuss the critical involvement of stem cells in the pathophysiology of aging and malignant conditions. Finally, we recapitulate current and future anti-aging, anti-cancer preventive strategies including diet and calorie-intake modification, exercise and pharmacologic interventions.

---

[1] Correspondence to: Panagiotis J. Vlachostergios MD, Department of Medical Oncology, University Hospital of Larissa, University of Thessaly School of Medicine Biopolis (Mezourlo) 41110, Larissa, Greece, Tel: +302413502785, Fax: +302410672467, Email: pvlacho@med.uth.gr.

Abbreviations

SSB,	single-strand break;
DSB,	double-strand break;
NHEJ,	non-homologous end joining;
HR,	homologous recombination;
FA,	Fanconi anaemia pathway;
MGMT,	O6-methylguanine DNA-methyltransferase;
NER,	nucleotide excision repair;
BER,	base excision repair;
MMR,	mismatch repair;
TLS,	translesion synthesis.

Introduction

The preservation of genome integrity in all organisms is of paramount importance for cell viability and defence against a continuous structural and functional deterioration imposed by primary exposure and secondary processing of extracellular and intracellular genotoxic stressors. The accumulation of emerging evidence on deregulation of the DNA damage response (DDR), which is considered the guardian of genomic health, places DDR at a crossroad between aging and cancer. Both these pathologies require the existence of aberrations in multiple steps of the inherent self-defence DDR system, which involves a plethora of recognition, signaling and DNA damage reversal pathways.

DDR: Definition/Stages

DDR is an essential component of a cell's defence machinery against the detrimental effects of continuous exposure to DNA damaging conditions. DDR consists of a wide variety of sensors, transducers and effectors orchestrating the appropriate repair of DNA damage and resolution of DNA replication blocks (Harper and Elledge, 2007). Although there is still a lot to learn about the extent of DDR and the complexity of its components, damage-sensing is composed of at least five independent sensing complexes of which ATM (Ataxia telangiectasia mutated) and ATR (Ataxia telangiectasia and Rad3 related) kinases are the best known (Harper and Elledge, 2007). ATM is recruited and activated by DNA double-strand breaks. It phosphorylates several key proteins that initiate activation of the DNA damage checkpoint, leading to cell cycle arrest, DNA repair or apoptosis. Several of these targets, including p53, Chk2, BRCA1, and H2AX, are tumor suppressors. ATR is activated in response to persistent single-stranded DNA that occurs at stalled replication forks. ATR works with a partner protein called ATRIP (ATR interacting protein) to recognize single-stranded DNA. Once ATR is activated, it phosphorylates Chk1, initiating a signal transduction cascade that culminates in cell cycle arrest (Harper and Elledge, 2007). Mediators are proteins that act directly downstream of the ATM and ATR kinases as substrates, playing a dual role as recruiters of additional substrates and as scaffolds to assemble the repair complex. Mdc1 (Mediator of DNA damage checkpoint protein 1), 53BP (p53 binding protein), the MRN (Mre11, Rad50 and Nbs1) complex and BRCA1 are important representatives of this DDR subgroup (Harper and Elledge, 2007). There exist five main DNA repair pathways which are involved in DNA damage reversal, depending on the nature of the induced DNA-lesions: direct or reversion repair by O^6-methylguanine DNA methyltransferase (MGMT), mismatch repair (MMR), base-excision repair (BER), nucleotide-excision repair (NER) and double-strand break (DSB) repair, which involves homologous recombination (HR) and non-homologous end joining (NHEJ) repair sub-pathways (Hansen and Kelley, 2000; Christmann et al., 2003; Wood et al., 2005; Pourquier, 2006). Additionally, there exist a DNA replication block bypassing repair mechanism termed post-replication repair (PRR), consisting of both error-prone translesion synthesis (TLS) and error-free damage avoidance (Barbour and Xiao, 2003) as well as a DNA-crosslink repair pathway, combining HR and TLS, the Fanconi Anaemia pathway (FA) (Kennedy and D'Andrea, 2005). These DNA repair pathways are illustrated in Chart 1.

DDR: Aging and Cancer

DNA damage produces two main effects on DNA-related processes: development of mutations and transcriptional stress. The first is a precursor of cancer, while the second leads to an impairment of replication, cellular senescence, and malfunction or death. Cellular senescence is the phenomenon by which normal diploid cells permanently lose the ability to divide. The word *senescence* is derived from the Latin word *senex*, meaning "old man" or "old age" or "advanced in age". Senescence contributes to aging via both attenuation of self-renewing cells causing decreased tissue regeneration and fostering of an inflammatory microenvironment, permissive for neoplastic growth (Chen et al., 2007). Eventually, a progressive loss of tissue homeostasis may result in tissue dysfunction and appearance of progeroid aging phenotypes (Schumacher et al., 2009a).

Similarly, there are two outcomes of genome maintenance effected by DDR, repair of lesions and activation of cellular responses that abort mitosis and execute apoptosis of cells when lesions are beyond repair. The initial recognition machinery involves proteins devoted to signaling, such as ATM and p53 or to chromatin remodeling such as histone H2AX, with cooperative roles in induction or/and facilitation of DNA repair (Vijg, 2008).

Chart 1. Most frequent DNA lesions and corresponding repair pathways

	DNA Lesions				
	SSB	DSB / crosslinks	DNA adducts / Base damage	Replication errors	Post-replication errors
DNA Repair Pathways	NER	NHEJ	MGMT	MMR	TLS
	BER	HR	NER	BER	Damage Avoidance
		FA	BER		

A molecular model of the aging process comprises two mechanisms, DNA damage and telomere shortening, acting together to induce additive generalized genomic instability, thus triggering p53 activation and senescence (Ding and Chen, 2008). This is also known as the "network" theory of aging, as it puts together two individual mechanisms, DDR and telomere maintenance. The latter has been proven an integral part of DDR machinery triggering because when telomeres become dysfunctional, DDR is activated. In addition, chromosomal repair may be required for DSB repair (Slijepcevic, 2008). Therefore, DNA damage is a central mediator in triggering replicative senescence, induced by telomere shortening and premature cellular senescence, induced by oncogenic or oxidative stress. In contrast, a cell that has secured maintenance of telomeres length may respond to DNA damage in a completely different way and the balance is tipped towards neoplasia.

Evidently, DDR abnormalities are also susceptibility factors for carcinogenesis and repair capacity tested with various DNA repair phenotype assays has been associated with increased cancer risk. Further investigation is needed to clarify whether genetic variations, in terms of single nucleotide polymorphisms (SNPs), may be correlated with variations in DNA repair capacity (DRC) phenotype in general population (Li et al., 2008). Furthermore, cells harboring DDR defects and failing to repair the damaged DNA cannot evade genomic instability and subsequent propensity to cellular transformation. Tumor cells have several defects in maintenance of genome integrity, which involves loss of appropriate DDR, affecting key proteins involved in checkpoints, histone modifications and DNA repair (Powell and Bindra, 2009). Interestingly, this property of cancer can be exploited for cancer therapy using the principle of synthetic lethality, based on a weakened DRC of tumor cells and a consequent selective sensitivity to specific inhibitors. A couple of clinically relevant paradigms include Chk1 inhibitors, which selectively target the tumor cells with p53 mutations and PARP-1 inhibitors, which showed selectively killing effects on the cells with defects of homologous recombination (HR) in the context of BRCA1/2 mutations (Liang et al., 2009). PARP inhibitors are in advanced stage of clinical development for the treatment of triple negative (ER, PR and HER2 negative) breast carcinoma.

The Tumor Suppressor Response

Tumor suppression mechanisms, acting downstream of DNA damage sensors and upstream of effectors of DNA repair, have a considerable contribution to the entire process of DDR, and comprise caretakers, which protect against mutations and gatekeepers, which are responsible for induction of apoptosis or cell cycle arrest and senescence. Signal transduction pathways activated by DNA damage converge on the p53-p21 pathway mediating senescence due primarily to telomere shortening and on the p16-Rb pathway mediating premature senescence (Chen et al., 2007).

Recent evidence has attributed an ambiguous role of tumor suppressors (p53, p16/p21) in DDR, given that the price for lifespan extention after suppression of oncogenesis is the loss of tissue regeneration as a result of cell death or senescence, thus predisposing to aging acceleration (Schumacher et al., 2008). The use of mouse models has enriched our knowledge on this role of caretakers in lifespan extention and cancer risk. In p53 knock-out mice (p53-/- or p53+/-), early death was observed due to cancer development, whereas p53-mutated mice (p53+/m), with increased p53 activity, disclosed a reduced incidence of cancer but acceleration of aging due to increased apoptosis and senescence (Van Heemst et al., 2007; Van Heemst, 2009). Furthermore, intriguing results have come to light from investigation of codon 72 polymorphism in human TP53 gene. In this position there are two alternative variants, either proline or arginine. The proline variant (Pro/Pro) has a higher p53 transcriptional activity while the arginine variant (Arg/Arg) displays a lower p53 activity. Persons homozygous for the proline variant with a higher activity demonstrated a higher rate of survival resulting from a higher potential of senescence, evidenced by increased expression of p21, beta-galactosidase and G1 arrest when compared to the arginine variant (Arg/Arg). However, increased cancer mortality was also observed in the Pro/Pro group, arguing for antagonistic effects of apoptosis and senescence on cancer and lifespan, which may be expressed by protection from cancer in early life but simultaneously promotion of aging phenotypes, including late-life cancer (Van Heemst et al., 2007; Van Heemst, 2009). Mechanisms responsible for cellular aging may have evolved as safeguards against carcinogenesis. Resistance to cancer early in life would convey a reproductive advantage while disease/dysfunction associated with advanced age would provide no selective disadvantage (Gilchrest and Bohr, 1997).

With respect to the role of p53 in carcinogenesis, the signal for tumor suppression is largely initiated by oncogene-activation and mediated by two distinct pathways, the induction of the tumor suppressor $p14^{ARF}$ and the promotion of DNA damage due to hyperproliferation. In both, there is a selective pressure for inactivation of checkpoint components to allow proliferating cells to progress to fully malignant tumors that trigger p53 activation (Meek, 2009).

The Hormonal Response: An Additional Variable

Studies in premature aging mice of NER -/- genotype demonstrated that their inherent genome instability induced an adaptive response, resembling the one observed after calorie restriction with up-regulation of stress- and defense-responses and suppression of growth hormone (GH), insulin growth factor-1 (IGF-1), thyrotroph and lactotroph axes and oxidative metabolism (Schumacher et al., 2009a). Consequently, a unifying concept of an evolutionarily conserved “survival” response has evolved, in which a shift of the organism’s resources from growth to maintenance is triggered as a response to stressors like DNA damage and starvation. This provides a plausible link between longevity-promoting suppression of the GH – IGF-1 somatotropic axis and pathogenesis of progeroid diseases (Schumacher et al., 2008).

The GH – IGF-1 axis is also involved in the development of neoplasia because IGF-1 is a known antiapoptotic, mitogenic and growth factor and there is an established association between its plasma levels and certain malignancies (Yu and Rohan, 2000). Important data have recently come to confirm this effect, based on a preclinical GH receptor/binding protein knock-out (GHR/BP KO) mouse model which demonstrated less overall incidence of neoplastic disease and less severe and fatal neoplastic lesions in knock-out mice compared with their wild type littermates (Ikeno et al., 2009).

The Role of Stem Cells

In an effort to link molecular events with evolution of progeroid and cancer phenotypes a deeper insight into the functional outcomes of DNA damage is required. It is established that even in tissues from younger organisms the renewal capacity of highly differentiated cells is limited. In tissues from older organisms the decline in the function and number of cells is more prominent than in younger tissues. Remaining cells, due to their extended life have accumulated hyperplastic or neoplasia-promoting genetic lesions (Vijg, 2008).

Adult tissue stem-cells play an important role in the long-term maintenance of tissues throughout life, by renewing the cell population to replace senescent, terminally differentiated cells. DNA damage in these cells may decrease their survival, with severe impact on tissue degeneration. If DNA damage is combined with loss of DRC, altered patterns of proliferation, quiescence and differentiation, these cells carry the risk of increased susceptibility to malignant transformation (Kenyon and Gerson, 2007). Cancer is increasingly being viewed as a stem cell disease. This is based on the self-renewal ability of a subpopulation of the tumor mass and the strong circumstantial evidence that cancer might originate in normal tissue stem cells or early progenitor cells. For normal or cancer cells to be defined as stem cells, it is necessary to demonstrate their self-renewing properties and show that they can reconstitute the normal tissue and cancer, respectively, over time. This has now been demonstrated in a wide range of carcinomas, such as teratocarcinoma, leukemia, brain cancer and breast cancer. Cancer development and growth is therefore driven by a small subpopulation of cancer stem cells, which can give rise to both more cancer stem cells as well as the bulk of cancer cells (Gudjonsson and Magnusson, 2006).

In a similar manner, stem cells have been demonstrated to undergo age-related changes, of genetic (e.g., telomere shortening and repair genes mutations), epigenetic (e.g. DNA methylation) and biochemical (e.g., oxidative stress and glycation) nature (Bassi and Sacco, 2009). These alterations constitute the driving force of stem cell aging which is ultimately responsible for all aging phenomena observed in an organism, given that the regenerative process of the organism is determined by the ability of its stem cells to replace damaged or terminally arrested cells inside each tissue (Bassi and Sacco, 2009).

Age-Related Changes in DDR

Another aspect that should not be overlooked is that DNA repair itself can be the subject of age-related changes and deterioration. Many DNA repair factors have been associated with progeroid mutant phenotypes, particularly proteins involved in NHEJ (Ku80, DNA-PKcs), HR (Rad50, BRCA1) and NER (XPD, CSA, CSB, ERCC1, XPF). Defective NHEJ, HR and TCR in association with dysfunctional telomeres produce genomic instability by allowing unrepaired DNA damage (Lombard et al., 2005). Moreover, as the network of interactions between damage and repair is not confined only to nuclear transcribed DNA but also includes telomeric and mitochondrial DNA, efficiency of other major elements of DNA repair in all these DNA types may also play a role in the aging process. Specifically, BER efficiency, although different in nuclear, mitochondrial and telomeric DNA, is relevant for the rate of aging. Poly (ADP-ribose) polymerase (PARP) activity also influences the aging rate via stimulation of BER activity and attenuation of mitochondrial formation of reactive-oxygen species (ROS) due to NAD^+ depletion and telomeric maintenance. Furthermore, replicative senescence induced by telomeric single-strand breaks (SSBs) is accompanied by profound changes in gene expression patterns and these contribute to age changes in tissues. Finally, damaged mitochondria, which can cause endogenous stress by ROS production, may multiply faster than intact ones, and by virtue of their ability to replicate independently of cell division, they will accumulate in some cells, causing cell loss (Zglinicki et al., 2001). On the other hand, inefficiency or low fidelity of MMR, BER and NER may lead to microsatellite instability (MMR), point mutations, even genomic rearrangements by use of NHEJ repair. Ultimately, the presence of fewer cells in G2 phase may cause changes in the DSB repair sub-pathway used, favoring the more mutagenic over the error free HR, causing genomic instability and loss of heterozygosity (LOH). Finally, a vicious cycle of genomic instability and aging becomes activated as the time-dependent accumulation of spontaneous mutations and rearrangements results in deregulation of transcription, impaired stress response and attenuated function of DNA repair genes, which in turn lead to more mutations, further exacerbating the age-related functional decline (Gorbunova et al., 2007). All these pathogenetic pathways converge to a "nucleocentric" view of aging where chromatin alterations constitute the downstream keypoint in the series of events following DNA damage as not only it may eventually lead to senescence but also renders cells more susceptible to further damage as well as epigenetic alterations, thus causing an amplification of the initial effect (Sinclair and Oberdoerffer, 2009).

CLINICAL EXAMPLES OF DNA REPAIR DEFICIENCIES

The biological significance of DNA repair mechanisms is highlighted by their evolutionarily conservation and by the phenotypes of patients with inherited defects in DNA repair genes. Strikingly, defects in DNA repair mechanisms can be associated with very different clinical phenotypes, including cancer and premature aging (Hoeijmakers, 2001). Xeroderma pigmentosum (XP) is an example of a human cancer-prone syndrome that is associated with inborn defects in global genome nucleotide excision repair (GG-NER) (de Boer and Hoeijmakers, 2000). XP is characterized by sun-induced pigmentation abnormalities, a risk of skin cancer more than 2000 times the risk in general population and an increased risk of internal tumors (Hoeijmakers, 2009). Due to the defects in GG-NER, XP patients show an accumulation of mutagenic lesions (Dolle et al., 2006). This accumulation is recapitulated in corresponding mouse models (Dolle et al., 2006). Mutagenic lesions may cause point mutations if not properly repaired and their persistence will put cells at a higher risk of neoplastic transformation (Hoeijmakers, 2007), which may explain the over 1000-fold increased risk of developing UV induced skin cancer observed in XP patients. Cockayne's syndrome (CS) and Trichothiodystrophy (TTD) are two examples of human accelerated ageing syndromes that are associated with inborn defects in transcription coupled repair (TCR or TC-NER). CS is characterized by early cessation of growth and development, severe and progressive neurodysfunction associated with demyelination, neural-sensori hearing loss, cataracts, cachexia and frailty. The average reported lifespan for patients with the disease is 12 years. TTD has features of CS in addition to brittle hair and nails and ichthyotic skin

(Hoeijmakers, 2009). Due to defects in TCR, CS and TTD patients and corresponding mouse models show an accumulation of cytotoxic lesions in actively transcribed genes (Dolle et al., 2006). The persistence of cytotoxic lesions will lead to the induction of apoptosis and/or senescence. An excessive induction of apoptosis and/or senescence may account for the absence of cancer as well as the presence of the segmented accelerated ageing phenotypes observed in CS and TTD patients (de Boer et al., 2002; van der Pluijm et al., 2007).

New Directions and Interventional Perspectives

The dramatic progress made over the last decade in the elucidation of key aspects within the complex mechanisms that propagate the aging process has enabled initial attempts towards implementation of rational preventive strategies. A reasonable way of delaying the onset of aging would be to avoid or at least limit exposure to exogenous genotoxic stressors. Moreover, the manipulation of endogenous genotoxins, such as ROS, or metabolic processing of environmental exposures could offer additional benefit. In this latter case, dietary interventions, for example calorie restriction, could favor tissue preservation by reducing ROS.

Diet and Calorie Restriction

Calorie restriction in adult men and women causes beneficial metabolic, hormonal, and functional changes, but the precise amount of calorie intake or body fat mass associated with optimal health and maximum longevity in humans is not known. It is also an open question whether calorie restriction extends lifespan in lean humans (Fontana and Klein, 2007). From the survival studies of overweight and obese people, it is estimated that long-term calorie restriction to prevent excessive weight gain could add 3–13 years to life expectancy. Thus the effects of calorie restriction on human life extention are significant but smaller than those achieved by medical and public health interventions, which have extended life by about 30 years in developed countries in the 20th century, by greatly reducing deaths from infections, accidents, and cardiovascular diseases (Everitt and Le Couteur, 2007). In addition, the absence of life-long studies and controlled randomized trials hinders the integration of calorie-restriction in a well established model and in clinical practice. It seems that a balanced diet of moderate proportions and exercise remain today the only proven fountain of youth (Dominguez et al., 2009).

Exercise

Regarding the relationship of exercise, longevity and DDR, some relevant data focus on the role of free radicals production (particularly oxygen- and nitrogen- radicals), and related reactive oxygen and nitrogen species in cells and tissues during exercise. The evidence strongly suggests that free radicals generated during mild to moderate endurance-type exercise act as intracellular signaling molecules to initiate exercise adaptation (Sachdev and

Davies, 2008). This mild oxidative stress possibly elicited by regular exercise appears to have beneficial outcomes by adaptively upregulating various antioxidant mechanisms, including antioxidative and repair-degradation enzymes for damaged molecules, like the proteasome (Goto et al., 2007). In a similar manner, exercise may be suppressing tumour initiation events by modifying carcinogen activation, specifically by enhancing the cytochrome P450 system and by enhancing selective enzymes in the carcinogen detoxification pathway, including, but not limited to, glutathione-S-transferases. Furthermore, exercise may enhance DNA repair systems and improve intracellular protein repair systems. In addition to altering processes related to tumour initiation, exercise may also exert a cancer-preventive effect by dampening the processes involved in the promotion and progression stages of carcinogenesis, including scavenging ROS, altering cell proliferation, apoptosis and differentiation, decreasing inflammation, enhancing immune function, and suppressing angiogenesis (Rogers et al., 2008). Thus, moderate exercise-mediated transient cell cycle arrest with a positive impact on self-defence activation via minimal ROS production, and longevity expansion remains currently speculative but worth testing in future studies.

The Concept of Hormesis

A view on retarding the onset of aging gaining in popularity in the scientific community is based on triggering innate adaptive responses by low dose of stress treatments that would otherwise in higher concentrations have adverse effects. This process is referred as hormesis and at the molecular level it works via accumulation of stalled RNA polymerase II (RNAPII) at sites of DNA damage, mediating the aging-related shift from growth to body resources preservation through somatotropic attenuation. Therefore, low amounts of persistent DNA damage in active genes might confer enhanced stress resistance and reduced cellular growth through RNAPII-dependent damage sensing. The underlying mechanisms responsible for this RNAPII-mediated upregulation of cellular responses are currently under investigation, but already available data from studies in DNA repair deficiency models, particularly TCR-defective diseases where RNAPII exerts its central role, may be of great significance in guiding this research (Schumacher, 2009b).

Pharmacologic Interventions

The use of pharmacological inhibitors of the mTOR (mammalian target of rapamycin) pathway, which is downstream of the GH-IGF-1 axis and therefore is involved in premature aging, has shown promising results in mice by extending their life span and delaying the development of conditions associated with aging, including cancer (Harrison et al., 2009; Hoeijmakers, 2009). These findings may have implications for further development of interventions targeting mTOR for the treatment and prevention of age-related diseases (Harrison et al., 2009).

Another important concept involving DDR in aging and cancer pathogenesis is the recently described "NEMO shuttle". Following genotoxic stress, the regulatory subunit of cytoplasmic IκB kinase complex, IKKγ or NEMO (NF-κB essential modulator) is translocated to the nucleus and consecutively subjected to SUMOylation, ATM-dependent phosphorylation, deconjugation of SUMO (small ubiquitin-like modifier) and ubiquitylation, thus facilitating its cytoplasmic re-localisation, in order to activate the transcription factor NF-κB (Nuclear factor-kappa B). This activation is effected through phosphorylation and ubiquitination of the NF-κB inhibitory proteins IκB-α and IκB-β. These proteins are substrates of 26S proteasome and their proteasomal degradation releases NF-κB from its complex, facilitating its transport to the nucleus (Huang et al., 2003; Janssens and Tschopp, 2006). Transcriptionally-active NF-κB evokes cellular responses such as inflammatory, anti-apoptotic and carcinogenic signals and tissue atrophy, which have been termed inflamm-aging (Salminen et al., 2009). Considering this emerging role of NF-κB in the pathogenesis of progeroid and cancer syndromes combined with its well-known regulation by the ubiquitin-proteasome system (UPS), the usefulness of proteasome inhibitors already used in cancer treatment (Hoeller and Dikic, 2009; Petroski, 2008, Yang et al., 2009) is currently tested in neurodegenerative diseases (Shah and Di Napoli, 2007). Furthermore, recent work on the differentiation of antibody secreting cells, revealed a regulatory circuit that integrates protein synthesis, secretion and degradation (proteostasis), into cell lifespan determination (Sensi et al., 2008). Even more intriguingly, a critical view of previous and ongoing research supports a direct role of UPS in the post-translational regulation of DNA repair proteins, as well as in transcriptional response of DNA repair after DNA damage and stress (Vlachostergios et al., 2009a; Vlachostergios et al., 2009b).

Conclusion

As the molecular knowledge about the pathogenesis of aging and cancer increases, more targets within the DDR machinery will be revealed, and it may be expected that interventions modulating environmental and epigenetic causative factors might be successful in ameliorating the effects of these processes and further extending life expectancy.

References

Barbour, L; Xiao, W. Regulation of alternative replication bypass pathways at stalled replication forks and its effects on genome stability: a yeast model. *Mutat Res.* 2003; 532: 137–55.

Bassi, P; Sacco, E. Cancer and aging: the molecular pathways. *Urol Oncol.* 2009;27: 620-7.

Chen, JH; Hales, CN; Ozanne, SE. DNA damage, cellular senescence and organismal ageing: causal or correlative? *Nucleic Acids Res.* 2007; 35: 7417-28.

Christmann, M; Tomicic, MT; Roos, WP; Kaina, B. Mechanisms of human DNA repair: an update. *Toxicology*. 2003; 193: 3-34.

de Boer, J; Hoeijmakers, JH. Nucleotide excision repair and human syndromes. *Carcinogenesis.* 2000; 21: 453-60.

de Boer, J; Andressoo, JO; de Wit, J; Huijmans, J; Beems, RB; van Steeg, H; Weeda, G; van der Horst, GT; van Leeuwen, W; Themmen, AP; Meradji, M; Hoeijmakers, JH. Premature aging in mice deficient in DNA repair and transcription. *Science.* 2002; 296: 1276-9.

Ding, SL; Shen, CY. Model of human aging: recent findings on Werner's and Hutchinson-Gilford progeria syndromes. *Clin. Interv. Aging.* 2008; 3: 431-44.

Dollé, ME; Busuttil, RA; Garcia, AM; Wijnhoven, S; van Drunen, E; Niedernhofer, LJ; van der Horst, G; Hoeijmakers, JH; van Steeg, H; Vijg, J. Increased genomic instability is not a prerequisite for shortened lifespan in DNA repair deficient mice. *Mutat. Res.* 2006; 596: 22-35.

Dominguez, LJ; Barbagallo, M; Morley, JE. Anti-aging medicine: pitfalls and hopes. *Aging Male.* 2009; 12: 13-20.

Everitt, AV ; Le Couteur, DG. Life extension by calorie restriction in humans. Ann *N Y Acad. Sci.* 2007; 1114: 428-33.

Fontana, L; Klein, S. Aging, adiposity, and calorie restriction. *JAMA.* 2007; 297: 986-94.

Gilchrest, BA; Bohr, VA. Aging processes, DNA damage, and repair. *FASEB J.* 1997; 11: 322-30.

Gorbunova, V; Seluanov, A; Mao, Z; Hine, C. Changes in DNA repair during aging. *Nucleic. Acids Res.* 2007; 35: 7466-74.

Goto, S; Naito, H; Kaneko, T; Chung, HY; Radák, Z. Hormetic effects of regular exercise in aging: correlation with oxidative stress. *Appl. Physiol. Nutr. Metab.* 2007; 32: 948-53.

Gudjonsson, T; Magnusson, MK. Stem cell biology and the cellular pathways of carcinogenesis. *APMIS.* 2005; 113: 922-9.

Hansen, WK; Kelley, MR. Review of mammalian DNA repair and translational implications. *J. Pharmacol. Exp. Ther.* 2000; 295: 1-9.

Harper, JW; Elledge, SJ. The DNA damage response: ten years after. *Mol. Cell.* 2007; 28: 739-45.

Harrison, DE; Strong, R; Sharp, ZD; Nelson, JF; Astle, CM; Flurkey, K; Nadon, NL; Wilkinson, JE; Frenkel, K; Carter, CS; Pahor, M; Javors, MA; Fernandez, E; Miller, RA. Rapamycin fed late in life extends lifespan in genetically heterogeneous mice. Rapamycin fed late in life extends lifespan in genetically heterogeneous mice. *Nature.* 2009; 460: 392-5.

Hoeijmakers, JH. DNA repair mechanisms. *Maturitas.* 2001; 38: 17–22.

Hoeijmakers, JH. Genome maintenance mechanisms are critical for preventing cancer as well as other aging-associated diseases. *Mech. Ageing Dev.* 2007; 128: 460–2.

Hoeijmakers, JH. DNA damage, aging, and cancer. *N Engl J Med.* 2009; 361: 1475-85.

Hoeller, D; Dikic, I. Targeting the ubiquitin system in cancer therapy. *Nature.* 2009; 458: 438-44.

Huang, TT; Wuerzberger-Davis, SM; Wu, ZH; Miyamoto, S. Sequential modification of NEMO/IKKγ by SUMO-1 and ubiquitin mediates NF-κB activation by genotoxic stress. *Cell.* 2003; 115: 565-76.

Ikeno, Y; Hubbard, GB; Lee, S; Cortez, LA; Lew, CM; Webb, CR; Berryman, DE; List, EO; Kopchick, JJ; Bartke, A. Reduced incidence and delayed occurrence of fatal neoplastic diseases in growth hormone receptor/binding protein knockout mice. *J Gerontol A Biol Sci. Med. Sci.* 2009; 64: 522-9.

Janssens, S; Tschopp, J. Signals from within: the DNA-damage-induced NF-κB response. *Cell Death Differ.* 2006; 13: 773-84.

Kennedy, RD; D' Andrea, AD. The Fanconi Anemia/BRCA pathway: new faces in the crowd. *Genes. Dev.* 2005; 19: 2925-40.

Kenyon, J; Gerson, SL. The role of DNA damage repair in aging of adult stem cells. *Nucleic Acids Res.* 2007; 35: 7557-65.

Li, C; Wang, LE; Wei, Q. DNA repair phenotype and cancer susceptibility--a mini review. *Int J. Cancer.* 2009; 124: 999-1007.

Liang, Y; Lin, SY; Brunicardi, FC; Goss, J; Li, K. DNA damage response pathways in tumor suppression and cancer treatment. *World J. Surg.* 2009; 33: 661-6.

Lombard, DB; Chua, KF; Mostoslavsky, R; Franco, S; Gostissa, M; Alt, FW. DNA repair, genome stability, and aging. *Cell.* 2005; 120: 497-512.

Meek, DW. Tumour suppression by p53: a role for the DNA damage response? *Nat. Rev. Cancer.* 2009; 9: 714-23.

Petroski, MD. The ubiquitin system, disease, and drug discovery. *BMC Biochem.* 2008; 9 Suppl 1:S7.

Pourquier, P. La réparation de l' ADN, cible potentielle d'un développement thérapeutique en cancérologie. *Bull. Cancer.* 2006; hors série: 124-44.

Powell, SN; Bindra, RS. Targeting the DNA damage response for cancer therapy. *DNA Repair (Amst).* 2009; 8: 1153-65.

Rogers, CJ; Colbert, LH; Greiner, JW; Perkins, SN; Hursting, SD. Physical activity and cancer prevention: pathways and targets for intervention. *Sports Med.* 2008; 38: 271-96.

Sachdev, S; Davies, KJ. Production, detection, and adaptive responses to free radicals in exercise. *Free Radic. Biol. Med.* 2008; 44: 215-23.

Salminen, A; Suuronen, T; Huuskonen, J; Kaarniranta, K. NEMO shuttle: a link between DNA damage and NF-kappaB activation in progeroid syndromes? *Biochem. Biophys. Res. Commun.* 2008; 367: 715-8.

Schumacher, B; Hoeijmakers, JH; Garinis, GA. Sealing the gap between nuclear DNA damage and longevity. *Mol. Cell Endocrinol.* 2009; 299: 112-7.

Schumacher, B. Transcription-blocking DNA damage in aging: a mechanism for hormesis. *Bioessays.* 2009; 31: 1347-56.

Schumacher, B; Garinis, GA; Hoeijmakers, JH. Age to survive: DNA damage and aging. *Trends Genet.* 2008; 24: 77-85.

Sensi, SL; Rapposelli, IG; Frazzini, V; Mascetra, N. Altered oxidant-mediated intraneuronal zinc mobilization in a triple transgeni c mouse model of Alzheimer's disease. *Exp. Gerontol.* 2008; 43: 488-92.

Shah, IM; Di Napoli, M. The ubiquitin-proteasome system and proteasome inhibitors in central nervous system diseases. *Cardiovasc Hematol Disord Drug Targets.* 2007; 7: 250-73.

Sinclair, DA; Oberdoerffer, P. The ageing epigenome: damaged beyond repair? *Ageing Res. Rev.* 2009; 8: 189-98.

Slijepcevic, P. DNA damage response, telomere maintenance and ageing in light of the integrative model. *Mech. Ageing Dev.* 2008; 129: 11-6.

van der Pluijm, I; Garinis, GA; Brandt, RM; Gorgels, TG; Wijnhoven, SW; Diderich, KE; de Wit, J; Mitchell, JR; van Oostrom, C; Beems, R; Niedernhofer, LJ; Velasco, S; Friedberg, EC; Tanaka, K; van Steeg, H; Hoeijmakers, JH; van der Horst, GT. Impaired genome maintenance suppresses the growth hormone–insulin-like growth factor 1 axis in mice with Cockayne syndrome. *PLoS. Biol.* 2007; 5(1):e2.

van Heemst, D. Biology of cancer and ageing. *Eur. J. Cancer.* 2009; 45 Suppl 1: 414-5.

van Heemst, D; den Reijer, PM; Westendorp, RG. Ageing or cancer: a review on the role of caretakers and gatekeepers. *Eur. J. Cancer.* 2007; 43: 2144-52.

Vijg, J. The role of DNA damage and repair in aging: new approaches to an old problem. *Mech. Ageing Dev.* 2008; 129: 498-502.

Vlachostergios, PJ; Patrikidou, A; Daliani, DD; Papandreou, CN. The Ubiquitin-Proteasome System in cancer, a major player in DNA Repair. Part 1: Post-translational regulation. *J Cell Mol. Med..* 2009; 13(9B): 3006-18.

Vlachostergios, PJ; Patrikidou, A; Daliani, DD; Papandreou, CN. The Ubiquitin-Proteasome System in cancer, a major player in DNA Repair. Part 2: Transcriptional regulation. *J Cell Mol. Med..* 2009; 13(9B): 3019-31.

von Zglinicki, T; Bürkle, A; Kirkwood, TB. Stress, DNA damage and ageing -- an integrative approach. *Exp. Gerontol.* 2001; 36: 1049-62.

Wood, RD; Mitchell, M; Lindahl, T. Human DNA repair genes. *Mutat. Res.* 2005; 577: 275-83.

Yang, H; Zonder, JA; Dou, QP. Clinical development of novel proteasome inhibitors for cancer treatment. *Expert Opin. Investig. Drugs.* 2009; 18: 957-71.

Yu, H; Rohan, T. Role of the insulin-like growth factor family in cancer development and progression. *J. Natl. Cancer Inst.* 2000; 92: 1472-89.

In: Encyclopedia of DNA Research (3 Volume Set)
Editors: S. J. Duncan and P. H. Wiley
ISBN 978-1-61324-305-3

Chapter 25

DNA Repair Gene XPC Codon 939 Polymorphism and AFB1-Related Hepatocellular Carcinoma Risk in Guangxi Population, China

Xi Dai Long[1 1], Yuan-Feng Zhou[1], Jin-Guang Yao[2], Pinhu Liao[2], Yu-Jin Tang[2], Zan-Song Huang[2], Bing-Pu Chen[3], Xiao-Ying Huang[1], Chao Wang[2], Ai-Ming Ma[4], Yun Yi[4], Qiu-Xiang Liang[4], and Yun-Long Lu[1 2]

[1] Department of Pathology, Affiliated Hospital of Youjiang Medical College for Nationalities, Baise, Guangxi Zhuang Autonomous Region, China
[2] Department of Medicine, Affiliated Hospital of Youjiang Medical College for Nationalities, Baise, Guangxi Zhuang Autonomous Region, China
[3] Department of Basic Medicine, Youjiang Medical College for Nationalities, Baise, Guangxi Zhuang Autonomous Region, China
[4] Department of Clinic Medicine, Youjiang Medical College for Nationalities, Baise, Guangxi Zhuang Autonomous Region, China

Abstract

Aim: Genetic polymorphisms in DNA repair genes may influence individual variation in DNA repair capacity, which may be associated with risk of hepatocellular carcinoma (HCC) related to the exposure of aflatoxin B1 (AFB1). In this study, we will focused on the polymorphism of xeroderma pigmentosum complementation group C (XPC) codon 939, involved in nucleotide excision repair.

[1] Corresponding author (X.D. Long): Department of Pathology, Youjiang Medical College for Nationalities, Chengxiang Road NO. 98, Baise City, Guangxi Zhuang Autonomous Region, China 533000. Email: xiaolonglong200166@yahoo.com.cn. Tel.: +86 776 2849481. Fax.: +86 776 2849481.
[2] These authors contributed equally to this work.

Methods: We conducted a case-control study including 579 HCC cases and 702 controls without any evidence of hepatic diseases to evaluate the associations between these two polymorphisms and HCC risk for Guangxi population by means of TaqMan-PCR.

Results: We did not found a significant difference in the frequency of XPC codon 939 genotypes between HCC and control groups. However, we found that the codon 939 Gln alleles, compared to Lys alleles, significantly increased the risk of HCC for individuals facing longer exposure years [adjusted odds ratios (ORs), 10.90 for the heterozygotes of XPC codon 939 Lys and Gln allele (XPC-LG) and long AFB1-exposure years; 19.10 for the homozygotes of XPC codon 939 Gln alleles (XPC-GG) and long AFB1-exposure years] or higher exposure levels (adjusted ORs, 6.87 for XPC-LG and high AFB1-exposure levels; 15.51 for XPC-GG and high AFB1-exposure levels) in the AFB1 exposure-genotypes interact analytical model.

Conclusions: These Results may suggest that the DNA repair gene XPC codon 939 polymorphism is associated with the increased risk of HCC correlated with AFB1 exposure among Guangxi Population.

ABBREVIATIONS

AFB1	aflatoxin B1;
CI	confidence interval;
HBsAg	hepatitis B surface antigen;
HBV	hepatic B virus;
HCC	hepatocellular carcinoma;
HCV	hepatic C virus;
OR	odds ratio;
NER	nucleotide excision repair;
PCR	polymerase chain reaction;
XPC	xeroderma pigmentosum complementation group D;
XPC-GG	the homozygotes of XPC codon 939 Gln alleles;
XPC-LG	the heterozygotes of XPC codon 939 Lys and Gln allele;
XPC-LL	the homozygote of XPC codon 939 Lys alleles.

BACKGROUND

Hepatocellular carcinoma (HCC) is the most common malignant tumor in Guangxi Zhuang Autonomous Region, the P.R. China [1, 2]. Epidemical evidences have showed that the exposure of aflatoxin B1 (AFB1), an important chemical carcinogen, is the most common cause of high incidence rates of HCC in this area [1, 2]. Many studies have suggested AFB1 can induce AFB1-DNA adducts formation and cause DNA strand break, DNA base damage, and oxidative damage that may ultimately lead to cancer [3]. AFB1-induced DNA damage can be repaired by base excision repair (BER), strand break repair pathway, and nucleotide excision repair (NER) [4, 5].

We have previously reported a relationship between HCC risk and allelic variations in the X-ray repair cross-complementing group 1 (XRCC1) [6, 7], XRCC3 [8-10], and the xeroderma pigmentosum complementation group D (XPD) genes [11] which are involved in BER , strand break repair, and nucleotide excision repair pathway [4, 12, 13]. In this study, we focused on the xeroderma pigmentosum complementation group C (XPC) gene, which is one of the seven genetic complementation groups encoding for proteins involved in the NER pathway [14-16]. The XPC codon 939 polymorphic locus has been of particular interest in molecular epidemiology studies [17-19]. Some studies have shown this polymorphism may be associated with decreased DNA repair capacity and increased tumor risk [16-29]. Therefore, we specifically conducted a hospital-based case-control study to examine whether this polymorphism modify the risk of HCC among Guangxi population from an AFB1 exposure area.

Materials and Methods

Study Population and Sample Collection

The protocol of the study was approved by the Ethic Committees of the hospitals involved in this study. HCC cases (n = 579) and control individuals (n = 702) were recruited from Affiliated Hospitals of the two main medical colleges in the Southwestern Guangxi, namely Guangxi Medical University and Youjiang Medical College for Nationalities, between January 2007 and November 2009. All cases and controls were residents of Guangxi Zhuang Autonomous Region from AFB1 exposure areas and accepted to be enrolled in this study. Cases included in present study, representing for a significant portion of HCCs from Guangxi population, were identified through hepato surgery, hepato pathology, oncology, hepatology centres and through cancer registries, and confirmed by histopathological diagnosis in 100% of the HCC cases. During the same period, controls without clinical evidence of liver diseases were randomly selected from a pool of healthy volunteers that visited the general health check-up center of the same hospitals because of their routine scheduled physical exams supported by local governments.

To control the effects of confounders which were likely risk factors for Guangxiese HCC, controls were frequently matched to cases based on sex, age (= 5 years), ethnicity (Han, Minority), and hepatic B virus (HBV) and HCV infection. These having hepatitis B surface antigen (HBsAg)-positive and anti-HCV-positive in their peripheral serum were defined as groups infected HBV and HCV. With informed consent, the characteristic information of all study subjects, including sex, age, ethnicity, HBsAg and anti-HCV, were ascertained as described previously, at the same time 4 mL of peripheral blood was also collected.

Nucleic Acid Isolation

Leukocytes from peripheral venous blood samples were isolated by standard procedures. DNA was then extracted from leukocyte samples by standard phenol-chloroform extraction and ethanol precipitation. DNA was stored at 4°C until additional analysis.

AFB1-Exposure Years

AFB1-exposure years were ascertained by our previously published methods [6-11]. In brief, AFB1-exposure years were defined as the length that each subject lived in AFB1exposure region and cumulative AFB1 exposure-years were calculated by the following formula: exposure-years = (age□ migration-years), where migration-years were defined as the years a person lived in non-exposure region of AFB1. To analyze, AFB1-exposure years were divided into three groups: short (< 40 years), medium (40 – 48 years), and long (> 48 years), according to the value of AFB1-exposure years, with two cutoff points of 40 and 48 years, the median exposure years among controls and cases.

AFB1-Exposure Levels

In this study, we evaluated the AFB1-exposure levels by means of AFB1-DNA adducts levels of DNA samples from all subjects' peripheral blood leukocytes resulting from the fact that the DNA samples of liver tissue were not obtained from the controls, which was described in our previously published articles [8, 11]. In order to analyze, AFB1-DNA adduct levels were divided into three groups: low level (≤ 1.00 μmol/mol DNA), medium level (1.01 – 2.00 μmol/mol DNA), and high level (≥ 2.01 μmol/mol DNA), according to the value of AFB1-DNA adduct levels, with two cutoff points of 1.00 and 2.00 μmol/mol DNA, the average adducts levels among controls and cases.

Genotyping Assays

Gene polymorphism analysis of XPC Lys939Gln was typed by using the TaqMan-PCR on iCycler iQ™ real-time PCR detection system (iQ5, Bio-Rad). The primers were: 5'-AGC AGC TTC CCA CCT GTT C-3' and 5'- GTG GGT GCC CCT CTA GTG-3'□ Applied Biosystems□ . The probes were: 5'-FAM-CAC AGC TGC TCA AAT-MGB-3' and 5'-Hex-CTC ACA GCT TCT CAA AT-MGB-3' □ Applied Biosystems□ . PCR reactions were run in a 25 μL final volume containing 1 × Premix Ex Taq™□ TaKaRa, catalog # DRR039A□ , 0.2 μM of each probe, 0.2 μM of each primer, and 50 – 100 ng of genomic DNA. Cycling conditions were 95 □ for 2 min, and 45 cycles of 95 □ for 10 s, 60 □ for 1 min and 72 □ 10 s. Controls were included in each run and repeated genotyping of a random 10% subset yielded 100% identical genotypes. Data analysis for allele discrimination was performed with the iCycler iQ software.

Statistical Analysis

The Pearson *Chi-square* test was used to test for the distributive differences of the frequency in gender, age, ethnicity, HBV and HCV infection, AFB1-exposure information, and XPC genotypes between the cases and the control subjects. The odds ratio (OR) and the 95% confidence interval (CI) were calculated as an estimate of relative risk. Multivariate logistic regression was used to control for possible confounding by age, gender, ethnicity, HBV and HCV infection, and AFB1-exposure information, where appropriate, and when examining interactions between the polymorphisms and AFB1-exposure information (including exposure years and levels).

Joint effects between genotypes and exposure information were assessed by using the full regression model including all possible cofounders. The interactive effects were evaluated according to the following formula (ascribed by *Brennan P* [30]): $OR_{eg} > OR_{eg'} \times OR_{e'g}$, where OR $_{eg}$, OR in presence of both high AFB1 exposure (namely long AFB1-exposure years or high AFB1-exposure levels) and risk genotypes of XPC codon 939 (adjusted OR > 1); OR$_{eg'}$, OR for short AFB1-exposure years or low AFB1-expsore levels in the presence of risk genotypes of XPC codon 939; OR$_{e'g}$, OR for non-risk genotype of XPC codon 939 in the presence of high AFB1 exposure. A *P*-value of < 0.05 was considered statistically significant. All statistical analyses were done using the statistical package for social science (SPSS) version 18.0 (SPSS Institute, Chicago, IL).

RESULTS

Demographic and Etiologic Characteristics for Subjects

There were no significant differences for sex, age, ethnicity, HBsAg and anti-HCV status ($P > 0.05$; Table 1), which suggested HCC patients' data are comparable with controls'.

AFB1-exposure Status and HCC Risk

Table 2 summarized AFB1-exposure information of all study population. We found that the HCC cases (48 years) had higher AFB1-exposure years than the controls (40 years) and the risk of HCC gradually increased with the increasing exposure years (adjusted OR was 3.30 to 9.89, $P < 0.01$). We also found that the levels of AFB1-DNA adducts were associated with increased risk of HCC (adjusted ORs were 2.02 and 6.65 for medium- and high-level adducts, respectively). These results were consistent with our previously published data [8, 10, 11].

XPC Codon 939 Polymorphism and HCC Risk

The genotypic distribution of XPC Lys939Gln for both cancer cases and controls were shown in Table 3. Genotypic distributions of this gene in controls were in Hardy-Weinberg equilibrium. The frequencies of the codon 939 Gln allele among cases (0.36) were higher than among controls (0.32). Logistic regression analyses showed that the adjusted OR for HCC for those individuals carrying the heterozygotes for codon 939 Lys and Gln allele (XPC-LG) compared with those exhibiting featuring the homozygote for Lys alleles (XPC-LL) was 1.24 (95% CI, 0.98 – 1.57), and the corresponding OR for those featuring the homozygote for Gln alleles (XPC-GG) was 1.35 (95% CI, 0.94 – 1.93), which showed the risk of HCC was associated with the number of codon 939 Gln alleles. However, there was not statistically significant relationship between HCC risk and this polymorphism.

Joint Effects of AFB1 Exposure and XPC Codon 939 Polymorphism on HCC Risk

To study the relationship between the polymorphism of XPC codon 751 and AFB1-exposure years in the risk for HCC, we analyzed the AFB1-exposure years-genotypes joint effect on HCC risk (Table 4). In this analysis, we used reference the lowest risk group: those who had the XPC-LL and short-term AFB1-exposure years. The results showed that increasing AFB1-exposure years consistently increased HCC risk, moreover, this effect was more pronounced among the XPC-LG or XPC-GG subjects.

Table 1. Demographic and etiologic characteristics of HCC cases and controls

	Controls	Cases	χ^2	Two-sided *P*
	No. (%)	No. (%)		
Sex			2.368	0.124
Male	531 (75.6)	416 (71.8)		
Female	171 (24.4)	163 (28.2)		
Age(years)[a]			1.167	0.558
<35	81 (11.5)	56 (9.7)		
35-65	551 (78.5)	456 (80.3)		
>65	70 (10.0)	58 (10.0)		
Ethnicity			1.081	0.298
Han	313 (44.6)	275 (47.5)		
Minority	389 (55.4)	304 (52.5)		
HBV infection			0.451	0.502
HBsAg (+)	501 (71.4)	423 (73.1)		
HBsAg (-)	201 (28.6)	156 (26.9)		
HCV infection			0.105	0.745
Anti-HCV (+)	126 (17.9)	108 (18.7)		
Anti-HCV (-)	576 (82.1)	471 (81.3)		

[a] The mean ± S.E. ages were 49.37 ± 0.47 and 49.23 ± 0.43 for cases and controls, respectively.

Table 2. AFB1 exposure and the risk of HCC

AFB1 exposure	Controls n	Controls %	HCCs n	HCCs %	OR (95% CI)	Adjusted OR (95% CI)	P_{trend}
AFB1-exposure years[a,b]							
Short	371	52.8	124	21.4	1[ref]	1[ref]	
Medium	195	27.8	181	31.3	2.78 (2.09-3.70)	3.30(2.37-4.58)[c]	<0.001
Long	136	19.4	274	47.3	6.03 (4.51-8.05)	9.89(6.84-14.31)[c]	<0.001
AFB1-DNA adducts levels[d,e]							
Low	374	53.3	139	24.0	1[ref]	1[ref]	
Medium	211	30.1	160	27.6	2.04(1.54-2.71)	2.02(1.49-2.74)[f]	<0.001
High	117	16.7	280	48.4	6.44(4.82-8.61)	6.65(4.84-9.14)[f]	<0.001

[a] The Median for AFB1-exposure years were 48 years and 40 years for cases and controls, respectively.

[b] AFB1-exposure years were: < 40 years for short exposure; 40 - 48 years for medium exposure; > 48 years for long exposure.

[c] Adjusted for age, sex, ethnicity, HBsAg, anti-HCV, and AFB1-DNA adducts levels.

[d] The mean ± S.E. level of AFB1-DNA adducts is 1.99 ± 0.05 and 1.03 ± 0.02 μmol/mol DNA for cases and controls, respectively.

[e] Adduct levels were: ≤ 1.00 μmol/mol DNA for low level; 1.01 – 2.00 μmol/mol DNA for medium; ≥ 2.01 μmol/mol DNA for high level.

[f] Adjusted for age, sex, ethnicity, HBsAg, anti-HCV, and AFB1-exposure years.

Table 3. The polymorphism of XPC and the risk of HCC

XPC genotype	Controls n	Controls %	HCCs n	HCCs %	OR (95% CI)	Adjusted OR (95% CI)[a]	P_{trend}
LL[b]	326	46.4	236	40.8	Reference	Reference	
LG[b]	301	42.9	270	46.6	1.18(0.92-1.51)	1.24(0.98-1.57)	0.073
GG[b]	75	10.7	73	12.6	1.35(0.93-1.97)	1.35(0.94-1.93)	0.110

[a] Adjusted for age, sex, ethnicity , HBsAg, anti-HCV, AFB1-exposure years, and AFB1-DNA adduct levels.

[b] LL, LG, and GG represented the homozygotes of XPC codon 939 Lys alleles, the heterozygotes of XPC codon 939 Lys and Gln allele, and the homozygotes of XPC codon 939 Gln alleles, respectively.

Table 4. Joint effects of AFB1-exposure years and XPC codon 939 polymorphism on HCC risk

AFB1-exposue years[a]	XPC genotypes	Controls n	Controls %	HCCs n	HCCs %	OR (95% CI)	Adjusted OR (95% CI)[b]
Short	LL	176	25.1	51	8.8	Reference	Reference
	LG	16	22.8	60	10.4	1.29(0.84-1.99)	1.03(0.64-1.64)
	GG	35	5.0	13	2.2	1.28(0.63-2.60)	1.45(0.68-3.09)
Medium	LL	83	11.8	71	12.3	2.95(1.89-4.60)	3.35(2.06-5.45)
	LG	85	12.1	85	14.7	3.45(2.24-5.32)	3.80 (2.35-6.13)
	GG	27	3.8	25	4.3	3.20(1.71-5.98)	2.97(1.50-5.91)
Long	LL	67	9.5	114	19.7	5.87(3.81-9.06)	8.80(5.35-14.46)
	LG	56	8.0	125	21.6	7.70(4.95-12.00)	10.90(6.54-18.19)
	GG	13	1.9	35	6.0	9.29(4.57-18.88)	19.10(8.63-42.27)

[a] AFB1-exposure years were: < 40 years for short exposure; 40 - 48 years for medium exposure; > 48 years for long exposure.

[b] Adjusted for age, sex, ethnicity, HBsAg, anti-HCV, and AFB1-DNA adducts levels.

Table 5. Joint effects of AFB1-DNA adducts levels and XPC codon 939 polymorphism on HCC risk

AFB1-DNA adducts levels[a,b]	XPD genotypes	Controls n	Controls %	HCCs n	HCCs %	OR (95% CI)	Adjusted OR (95% CI)[c]
Low	LL	172	24.5	61	10.5	Reference	Reference
	LG	155	22.1	59	10.2	1.07(0.71-1.63)	1.04(0.66-1.63)
	GG	47	6.7	19	3.3	1.14(0.62-2.09)	1.09(0.57-2.10)
Medium	LL	104	14.8	70	12.1	1.90(1.25-2.89)	1.86(1.18-2.94)
	LG	88	12.5	71	12.3	2.78(1.48-3.49)	2.23(1.41-3.55)
	GG	19	2.7	19	3.3	2.80(1.40-5.68)	2.43(1.15-5.14)
High	LL	50	7.1	105	18.1	5.92(3.79-9.25)	5.92(3.65-9.61)
	LG	58	8.3	140	24.2	6.81(4.46-10.39)	6.87(4.34-10.87)
	GG	9	1.3	35	6.0	10.97(4.98-24.13)	11.51(5.00-26.49)

[a] AFB1-DNA adducts levels were: ≤ 1.00 μmol/mol DNA for low; 1.01 – 2.00 fμmol/mol DNA for medium; ≥ 2.01 μmol/mol DNA for high.

[b] Adjusted for age, sex, ethnicity, HBsAg, anti-HCV, and AFB1-exposure years.

For example, individuals who featured long-term AFB1-exposure years and carried the XPC-LG had an adjusted OR of 10.90, whereas those who had the XPC-GG had an adjusted OR of 19.10 relevant to the reference group (Table 4). Interestingly, we found there was interactive effects between the genotypes of this loca and AFB1-exposure years on the risk of HCC ($19.10 > 1.45 \times 8.80$). Similar increasing-risk trend was also found in the joint effective analysis between XPC codon 939 genotypes and AFB1-DNA adducts levels for the risk of HCC ($11.51 > 1.09 \times 5.92$, Table 5).

DISCUSSION

To the best of our knowledge, no studies have investigated the role of DNA-repair gene XPC for patients suffering HCC, especially from AFB1 exposure areas. In this study, we analyzed the association between XPC Lys939Gln polymorphism and HCC risk among Guangxi population, from a high AFB1-exposure area. The results showed XPC codon 939 Gln alleles increased the risk of HCC in Guangxi population (adjusted ORs = 1.24 and 1.35 for XPC-LG and –GG, respectively), although there was not statistical significance. Interestingly, a gene-environment interactive effect was found for those populations who had XPC codon 939 Gln alleles, and experienced dissimilar AFB1-exposure years or levels. These results may suggest that there were interactions between the genetic polymorphism of XPC codon 939 and AFB1 exposure, and imply that this polymorphism may have functional significance in HCC related to AFB1.

HCC, the most common cause of which is AFB1 exposure via corn and groundnut consumption, main food consumption types for Guangxiese, is one of major cancer types in Guangxi Zhuang Autonomous Region, P.R. China [1, 2]. Our previous studies [6-11] and present study also showed HCC patients featured longer AFB1-exposure years and higher AFB1-exposure levels, moreover, we found the risk of HCC was associated with different AFB1-exposure status. AFB1 is an important chemical carcinogen, which is mainly metabolized by cytochrome P450 into the genotoxic metabolic AFB1-epoxide that can bind to DNA, causing the formation of AFB1-guanine adducts and inducing DNA damage including base damage and oxidative DNA damage that can be repair by NER pathway [3-5].

While XPC, a 940-amino acid protein encoded by XPC gene which spans 33kb on chromosome 3p25 and contains 16 exons and 15 introns (Genbank accession no. AC090645) [31], is an important DNA damage recognition molecule involved in global genome DNA repair pathway, NER [14-16]. XPC recognizes a variety of DNA adducts formed by exogenous carcinogens such as AFB1 [14-16], which may play a role in the pathogenesis of HCC. It binds tightly with HR23B to form a stable XPC-HR23B complex, the first protein component that recognizes and binds to the DNA damage sites [14-16]. Some recent studies have showed that defects in XPC have been related to many types of malignant tumors [27, 32]. For example, Takebayashi Y, *et al.* [27] analyzed the loss of heterozygosity of XPC in sporadic ovarian, colon and lung carcinomas. Their results showed the deficiency of XPC resulted in the carcinogenesis of human tumors. Transgenic mice studies also revealed predisposition to many types of tumors in XPC gene knockout mice [33]. Our data also implied the deficiency of XPC function promoted the carcinogenesis of HCC induced by

AFB1 exposure. These results might suggest that XPC gene plays an important role in the prevention of DNA damage-mediated malignant tumors occurrence.

More than a hundred polymorphisms in the XPC gene have been identified and more and more evidence has expressed that the polymorphisms of this gene are associated with the function of DNA repair capacity, which might modify the risk of malignant tumors such as lung cancer, breast cancer, esophagus cancer, skin cancer, oral cavity cancer, gastric cancer, and head and neck cancer [17, 18]. In present study, we only analyzed XPC Lys939Gln polymorphism because this polymorphism loci localize at conserved sites of this gene [31] and change the amino acids coded, which may be associated with decreased DNA repair capacity [18, 19, 21-26, 28, 29, 34], increased frequency of p53 mutations [35, 36], and increased tumor risk [17, 23]. Our results provided some evidence for the effects of XPC codon 939 polymorphism on HCC risk, although this increased risk of HCC had not statistical significance which was correlated with small samples.

In this study, we stratified the analysis of XPC codon 939 genotypes by AFB1-exposure. This was done primarily because previous several studies had provided evidence that there might be interactive effects between this polymorphism and carcinogens on cancer risk [25, 29]. For example, Mechanic LE, *et al.* [25] conducted a hospital-based case-control study (including 2311 cases and 2022 controls) to elucidate whether XPC codon 939 Gln alleles modify the risk breast cancer associated with smoking. They found some multiplicatively interactive effects between XPC polymorphism and smoking status on the risk of breast risk. While Zhou RM, *et al.* [29] also statistically significant interaction between this polymorphism and environmental carcinogens exposure on esophagus cancer among another Chinese population, Hebei population (OR=2.05, 95% CI=1.15-3.66). Our data not only supported aforementioned studies, but also we also found a positively modified effects of XPC codon 939 Gln alleles on HCC carcinogenesis induced by AFB1 exposure.

Those results suggest that this polymorphism may alter the normal protein function, and consequently may be associated with a reduction in DNA repair capacity. Thus, the DNA damage induced by AFB1 cannot be repaired effectively and duly, consequently may induce genic mutation such as p53, and hepatocellular canceration. Thus, XPC Lys939Gln polymorphism may play a role in the carcinogenetic pathway of Guangxiese HCC related to AFB1 exposure.

Overall, to the best of our knowledge, this is the first report to investigate association between XPC codon 939 polymorphisms and the risk of HCC for Guangxi population from an high AFB1-exposure area. We find evidence to suggest that the genotypes of XPC with codon939 Gln alleles may increase the risk of AFB1-related HCC and the NER pathway may play an important role in the mechanism of action of this genotoxin. However, there were several limitations to our study. Despite XPC codon 939 polymorphism analyzed, we did not analyze other polymorphisms of this gene [17, 18], possibly being able to modify the risk of AFB1-relevant to HCC. Our findings were based on relatively small numbers and limited by small number subjects in part of the genotype strata. Therefore, the more genes deserve further elucidation based on a large sample and the combination of genes and AFB1 exposure.

ACKNOWLEDGMENTS

We thank Dr. Yun Ma, Dr. Zhou-Lin Deng, Dr. Yi-Ping Wei, and Dr. Ni-Bo Wei for DNA sample extraction and management; Dr. Feng Liang, and Dr. Qun Qing Yu for molecular biochemical technique. We also thank all members of Department of Medical Test and Infective Control, Affiliated Hospital of Youjiang Medical College for Nationalities for their help.

This study is supported by the Youth Science Foundation of Guangxi (NO. 0833097), and the Science Foundation of Youjiang Medical College for Nationalities (NO. 2005 and 2008).

REFERENCES

[1] Wang JS, Huang T, Su J, Liang F, Wei Z, Liang Y, Luo H, Kuang SY, Qian GS, Sun G *et al*: Hepatocellular carcinoma and aflatoxin exposure in Zhuqing Village, Fusui County, People's Republic of China. *Cancer Epidemiol. Biomarkers Prev.* 2001, 10(2):143-146.

[2] Yeh FS, Mo CC, Yen RC: Risk factors for hepatocellular carcinoma in Guangxi, People's Republic of China. *National Cancer Institute monograph.* 1985, 69:47-48.

[3] Wang JS, Groopman JD: DNA damage by mycotoxins. *Mutation research* 1999, 424(1-2):167-181.

[4] Wood RD: DNA damage recognition during nucleotide excision repair in mammalian cells. *Biochimie* 1999, 81(1-2):39-44.

[5] Wilson DM, 3rd, Thompson LH: Life without DNA repair. *Proceedings of the National Academy of Sciences of the United States of America* 1997, 94(24):12754-12757.

[6] Long XD, Ma Y, Huang HD, Yao JG, Qu de Y, Lu YL: Polymorphism of XRCC1 and the frequency of mutation in codon 249 of the p53 gene in hepatocellular carcinoma among Guangxi population, China. *Mol. Carcinog.* 2008, 47(4):295-300.

[7] Long XD, Ma Y, Wei YP, Deng ZL: The polymorphisms of GSTM1, GSTT1, HYL1*2, and XRCC1, and aflatoxin B1-related hepatocellular carcinoma in Guangxi population, China. *Hepatol. Res.* 2006, 36(1):48-55.

[8] Long XD, Ma Y, Deng ZL: GSTM1 and XRCC3 polymorphisms: Effects on levels of aflatoxin B1-DNA adducts. *Chinese J. Cancer Res.* 2009, 21(3):177-184.

[9] Long XD, Ma Y, Deng ZL, Huang YZ, Wei NB: [Association of the Thr241Met polymorphism of DNA repair gene XRCC3 with genetic susceptibility to AFB1-related hepatocellular carcinoma in Guangxi population]. *Zhonghua yi xue yi chuan xue za zhi = Zhonghua yixue yichuanxue zazhi = Chinese journal of medical genetics* 2008, 25(3):268-271.

[10] Long XD, Ma Y, Qu de Y, Liu YG, Huang ZQ, Huang YZ, Lin ZH, Wei NB, Zhou SC: The polymorphism of XRCC3 codon 241 and AFB1-related hepatocellular carcinoma in Guangxi population, China. *Ann. Epidemiol.* 2008, 18(7):572-578.

[11] Long XD, Ma Y, Zhou YF, Yao JG, Ban FZ, Huang YZ, Huang BC: XPD Codon 312 and 751 Polymorphisms, and AFB1 Exposure, and Hepatocellular Carcinoma Risk. *BMC cancer* 2009, 9(1):400.

[12] Thompson LH, West MG: XRCC1 keeps DNA from getting stranded. *Mutation research* 2000, 459(1):1-18.

[13] Griffin CS, Simpson PJ, Wilson CR, Thacker J: Mammalian recombination-repair genes XRCC2 and XRCC3 promote correct chromosome segregation. *Nat. Cell Biol.* 2000, 2(10):757-761.

[14] Araujo SJ, Wood RD: Protein complexes in nucleotide excision repair. *Mutation research* 1999, 435(1):23-33.

[15] Batty DP, Wood RD: Damage recognition in nucleotide excision repair of DNA. *Gene* 2000, 241(2):193-204.

[16] Sugasawa K: XPC: its product and biological roles. *Advances in experimental medicine and biology* 2008, 637:47-56.

[17] Qiu L, Wang Z, Shi X, Wang Z: Associations between XPC polymorphisms and risk of cancers: A meta-analysis. *Eur. J. Cancer* 2008, 44(15):2241-2253.

[18] Zhang D, Chen C, Fu X, Gu S, Mao Y, Xie Y, Huang Y, Li Y: A meta-analysis of DNA repair gene XPC polymorphisms and cancer risk. *Journal of human genetics* 2008, 53(1):18-33.

[19] Liang J, Gu A, Xia Y, Wu B, Lu N, Wang W, Lu C, Zheng Q, Wang S, Wang X: XPC gene polymorphisms and risk of idiopathic azoospermia or oligozoospermia in a Chinese population. *International journal of andrology* 2009, 32(3):235-241.

[20] Caronia D, Patino-Garcia A, Milne RL, Zalacain-Diez M, Pita G, Alonso MR, Moreno LT, Sierrasesumaga-Ariznabarreta L, Benitez J, Gonzalez-Neira A: Common variations in ERCC2 are associated with response to cisplatin chemotherapy and clinical outcome in osteosarcoma patients. *The pharmacogenomics journal* 2009, 9(5):347-353.

[21] De Ruyck K, Szaumkessel M, De Rudder I, Dehoorne A, Vral A, Claes K, Velghe A, Van Meerbeeck J, Thierens H: Polymorphisms in base-excision repair and nucleotide-excision repair genes in relation to lung cancer risk. *Mutation research* 2007, 631(2):101-110.

[22] Dong Z, Guo W, Zhou R, Wan L, Li Y, Wang N, Kuang G, Wang S: Polymorphisms of the DNA repair gene XPA and XPC and its correlation with gastric cardiac adenocarcinoma in a high incidence population in North China. *Journal of clinical gastroenterology* 2008, 42(8):910-915.

[23] Francisco G, Menezes PR, Eluf-Neto J, Chammas R: XPC polymorphisms play a role in tissue-specific carcinogenesis: a meta-analysis. *Eur. J. Hum. Genet.* 2008, 16(6):724-734.

[24] Hansen RD, Sorensen M, Tjonneland A, Overvad K, Wallin H, Raaschou-Nielsen O, Vogel U: XPA A23G, XPC Lys939Gln, XPD Lys751Gln and XPD Asp312Asn polymorphisms, interactions with smoking, alcohol and dietary factors, and risk of colorectal cancer. *Mutation research* 2007, 619(1-2):68-80.

[25] Mechanic LE, Millikan RC, Player J, de Cotret AR, Winkel S, Worley K, Heard K, Tse CK, Keku T: Polymorphisms in nucleotide excision repair genes, smoking and breast cancer in African Americans and whites: a population-based case-control study. *Carcinogenesis* 2006, 27(7):1377-1385.

[26] Shore RE, Zeleniuch-Jacquotte A, Currie D, Mohrenweiser H, Afanasyeva Y, Koenig KL, Arslan AA, Toniolo P, Wirgin I: Polymorphisms in XPC and ERCC2 genes, smoking and breast cancer risk. *International journal of cancer* 2008, 122(9):2101-2105.

[27] Takebayashi Y, Nakayama K, Kanzaki A, Miyashita H, Ogura O, Mori S, Mutoh M, Miyazaki K, Fukumoto M, Pommier Y: Loss of heterozygosity of nucleotide excision repair factors in sporadic ovarian, colon and lung carcinomas: implication for their roles of carcinogenesis in human solid tumors. *Cancer letters* 2001, 174(2):115-125.

[28] Zhang L, Zhang Z, Yan W: Single nucleotide polymorphisms for DNA repair genes in breast cancer patients. *Clinica chimica acta; international journal of clinical chemistry* 2005, 359(1-2):150-155.

[29] Zhou RM, Li Y, Wang N, Zhang XJ, Dong XJ, Guo W: [Correlation of XPC Ala499Val and Lys939Gln polymorphisms to risks of esophageal squamous cell carcinoma and gastric cardiac adenocarcinoma]. *Ai zheng = Aizheng = Chinese journal of cancer* 2006, 25(9):1113-1119.

[30] Brennan P: Gene-environment interaction and aetiology of cancer: what does it mean and how can we measure it? *Carcinogenesis* 2002, 23(3):381-387.

[31] Li L, Peterson C, Legerski R: Sequence of the mouse XPC cDNA and genomic structure of the human XPC gene. *Nucleic Acids Res.* 1996, 24(6):1026-1028.

[32] Miyashita H, Mori S, Tanda N, Nakayama K, Kanzaki A, Sato A, Morikawa H, Motegi K, Takebayashi Y, Fukumoto M: Loss of heterozygosity of nucleotide excision repair factors in sporadic oral squamous cell carcinoma using microdissected tissue. *Oncology reports* 2001, 8(5):1133-1138.

[33] Cheo DL, Burns DK, Meira LB, Houle JF, Friedberg EC: Mutational inactivation of the xeroderma pigmentosum group C gene confers predisposition to 2-acetylaminofluorene-induced liver and lung cancer and to spontaneous testicular cancer in Trp53-/- mice. *Cancer research* 1999, 59(4):771-775.

[34] Laczmanska I, Gil J, Karpinski P, Stembalska A, Trusewicz A, Pesz K, Ramsey D, Schlade-Bartusiak K, Blin N, Sasiadek MM: Polymorphism in nucleotide excision repair gene XPC correlates with bleomycin-induced chromosomal aberrations. *Environmental and molecular mutagenesis* 2007, 48(8):666-671.

[35] Ryk C, Kumar R, Sanyal S, de Verdier PJ, Hemminki K, Larsson P, Steineck G, Hou SM: Influence of polymorphism in DNA repair and defence genes on p53 mutations in bladder tumours. *Cancer letters* 2006, 241(1):142-149.

[36] Sakano S, Matsumoto H, Yamamoto Y, Kawai Y, Eguchi S, Ohmi C, Matsuyama H, Naito K: Association between DNA repair gene polymorphisms and p53 alterations in Japanese patients with muscle-invasive bladder cancer. *Pathobiology* 2006, 73(6):295-303.

In: Encyclopedia of DNA Research (3 Volume Set)
Editors: S. J. Duncan and P. H. Wiley

ISBN 978-1-61324-305-3

Chapter 26

INFLUENCE OF AGE ON EXPRESSION OF DNA REPAIR GENES IN ARABIDOPSIS

***Megan J. Osmond*[1,2] *and Bernard A. Kunz*[1]**
[1] School of Life and Environmental Sciences, Deakin University, Geelong, Victoria 3217, Australia
[2] CSIRO Food and Nutritional Sciences, North Ryde, New South Wales, 2112, Australia

ABSTRACT

Numerous studies suggest that ageing in mammals may be associated with a reduction in DNA repair, whereas little is known about the DNA repair capacity of plants as they age. In this study we examined the effects of ageing on the expression of genes thought to be involved in nucleotide excision repair (*AtERCC1*, *AtGTF2H2*, *AtGTF2H5*, *AtXPB1*, *AtXPD*, *AtXPF*) or translesion replication (*AtPOLH*, *AtREV1*, *AtREV3*, *AtUBC2*) of UV photoproducts in Arabidopsis (*Arabidopsis thaliana*). Two- or four-week old plants were unirradiated or treated with 254 nm ultraviolet (UV) radiation (0.75 or 1.5 kJm^{-2}), incubated for 3 or 9 hr, and gene expression was analysed via quantitative PCR. With the exception of *AtPOLH*, transcript levels for all genes investigated were at least four-fold greater in unirradiated four-week old plants than unirradiated two-week old plants. Compared to unirradiated age-matched plants, two-week old plants generally showed no consistent change in transcript levels for either UV dose or post-irradiation incubation period. On the other hand, transcript levels in four-week old plants were increased over those in two-week old plants for the majority of genes by 9 hr post-irradiation with 0.75 or 1.5 kJm^{-2} UV. No other consistent responses were observed for UV treatment. Collectively, our results are consistent with the possibility that ageing may be associated with increased DNA repair capacity in plants.

INTRODUCTION

Nearly all eukaryotic organisms age (Lombard et al., 2005), and an accumulation of DNA damage and loss of genomic stability are associated with the ageing process in mammals

(Vijg, 2000; Karanjawala and Lieber, 2004; Collado et al., 2007; Rass et al., 2007). To eliminate or cope with DNA lesions organisms have evolved mechanisms for damage reversal, repair or tolerance, depending on the type of damage present (Friedberg et al., 2006). These processes are not always performed with complete fidelity, however, and an accruement of DNA mutations resulting from imprecise repair has been linked to a number of age-related pathologies (Karanjawala and Lieber, 2004). There is also suggestive evidence that a decline in repair efficiency may contribute to a cumulative increase in DNA damage with age (Cabelof, 2007; Gorbunova et al., 2007). A decrease in DNA damage processing capacity could be due to reduced expression of DNA repair genes, decreased stability and activity of DNA repair proteins, or a combination of these factors (Cabelof, 2007). For example, ultraviolet (UV) irradiated fibroblasts from different aged donors exhibit age-related decreases in DNA photoproduct removal as well as the levels of transcripts and proteins encoded by the nucleotide excision repair (NER) genes *XPA* and *XPB* (Goukassian et al., 2000). However, the extent to which repair diminishes with age in mammals is not yet clear, and similar phenomena have yet to be investigated in plants.

In plants senescence occurs in leaf tissue after a given photosynthetic lifespan. Leaf senescence involves the catabolism of chlorophyll, proteins, lipids and polynucleotides, and is the final stage of leaf development (Quirino et al., 2000). Nutrients and organic molecules are subsequently incorporated into growing areas of the plant, such as buds, fruits and seeds. Chloroplasts are among the earliest organelles to undergo catabolism, leading to the yellowing of leaf tissue with age, similar to that caused by UV radiation, whereas the nucleus and mitochondria are the last (Quirino et al., 2000; Smart, 2005). Total protein and mRNA content of a cell decreases as it enters the latter stages of senescence, and while some RNA synthesis occurs, it does so at a lower rate (Quirino et al., 2000). It is not surprising then that many plant genes undergo decreased expression during senescence (Casati and Walbot, 2004). Yet certain genes are up-regulated in a senescence-specific manner (Smart, 2005). For example, defence-related genes are expressed during senescence, and older plants can show an enhanced resistance to pathogens (Quirino et al., 1999; Kus et al., 2002). Recently, we demonstrated links between pathogen resistance and DNA repair capacity (Kunz et al., 2008), but whether expression of DNA damage processing genes in plants is also affected by ageing is not known. However, the fact that plants form reproductive tissue in the later stages of development suggests there might be a continuing or increased need for DNA repair in older plants. In this chapter, we examine the effects of age alone and in combination with a DNA damage inducer (UV radiation) on transcript levels for a number of genes required for NER or tolerance of UV photoproducts in the model plant Arabidopsis (*Arabidopsis thaliana*).

Repair Gene Transcript Levels in Two-Week Versus Four-Week Old Plants

We first examined the influence of plant age on expression of genes suggested to be involved in NER (*AtERCC1*, *AtGTF2H2*, *AtGTF2H5*, *AtXPB1*, *AtXPD*, *AtXPF*) or translesion replication (*AtPOLH*, *AtREV1*, *AtREV3*, *AtUBC2*) of UV photoproducts on the basis of functional complementation data or sequence homology (Kunz et al., 2005, 2008; Grice et al., 2007; Kunz and Xiao, 2007). Transcripts representing these genes were quantitatively amplified using RNA samples isolated from two- or four-week old plants. These ages were selected because one- to two-week old seedlings are most often used for investigating DNA repair in *Arabidopsis* (eg, Quaite et al. 1994b; Fidantsef et al. 2000; Tanaka et al. 2002). Furthermore, since this plant flowers at four weeks, expression of repair genes might change in somatic cells at this time, as suggested by the reduced cyclobutane pyrimidine dimer photolyase activity and transcript levels in the stem and leaves of four-week old plants (Pang and Hays 1991; Ahmad et al. 1997).

Transcript levels for the ten genes varied considerably among the two-week and four-week old plants, but they were consistently higher in the older plants (Figure 1). Although the transcript level for *AtPOLH*, which encodes a translesion synthesis polymerase (Kunz et al., 2008), was three-fold greater in the four-week old plants the difference was not significant. Analysis of the data involved normalisation of the gene transcript levels to the 18S rRNA levels present in the same samples. The immediate concern would be that if the 18S gene was differentially expressed at the two ages, the rRNA levels might be lower in the four-week old plants thereby falsely normalising the transcript levels for the repair genes to higher than their true values. However, there was no significant difference between the 18S rRNA transcript levels in two- and four-week old plants under any conditions used in this study. Thus, the normalisation procedure did not artificially elevate the levels of transcripts calculated for damage processing genes in the four-week old plants.

The apparent age-specific increase in transcript levels for almost all of the NER and damage tolerance genes examined here contrasts with mammalian studies that point to an age-related decrease in DNA repair capacity (Venema et al., 1990; Goukassian et al., 2000; Takahashi et al., 2004; Cabelof, 2007). At four-weeks of age Arabidopsis begins to flower and produce seeds. Although cells in the mature leaf tissue are no longer dividing, it is important that the plant overall survives through this phase, having invested so much in seed production. Unrepaired transcription-blocking DNA damage in the mature cells might still be detrimental if, for example, the photosynthetic capacity previously provided by now dead or critically damaged cells was not recovered. Furthermore, previous reports indicate that older plants can show elevated levels of transcripts for stress- or defence-related genes and increased resistance to pathogens (Quirino et al., 1999; Kus et al., 2002; Casati and Walbot, 2004).

Effects of UV-C Plus Age on Repair Gene Transcript Levels

We next determined whether UV radiation can influence the age-related difference in the expression of genes whose products act to counter UV-induced DNA damage.

Biologically relevant doses of UV-B (280-320 nm) and sunlight induce cyclobutane pyrimidine dimers in seedling leaves (Quaite et al., 1992b). These photoproducts are also induced by 254 nm UV (Jenkins et al., 1995; Jiang et al., 1997; Gallego et al., 2000), but this wavelength is close to the DNA absorption peak of 260 nm and is not absorbed well by proteins. Thus, 254 nm UV induces DNA-specific damage more efficiently than UV-B (Mitchell et al., 1991; Quaite et al., 1992a; Perdiz et al., 2000; Birrell et al., 2001). So to more conveniently produce UV photoproducts we used a lamp emitting primarily 254 nm radiation (henceforth designated UV).

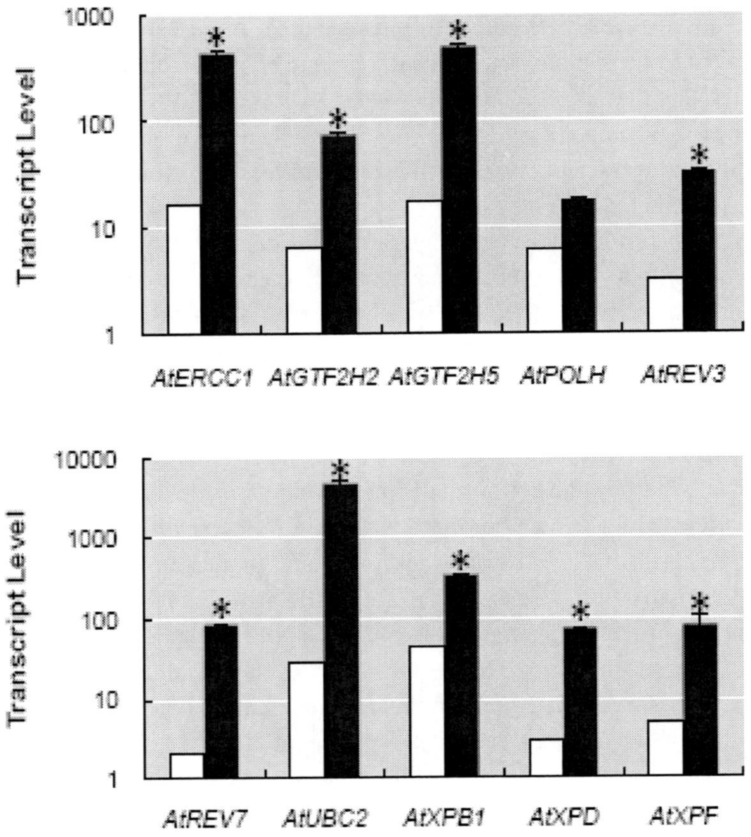

Figure 1. Effect of age on transcript levels. *A. thaliana* Landsberg *erecta* was grown at 21°C and 73% relative humidity under continuous light using eight 60 W cool-white fluorescent lamps and four tungsten 60 W bulbs emitting a total average light intensity of 209 μmol m^{-2} s^{-1}. Other procedures were performed as described previously (Vonarx et al., 2006). Leaf tissue was harvested from two-week old or four-week old plants and frozen in liquid nitrogen. RNA was isolated, assessed for concentration, purity and integrity, and treated with RNase-free DNase I. Repair gene and 18S rRNA cDNA fragments were quantitatively amplified from reverse-transcribed RNA using primers shown in Table 1. Amplicon lengths and identities were confirmed by non-quantitative PCR, size-evaluation and sequencing. Appropriate linearity tests were performed to ensure that differences in the threshold cycle values for the cDNAs remained constant despite variation in the amount of starting material. Primer concentrations used gave the lowest threshold cycle values and no amplification in no-template controls. Transcript levels are the means of three independent replicates, each normalised to the level of 18S rRNA in the same RNA preparation. Statistical significance ($p<0.05$) of differences between transcript levels for two-week old (□) and four-week old plants (■) was assessed using a two-tailed unpaired t-test. Significant differences are denoted by asterisks above the columns for the four-week old plants.

It was then necessary to select a UV dose that would not critically injure the plants but could influence gene expression. Previously, 0.6 kJm^{-2} UV was shown to modify gene expression in the Columbia ecotype by six hr post-irradiation (Molinier et al., 2005), and is in the range that causes non-lethal yellowing or curling of leaves within ten days of UV treatment (Jenkins et al., 1995). The Landsberg *erecta* ecotype we used is more resistant to UV than Columbia (Curtis and Hays, 2007).

Table 1. Primers used in this study

Gene	Locus	Amplicon Length (bp)	Forward Primer (5′ → 3′)	Reverse Primer (5′ → 3′)
AtERCC1	At3g05210	127	AACGCAAGAAGAAGGAGCC	CGCCGTTGTTGTATCTTTCT
AtGTF2H2	At1g05055	109	TTGGGTGACTCCTCTTTGC	GTGCATAAAGCTGAGTACAAT
AtGTF2H5	At1g12400	145	GCAATCAAAGGAGTGTTCGT	ATCTGCTCTACATGTGGTTG
AtPOLH	At5g44740	120	ATGCCTGAAGTAGACACCG	ACATCCGTTGCTTCGTTTGA
AtREV3	At1g67500	144	ACACCTAGCCACGATATGC	AGTAGCAATGGAGGAGAGG
AtREV7	At1g16590	133	GATTCCAACAGACACAAAGC	GGTTGACAAGGTTGAGTGG
AtUBC2	At2g02760	139	CCTCCATTCAGTCATTGCTA	CTAGTCGGCAGTCCAGCT
AtXPB1	At5g41370	118	ATGATTAAACCAGTGATACCC	CTGAGAAGAACAAAGAAGGAA
AtXPD	At1g03190	143	GCAAGATATTACGAGCAAGAT	TCATCATCCCATAATCAGCC
AtXPF	At5g41150	140	GGTCTCGAAGTCTACATGC	AGTTCTCGGCTCTGATGTC
At18S rRNA	X16077	123	AGTATGGTCGCAAGGCTGA	TGTCAATCCTTACTATGTCTG

However, we inferred from the Columbia data that a UV dose which provoked a similar phenotypic response in Landsberg *erecta* would be sufficient to affect gene expression. The lowest dose tested at which nearly all irradiated plants showed modest yellowing or curling of leaves under non-photoreactivable conditions (incubation under gold light which lacks wavelengths necessary for photoreactivation) was 1.5 kJm^{-2} (Figure 2, Table 2). A dose of 0.75 kJm^{-2} was used as intermediate between 0 and 1.5 kJm^{-2}. Neither UV dose had any phenotypic effect on the plants within the first 24 hr of post-irradiation incubation.

Plants were UV-irradiated, incubated for three or nine hr under gold light, and RNA samples were isolated for quantitative amplification of transcripts. Regardless of the UV dose or post-irradiation incubation period, there was no consistent increase or decrease in the transcript levels in two-week old plants compared to unirradiated age- and incubation period-matched controls (Figure 3). Indeed, for the most part transcript levels remained close to those for the unirradiated controls.

There also was no consistent response among the four-week old plants incubated for three hours following irradiation. In this case, however, the transcript levels for four of the genes (*AtREV3*, *AtUBC2*, *AtXPB1*, *AtXPD*) were significantly higher at one or both UV doses in the four-week old than the two-week old plants, and substantially greater than the unirradiated age-matched control values. After incubation for nine hr, transcript levels at one or both UV doses were significantly higher for nine of the ten genes in the four-week old plants compared to the two-week old plants.

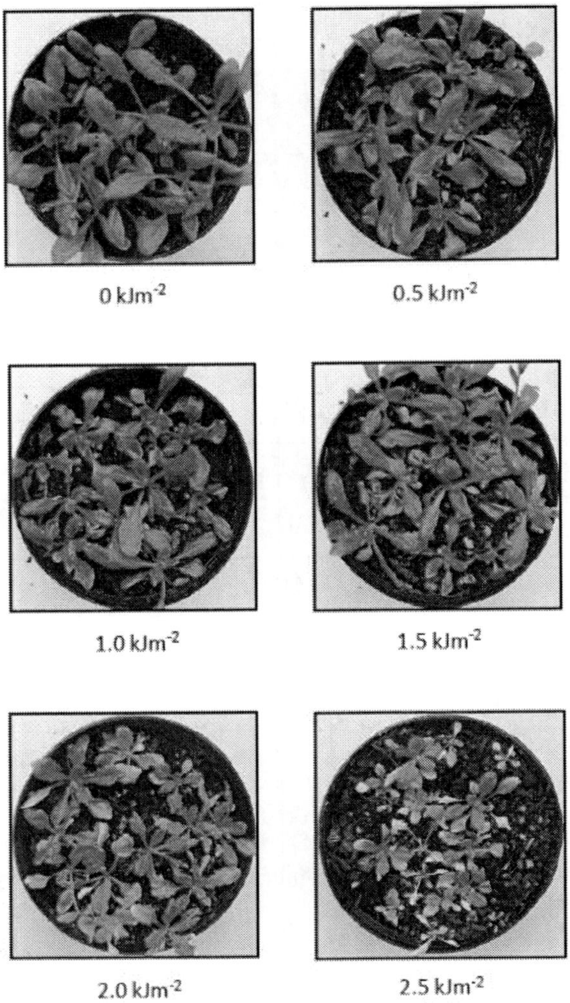

Figure 2. Phenotypic response of Arabidopsis to increasing UV doses. *A. thaliana* Landsberg *erecta* was grown as in the legend to Figure 1. Fourteen day-old plants in pots were exposed to 254 nm UV radiation. The UV source was a 60 cm germicidal tube (model G36T6L, Davis Ultraviolet) emitting 94% of its radiant energy at 254 nm and set to an incident dose rate of 6.36 $Jm^{-2}s^{-1}$ using a UVX Digital Radiometer fitted with a UVX-25 sensor. Following irradiation the pots were immediately transferred to a growth chamber and incubated under gold light with conditions otherwise the same as those used for routine growth. Plants were photographed after five days incubation.

The transcript levels also were greater than the unirradiated control values for eight of those nine genes. These results suggest that UV treatment may stimulate expression in the older plants of the majority of damage processing genes examined, and influence the age-related increase in transcript levels in a gene- and post-irradiation incubation period-dependent manner.

Table 2. UV-induced leaf yellowing/curling

Dose (kJm^{-2})	Number of Pots	Total Number of Plants Inspected	Affected Plants Per Pot (Mean % ± SEM)	Significance of Difference Compared to Control
0	13	91	34 ± 8.6	$p > 0.05$
0.5	12	85	41 ± 8.9	$p < 0.001$
1	13	84	85 ± 3.9	$p < 0.0001$
1.5	11	82	97 ± 1.2	$p < 0.0001$
2	10	56	100 ± 0	$p < 0.0001$
2.5	10	63	100 ± 0	$p < 0.0001$

SEM: standard error of the mean. Differences in the mean percentage of affected UV irradiated plants compared to the unirradiated control were analysed using a Students *t*-Test with the degrees of freedom = $(N_1-1) + (N_2-1)$ where N_1 = the number of unirradiated pots and N_2 = the number of irradiated pots for each comparison.

CONCLUSIONS

DNA repair capacity has been found to decline with age in mammalian studies. Thus, it seemed reasonable to think that transcripts encoding damage processing proteins in Arabidopsis, which has a six-week life cycle, might be less abundant in four-week old plants than in two-week old plants. Especially if endoreduplication, where chromosomal DNA is replicated in the absence of cell division, attenuates the impact of DNA damage in mature plant cells by providing multiple copies of each gene (Edgar and Orr-Weaver, 2001). Instead, transcript levels for the genes examined here were almost uniformly higher in the four-week old than two-week old plants. It is possible that because four-week old Arabidopsis is in the process of flowering, it might be at the most critical phase of its life-cycle. This may explain the up-regulation of defence-related genes and enhanced pathogen resistance in older plants (Quirino et al., 1999; Kus et al., 2002). A similar rationale might account for the higher levels of DNA repair gene transcripts in the four-week old plants. It also would be consistent with the hypothesis that damage processing mechanisms tend to decline in mammals after the peak reproductive years (Kirkwood and Holliday, 1979; Lombard et al., 2005; Rattan, 2005). However, we did not measure repair and so cannot say whether the increased transcript levels in the four-week old plants were associated with enhanced repair capacity. Furthermore, we did not examine mRNA stabilities and translation rates, or the locations, stabilities and activities of the proteins encoded by the genes we studied. Consequently, if the greater transcript levels do contribute to an increase in repair as plants age, post-transcriptional changes in gene expression as well as protein redistribution also may play important roles.

Following UV treatment transcript levels in the two-week old plants remained at about the levels measured for the corresponding unirradiated controls. Clearly, the genes examined did not display any coordinated response to UV in the younger plants under the conditions we used.

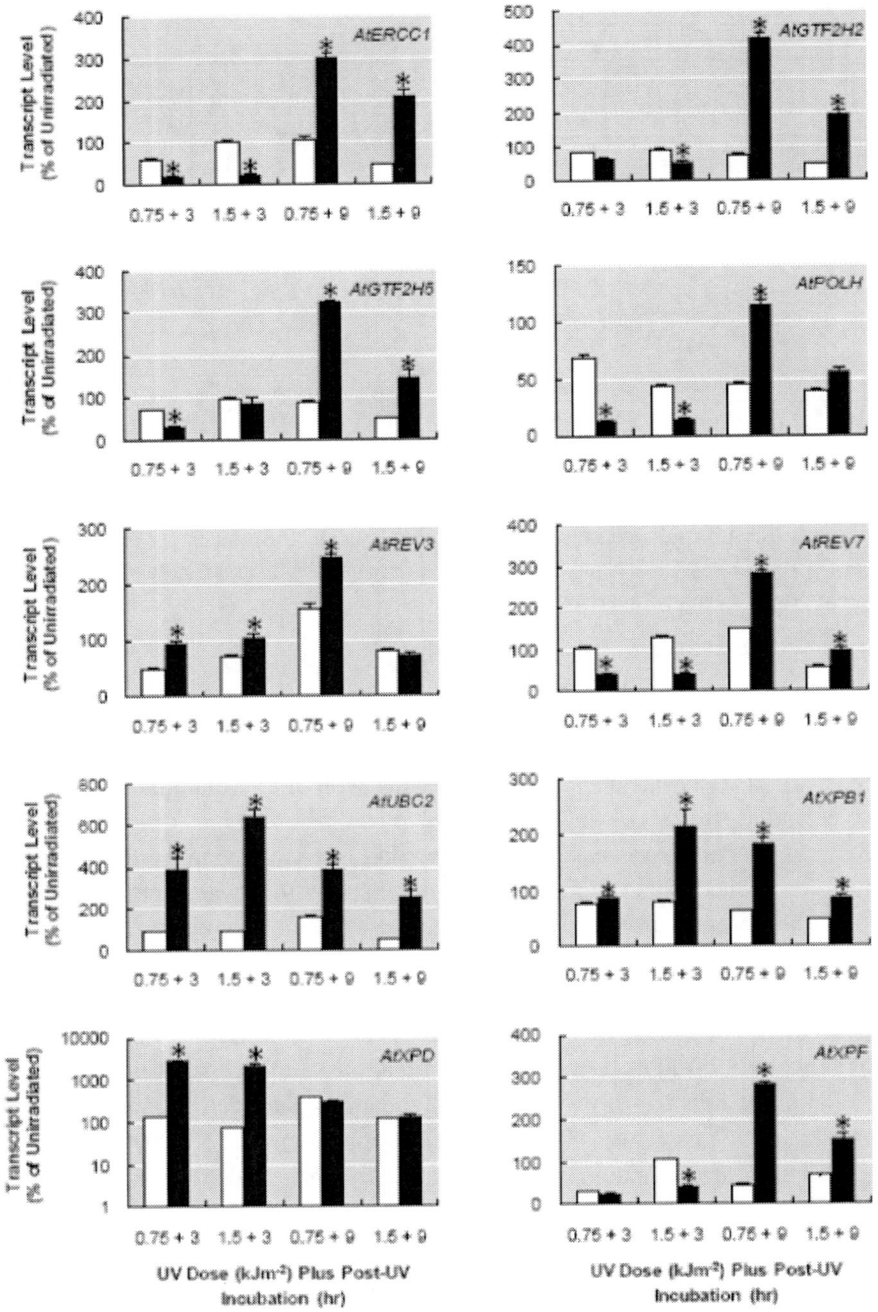

Figure 3. Transcript levels in plants UV-irradiated at two or four weeks of age. *A. thaliana* Landsberg *erecta* was grown as in the legend to Figure 1. Plants were UV-irradiated and incubated as described in the legend to Figure 2. Following incubation for three or nine hr leaf tissue was harvested and transcript quantification was carried out as described in the legend to Figure 1. Transcript levels are the means of three independent replicates, each normalised to the level of 18S rRNA in the same RNA preparation. Transcript levels in treated plants are expressed as the percentages of those in age-matched untreated plants that had been incubated under gold light for 3 hr or 9 hr alongside the irradiated plants. Differences in transcript levels were assessed for statistical significance ($p<0.05$) by two-way ANOVA with Bonferroni post-tests.

In contrast, by nine hr after UV irradiation most of the transcripts examined here were present at higher levels in the older plants than observed for the treatment-matched, two-week old plants or the unirradiated controls.

Thus, it appeared that UV radiation might be able to potentiate the age-related increase in the transcript levels. Given their daily exposure to large doses of incident solar UV wavelengths, it would not be surprising if some genes required for processing UV photoproducts were UV-inducible in plants. As mentioned above, however, we do not know whether repair actually is more effective in the older plants. Furthermore, our results must be interpreted cautiously. This study involved incubating irradiated plants under gold light to minimise the contribution of photoreactivation to DNA damage repair. Consequently, the NER and damage tolerance genes examined may have been expressed to a greater extent in the four-week old plants than they would have been had photoreactivation been available to reduce the damage load. Still, it remains possible that enhanced processing of UV-induced DNA damage in leaf cells might be important at this stage in the plant's life cycle, particularly if levels of UV-absorbing flavonoid pigments decline with age in Arabidopsis as they do in other members of the *Brassicaceae* (Reifenrath and Muller, 2007).

REFERENCES

Ahmad, M., Jarillo, J. A., Klimczak, L. J., Landry, L. G., Peng, T., Last, R. L. and Cashmore, A. R. (1997). An enzyme similar to animal type II photolyases mediates photoreactivation in *Arabidopsis*. *Plant Cell.* 9. 199-207.

Birrell, G. W., Brown, J. A., Wu, H. I., Giaever, G., Chu, A. M., Davis, R. W. and Brown, J. M. (2002). Transcriptional response of *Saccharomyces cerevisiae* to DNA-damaging agents does not identify the genes that protect against these agents. *Proc. Natl. Acad. Sci. USA* 99. 8778-8783.

Cabelof, D. C. (2007). Aging and base excision repair: In need of a comprehensive approach. DNA Repair. 6. 1399–1402.

Casati, P. and Walbot, V. (2004). Rapid transcriptome responses of maize (*Zea mays*) to UV-B in irradiated and shielded tissues. *Genome Biol.* 5. R16.1-R16.17.

Collado, M., Blasco, M. and Serrano, M. (2007). Cellular senescence in cancer and aging. *Cell.* 130. 223-233.

Curtis, M. J. and Hays, J. B. (2007). Tolerance of dividing cells to replication stress in UVB-irradiated Arabidopsis roots: requirement for DNA translesion polymerases eta and zeta. *DNA Repair*. 6. 1341-58.

Edgar, B. A. and Orr-Weaver, T. L. (2001). Endoreplication cell cycles: more for less. *Cell.* 105. 297-306.

Fidantsef, A. L., Mitchell, D. L. and Britt, A. B. (2000). The *Arabidopsis UVH1* gene is a homolog of the yeast repair endonuclease *RAD1*. *Plant Physiol.* 124. 579-586.

Friedberg, E. C., Walker, G. C., Siede, W., Wood, R. D., Schultz, R. A. and Ellenberger, T. (2006). DNA Repair and Mutagenesis. Washington, DC: American Society for Microbiology.

Gallego, F., Fleck, O., Wyrzykowska, J. and Tinland, B. (2000). AtRAD1, a plant homologue of human and yeast nucleotide excision repair endonucleases, is involved in dark repair of UV damage and recombination. *The Plant Journal*. 21. 507-518.

Gorbunova, V., Seluanov, A., Mao, Z. and Hine, C. (2007). Changes in DNA repair during aging. *Nucleic Acids Res*. 35. 7466-7474.

Goukassian, D., Gad, F., Yaar, M., Eller, M. S., Nehal, U. S. and Gilchrest, B. A. (2000). Mechanisms and implications of the age-associated decrease in DNA repair capacity. *J. Fed. Am. Soc. Exptl. Biol*. 14. 1325-1334.

Grice D. M., Osmond M. J., Stewart M., Vonarx E. J. and Kunz B. A. (2007) Roles of plant transcription factor IIH subunit homologs in nucleotide excision repair and transcription. In B.R. Landseer (Ed.), New Research on DNA Repair (pp. 251-272). New York, New York: Nova Science Publishers.

Jenkins, M. E., Harlow, G. R., Liu, Z., Shotwell, M. A., Ma, J. and Mount, D. W. (1995). Radiation-sensitive mutants of *Arabidopsis thaliana. Genetics*. 140. 725-732.

Jiang, C. Z., Yen, C.-N., Cronin, K., Mitchell, D. and Britt, A. B. (1997). UV- and gamma-radiation sensitive mutants of *Arabidopsis thaliana. Genetics*. 147. 1401-1409.

Karanjawala, Z. E. and Lieber, M. R. (2004). DNA damage and aging. *Mech. Ageing Dev*. 125. 405-416.

Kirkwood, T. B. and Holliday, R. (1979). The evolution of ageing and longevity. Proc. Royal Soc. Biol. Sci. 205. 531-546.

Kunz, B. A., Anderson, H. J., Osmond, M. J. and Vonarx, E. J. (2005). Components of nucleotide excision repair and DNA damage tolerance in *Arabidopsis thaliana*. Environ. Mol. Mutagenesis. 45. 115-127.

Kunz B. A., Dando, P. K., Grice, D. M., Mohr, P. G., Schenck, P. M. and Cahill D. M. (2008). UV-induced DNA damage promotes resistance to the biotrophic pathogen *Hyaloperonospora parasitica* in Arabidopsis. *Plant Physiol*. 148. 1021-1031.

Kunz, B. A. and Xiao, W. (2007). DNA damage tolerance in plants via translesion synthesis. Genes, Genomes, *Genomics*. 1. 89-99.

Kus, J. V., Zaton, K., Sarkar, R. and Cameron, R. K. (2002). Age-related resistance in *Arabidopsis* is a developmentally regulated defense response to *Pseudomonas syringae*. Plant Cell. 14. 479-490.

Lombard, D. B., Chua, K. F., Mostoslavsky, R., Franco, S., Gostissa, M. and Alt, F. W. (2005). DNA repair, genome stability, and aging. *Cell*. 120. 497-512.

Mitchell, D. L., Jin, J. and Cleaver, J. E. (1991). Relative induction of cyclobutane dimers and cytosine photohydrates in DNA irradiated in vitro and in vivo with ultraviolet-C and ultraviolet-B light. *Photochem. Photobiol*. 54. 741-746.

Molinier, J., Oakeley, E. J., Niederhauser, O., Kovalchuk, I. and Hohn, B. (2005). Dynamic response of plant genome to ultraviolet radiation and other genotoxic stresses. *Mutat. Res*. 571. 235-247.

Pang, Q. and Hays, J. B. (1991). UV-B-inducible and temperature sensitive photoreactivation of cyclobutane pyrimidine dimers in *Arabidopsis thaliana. Plant Physiol*. 95. 536-543.

Perdiz, D., Grof, P., Mezzina, M., Nikaido, O., Moustacchi, E. and Sage, E. (2000). Distribution and repair of bipyrimidine photoproducts in solar UV-irradiated mammalian cells. *J. Biol. Chem*. 275. 26732-26742.

Quaite, F. E., Sutherland, B.M. and Sutherland, J. C. (1992a) Action spectrum for DNA damage in alfalfa lowers predicted impact of ozone depletion. *Nature*. 358. 576-578.

Quaite F. E., Sutherland, B. M. and Sutherland, J. C. (1992b) Quantitation of pyrimidine dimers in DNA from UVB-irradiated alfalfa (*Medicago sativa* L.) seedlings. *Appl. Theoret. Electrophor.* 2. 171-175.

Quaite, F. E., Takayanagi, S., Ruffini, J., Sutherland, J. C. and Sutherland, B. M. (1994). DNA damage levels determine cyclobutyl pyrimidine dimer repair mechanisms in alfalfa seedlings. *Plant Cell.* 6. 1635-1641.

Quirino, B. F., Normanly, J. and Amasino, R. M. (1999). Diverse range of gene activity during *Arabidopsis thaliana* leaf senescence includes pathogen-independent induction of defense-related genes. *Plant Mol. Biol.* 40. 267-278.

Quirino, B. F., Noh, Y.-S., Himelblau, E. and Amasino, R. M. (2000). Molecular aspects of leaf senescence. *Trends Plant Sci.* 5. 278-282.

Rass, U., Ahel, I. and West, S. C. (2007). Defective DNA repair and neurodegenerative disease. *Cell.* 130. 991-1004.

Rattan, S. I. S. (2006). Theories of biological aging: Genes, proteins, and free radicals. *Free Radical Res.* 40. 1230-1238.

Reifenrath, K. and Muller, C. (2007). Species-specific and leaf-age dependent effects of ultraviolet radiation on two Brassicaceae. *Phytochem.* 68. 875-885.

Smart, C. M. (2005). Gene expression during leaf senescence. New Phytol. 126. 419-448.

Takahashi, Y., Moriwaki, S.-I., Sugiyama, Y., Endo, Y., Yamazaki, K., Mori, T., Takigawa, M. and Inoue, S. (2004). Decreased gene expression responsible for post-ultraviolet DNA repair synthesis in aging: a possible mechanism of age-related reduction in DNA repair capacity. *J. Invest. Dermatol.* 124. 435-442.

Tanaka, A., Sakamoto, A., Ishigaki, Y., Nikaido, O., Sun, G., Hase, Y., Shikazono, N., Tano, S. and Watanabe, H. (2002). An ultraviolet-B-resistant mutant with enhanced DNA repair in *Arabidopsis*. *Plant Physiol.* 129. 64-71.

Venema, J., Mullenders, L. H. F., Natarajan, A. T., van Zeeland, A. A. and Mayne, I. V. (1990). The genetic defect in Cockayne syndrome is associated with a defect in repair of UV-induced DNA damage in transcriptionally active DNA. *Proc. Natl. Acad. Sci. USA.* 87. 4707-4711.

Vijg, J. (2000). Somatic mutations and aging: a re-evaluation. *Mutat. Res.* 447. 117-135.

In: Encyclopedia of DNA Research (3 Volume Set)
Editors: S. J. Duncan and P. H. Wiley
ISBN 978-1-61324-305-3

Chapter 27

Acquired Antibiotic Resistance in Starters and Probiotics Strains of Lactic Acid Bacteria and Bifidobacteria Species

Baltasar Mayo**, Ana B. Flórez, Mohammed S. Ammor and Susana Delgado
Department of Microbiology and Biochemistry, Instituto de Productos Lácteos de Asturias (CSIC), Carretera de Infiesto s/n, 33300 Villaviciosa, Asturias, Spain

1. Abstract

Lactic acid bacteria (LAB) are non-pathogenic microorganisms found naturally in raw plant and animal materials or added deliberately as functional starter cultures to promote desirable biochemical changes in food and feed fermentations. LAB and bifidobacteria are also natural inhabitants of the gastrointestinal tract (GIT) of human and animals where they exert an array of protective and beneficial effects; a property that has led to the use of members of these two microbial types as probiotics. Antibiotics have greatly contributed over the last 65 years to human life expectancy and quality of life in the fight against infectious diseases. However, their efficacy is being severely threatened by the appearance and worldwide spread of antibiotic resistance. At present, there is great concern that commensal and beneficial bacterial populations, such as the food-borne and GIT-associated bacteria, become a reservoir for resistances, from which they could ultimately be transferred to human pathogens and opportunistic bacteria. Transferable resistances carried by starters and probiotics could be spread either during food manufacture or during passage through the GIT. Consequently, discriminating between resistant and susceptible strains, and distinguishing between intrinsic, non-transferable and acquired, transferable resistances is vital in order of not to disseminate antibiotic resistances via the food chain. Genes conferring resistance to several antibiotics have already been identified and characterized in strains of LAB and bifidobacteria. In particular, *tet*(W) and *erm*(B), encoding tetracycline and erythromycin resistance,

* Corresponding Author: Baltasar Mayo, Department of Microbiology and Biochemistry, Instituto de Productos Lácteos de Asturias (CSIC), Carretera de Infiesto s/n, 33300 Villaviciosa, Asturias, Spain Phone number: +34 985 89 12 31 Fax number: +34 985 89 22 33 E-mail address: baltasar.mayo@ipla.csic.es.

respectively, seem to be widely spread among species of these bacterial groups. Some of these genes have proved to be transferable in laboratory conditions, but also in food and in animal GIT systems.

Keywords: antibiotic resistance, acquired resistance, lactic acid bacteria, bifidobacteria, starters, probiotics

2. General Introduction

2.1. Antibiotics and Antibiotic Resistance

Antibiotics are arguably the most successful form of chemotherapy developed over the entire history of medicine. Their clinical use has achieved a significant reduction in the morbidity and mortality associated with infectious diseases, and has had a profound impact on the success of surgery and cancer therapy (Levy and Marshall, 2004). Antibiotics were soon extended to veterinary medicine, where they have been employed as therapeutic and prophylactic agents, and as animal growth promoters (Wegener, 2003). Furthermore, they have been extensively used in aquaculture and agriculture practices (McManus et al., 2002). The presence of antibiotics in an ecosystem exerts a strong selective pressure for the appearance and spread of resistant bacteria, because they kill the susceptible bacteria, raises the frequency of mutation, and, more importantly, allow the resistant bacteria to grow and dominate the niche (Baquero et al., 2002; Barbosa and Levy, 2000). Antibiotic resistance has been a continuing problem since antibiotics were introduced, but it is the increase in the number, diversity and range of resistant organisms that has become a huge clinical problem worldwide (Tenover, 2006). The phenomenon has been recently defined by the Alliance for the Prudent Use of Antibiotics as a shadow epidemic (http://www.apua.org). It has been estimated that some 1-10 million tons of antibiotics have already been released into the biosphere (European Commission, 2005). Not surprisingly, a clear correlation between antibiotic use and level of resistant bacteria has been repeatedly reported (Bronzwaer et al., 2002). Antibiotic resistance complicates and put up the price of the treatment of infections (Levy and Marshall, 2004). The real human and economic cost of the resistances is difficult to estimate, but it becomes clear that infections by resistant microorganisms duplicate the average stay at hospitals and double both mortality and morbidity rates (Cosgrove and Carmelli, 2003; Baquero et al., 2002). The spread of antibiotic resistance is conjugated with a discontinued research on new antimicrobials, thought to be a reflection of the industrial perception of the ever-increasing stringent criteria of regulatory agencies for new drugs and the short time protection of the discoveries (Shales et al., 2004).

Resistances can be inherent to a bacterial species or genus (referred to as intrinsic or natural resistance) or can be acquired (European Commission, 2005). Intrinsic or "natural" resistance involves absence of the target or presence of low affinity targets, low cell permeability, enzymatic inactivation of the antibiotics, and activity of efflux mechanisms (Tenover, 2006). The acquisition of antibiotic resistance occurs via mutation of pre-existing genes or, more frequently, by horizontal transmission. With some exceptions, intrinsic resistance and resistance by mutation are unlikely to be disseminated; horizontally transferred genes, particularly those carried on mobile genetic elements, are those most likely to be transmitted (Normak and Normak, 2002).

Acquired antibiotic resistance is a process intimately associated to the genuine evolution processes; thus being predictable and anticipated. The appearance of resistances mostly depends on the volume of antimicrobial utilized (Bronzwaer et al., 2002), the frequency of mutation of the microorganisms (Miller, 1996) and of the biological cost of the resistance (Bjorkman and Anderson, 2000). The biological cost seems to be not too strong, and occasionally compensatory mutations can reverse any deleterious effect of the resistance mutation (Andersson and Levin, 1999). In fact, the rapid appearance of resistances contrasts with a slow disappearance, in the absence of the selective agent (Andersson, 2003). Resistances could be maintained by a continued exposure to residual concentration of antibiotics into the environment (Donoghue, 2003), but resistance genes are frequently associated with resistance determinants to other toxic compounds, in such a way that the resistances could be maintained by co-selection (Borgen et al., 2002; Aarestrup, 2000). Finally, from an ecological point of view, the resistant microorganisms selected by the treatment of an individual would be diluted in an environment colonized by susceptible microorganisms; however, the treatment of whole populations with the same antimicrobial agent increases the selective pressure for resistant organisms and reduces the chances of re-colonization by those susceptible (Levy, 1997).

2.2. Emergence and Spread of Antibiotic Resistance

The first resistant bacteria were detected in hospital environments soon after the introduction of the antimicrobials (for a review, Levy and Marshall, 2004). During the 50s and 60s, several enteric strains of *Salmonella, Shigella*, and *Escherichia coli* were isolated harbouring multiple resistances. However, the resistance problem was not evident until the 70s when pathogenic strains of *Haemophilus influenzae* and *Neisseria gonorrhoeae* resistant to ampicillin were firstly identified (Elwell et al., 1977; de Graaf et al., 1976).

Resistance genes are usually located in mobile genetic elements (such as plasmids and transposons) having a high transference capacity (Normak and Normak, 2002). Furthermore, some transposons contain integrons: complex transposons containing dedicated places for insertion of resistance and other adaptive genes, which are all expressed from a unique promoter (Nandi et al., 2004; Hall et al., 1999). These elements pass from one cell to another, jumping in some cases the species and genus barriers. Indeed, some determinants have even been found to be transferred into the GIT of mammals (Netherwood et al., 1999). As a result, at present, some strains of species of *Acinetobacter bumanii, Enterobacter cloacae, Enterococcus faecium, Klebsiella pneumoniae, Mycobacterium tuberculosis, Pseudomonas aeruginosa, Staphylococcus aureus,* and *Streptococcus pneumoniae* are resistant to all antimicrobials available for the treatment of the infections they cause (Levy and Marshall, 2004; Wright, 2003). The collection of all genetic determinants having the potential to function as resistance genes has been recently called the antibiotic resistome (D'Acosta et al., 2006). The resistome consists in an impressive long list of genes, including the antibiotic resistance genes of pathogenic and non-pathogenic microorganisms, but also resistance genes from antibiotic producers, cryptic and silent antibiotic resistance genes, and resistance precursors genes that, under appropriated selection pressure, might evolve into genuine resistance genes. However, the clonal character of infectious strains (Levy and Marshall, 2004) suggests that more other factors than antibiotic resistances are contributing to the great success of pathogenic strains.

2.3. Commensal Microorganisms and Antibiotic Resistance

Commensal bacteria can act as a reservoir of resistant elements (plasmids, transposons) able to be transferred to pathogen microorganisms, a fact that has been firmly stated in some bacterial groups (DeFlaun and Levy, 1989). As an example, the commensal *Haemophilus parainfluenzae* supplies plasmids encoding β-lactamases to *H. influenzae* strains (Levy, 1985). Similarly, strains of *Staphylococcus epidermidis* serve as a reservoir of resistance genes and plasmids for the pathogenic *S. aureus* strains (Cohen et al., 1982). Although not encoded on mobile elements, the origin of penicillin resistance in *S. pneumoniae* has been found into the commensal species *Streptococcus viridans*, naturally resistant to this antibiotic (Balsalobre et al., 2003). Thus, the public health concern of commensal populations acting as a reservoir for antibiotic resistance that could be cross-species transferred to pathogenic organisms is well supported by an ever growing body of scientific evidence (Levy and Marshall, 2004; Salyers et al., 2004; http://www.roarproject.org). However, little is known at present about the abundance, diversity, and distribution of resistance genes in many commensal bacterial groups, because, until recently, bacterial pathogens have been the primary focus of studies on antibiotic resistance. Moreover, the mechanisms involved in horizontal gene transfer from commensals to pathogens are equally poorly understood.

3. Antibiotic Resistance in LAB and Bifidobacteria

LAB are a taxonomically diverse group of microorganisms, whose common characteristic is the production of lactic acid from the fermentation of carbohydrates (Leroy and de Vuyst, 2004). The most typical LAB members are Gram-positive, non spore-forming, catalase-negative organisms of the low GC branch, belonging to the order *Lactobacillales. Lactobacillus, Lactococcus, Leuconostoc* and *Pediococcus* are among the typical genera (Carr et al., 2002). Important pathogens and opportunistic microorganisms, such as several *Streptococcus* and *Enterococcus* species, also fall into the functional LAB definition. Of these two genera, only the non-pathogenic species *Streptococcus thermophilus* is of industrial relevance. For practical and ecological reasons, *Propionibacterium* and *Bifidobacterium* (both belonging to the order *Actinobacteria*, within the high GC branch of Gram-positives) have traditionally been included among the typical LAB genera (Wood and Warner, 2003). Owing to their aerotolerant anaerobic nature, the members of the LAB community occupy a wide range of nutrient-rich environments, which includes derived-plant material, milk and dairy products, and meat and meat products (Carr et al., 2002). Many LAB species are involved in the manufacture and preservation of fermented foods from raw agricultural materials in which they are present as contaminants, contributing to enhance the organoleptic, rheological and nutritional properties of fermented foods and animal feeds (Leroy and de Vuyst, 2004). In large-scale industrial fermentations, their accidental presence has been substituted by a deliberate addition in the form of functional starter cultures. Therefore, they are of great economic importance. LAB species are also commonly found among the resident microbiota of the GIT and genital-urinary tract (GUT) of vertebrates (Carr et al., 2002), where they are thought to play a key role in the microbial balance associated with health (Ouwehand et al., 2002). In this context, probiotic organisms are defined as "non-pathogenic microorganisms that, when ingested in certain numbers exert a positive influence on host physiology and health beyond inherent general nutrition" (Ouwehand et al., 2002). Large numbers of these bacteria are consumed to maintain and restore the microbial balance in the intestines, to increase their beneficial activities, and/or to counteract the action of harmful populations. Given their long history of safe consumption in traditional and industrial fermented foods, and their beneficial presence in GIT and GUT from the very beginning of life, LAB have been awarded a generally regarded as safe (GRAS) status by the American Food and Drug Association (FDA, 1997). Over the past two decades, interest in LAB physiology and genetics has greatly increased, reflecting their growing importance as starters in fermentation processes and an increasing potential of the probiotics market. Strains used as starters and as probiotics must present desirable technological and/or probiotic properties for which they should, therefore, be carefully selected. Besides desirable technological properties, the safety aspects of these bacteria are of great concern, including the absence of potentially transferable antibiotic resistances (Saarela et al., 2002).

Enterococci are true LAB species commonly present in fermented products and found as resident microorganisms in human and animal GITs. However, they will not be covered in the present review. These bacteria remain controversial, because they harbour a series of virulence factors and some species have been associated with a number of human infections (Foulquié Moreno et al., 2006). Moreover, antibiotic resistance in enterococci has received much attention and readers are addressed to some excellent recent reviews (Foulquié Moreno et al., 2006; Mathur and Shing, 2005; Klein, 2003).

As already mentioned, LAB species from foods and from the GIT and GUT may also act as reservoirs of antimicrobial resistance genes that could ultimately be horizontally transferred to pathogens, either in food and feed matrices or during passage though the GIT (Teuber et al., 1999). Moreover, the food chain has been recognized as one of the main routes of transmission of antibiotic resistance from animal to human bacterial populations (Gevers et al., 2003a; White, 2000; Teuber et al., 1999). It is the very large numbers of LAB in fermented products and the GIT that helps in the appearance of different resistant mechanisms via mutations. In addition, LAB can also acquire resistance from other bacteria in these environments, which, after amplification, may be transmitted to a further host. Then, fermented products provide a vehicle for transmission from the indigenous microbiota of animals to the bacteria of the human GIT (Salyers et al., 2004; Salyers et al., 2002; White, 2000). In fact, the presence of several resistance genes in many LAB strains from foods and from the human GIT has been firmly stated; for recent reviews see (Ammor et al., 2007a; Mathur and Singh, 2005).

3.1. Assays for Antibiotic Resistance in LAB

Traditionally, LAB strains have been examined for antibiotic resistance as a negative criterion during their physiological and technological characterization in the search for new starter and probiotic cultures (Masco et al., 2006; Delgado et al., 2005; Katla et al., 2001; Yazid et al., 2000; Orberg and Sandine, 1985; Cogan, 1972). But most LAB species grow relatively poorly, if at all, on well-known antibiotic testing media such as Mueller-Hinton (Difco) and Iso-Sensitest (Oxoid). Furthermore, de Man, Rogosa, and Sharpe (MRS) medium can inactivate some antibiotics (e.g., imipenem). Thus, a suitable culture medium for assaying the majority of LAB species is still required (Tosi et al., 2007; Mättö et al., 2006; Klare et al., 2005; Zarazaga et al., 1999). Very recently, Klare and co-workers developed the LSM testing medium, a mixed formulation of Iso-Sensitest (90%) and MRS (10%), complemented with L-cysteine for anaerobic bacteria (Klare et al., 2005). This medium provides optimal growth for most species. The MLS medium has been extensively evaluated and validated in a series of recent antibiotic surveys, involving *Lactococcus lactis* (Flórez et al., 2005; Flórez et al., 2006a), *Lactobacillus plantarum* (Flórez et al., 2006b), and *Lactobacillus* spp. (Danielsen et al., 2007a; Egervärn et al., 2007a; Danielsen and Wind, 2003), *Bifidobacterium* spp. (Mättö et al., 2007), and *Pediococcus* spp. (Danielsen et al., 2007b) species. The LSM medium has to be supplemented with lactose (0.5% w/v) for sufficient growth of *Streptococcus thermophilus* strains (Tosi et al., 2007).

The Minimum Inhibitory Concentration (MIC) of a given antibiotic is the lowest antibiotic concentration suppressing growth of the strain under study. For clinically-important microbial species, MICs can either be measured in liquid or solid media by well standardized procedures (CLSI, 2004). Standardization of media and procedures for testing beneficial and commensal organisms has only recently been addressed (Danielsen et al., 2004; Klare et al., 2005; Egervärn et al., 2007b; Mayrhofer et al., 2008).

Traditionally, several methods have been reported for antimicrobial susceptibility testing of these bacteria, including the Etest (Danielsen and Wind, 2003; Charteris et al., 2001; Felten et al., 1999), agar dilution (Herrero et al., 1996), disk diffusion (Coppola et al., 2005; Herreros et al., 2005; Felten et al., 1999; Chateris et al., 1998a,b), and microbroth culture (Delgado et al., 2005; Flórez et al., 2005). Except for some specific agent-related effects, agreement between MICs resulting from the broth microdilution, Etest and disk diffusion has been recently reported (Mayrhofer et al., 2008). However, some assays seem to be not reliable for particular antibiotics. For instance, disk diffusion is not reliable for detecting low-level vancomycin resistant enterococci (MICs from 8 to 32 μg/ml) (Swenson et al., 1992). Variations in the cation content and the concentration of critical compounds (such as thymine) can modify the results, as can the inoculum size, the temperature and the incubation time, etc. (Egervärn et al., 2007b; Danielsen et al., 2004). Because of all these factors, there still is a lack of agreement regarding the resistance-susceptibility breakpoints for most antibiotics in LAB species (European Commission, 2005; Danielsen and Wind, 2003; Katla et al., 2001; Charteris et al., 1998a,b). Clear definition of the breakpoints discriminating between resistant and susceptible strains is vital, as it is the distinguishing between intrinsic, non-specific, non-transferable and acquired, transferable resistances, for which comparison of antimicrobial-resistance patterns in many LAB species from different sources may be required (European Commission, 2005; Teuber et al., 1999). Distinguishing is usually undertaken by comparing the MICs obtained in different surveys (Danielsen and Wind, 2003; Charteris et al., 2001). The study of the MIC distribution across species and strains is also of great help when attempting to separate resistant strains from a majority of susceptible bacteria (Ammor et al., 2008a; Danielsen and Wind, 2003) (Figure 1).

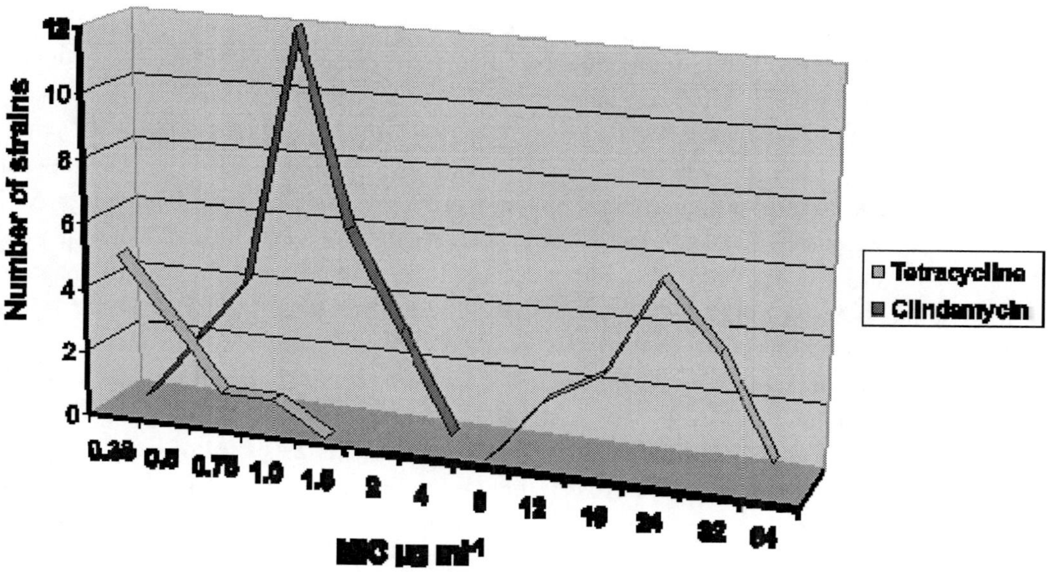

Figure 1. Distribution of tetracycline (in green) and clindamycin MICs (in red) in a series of human *Bifidobacterium* strains. The unimodal distribution of clindamycin MICs suggests the presence of no resistant strains, while the clear bimodal distribution of tetracycline MICs indicates the presence of qualitatively-distinct populations.

Phenotypic tests alone can not confirm the presence or absence of transferable resistance genes. Phenotypic assays, then, have to be complemented by molecular methods in which bacterial strains are directly screened for the presence of antibiotic resistance determinants. These methods include amplification by PCR with specific primers of single (Aquilanti et al., 2007; Hummel et al., 2007; Rojo-Bezares et al., 2006) or multiplex (Strommenger et al., 2003) antibiotic resistance genes, real time quantitative PCR (Volkmann et al., 2004), or the use of DNA microarrays containing large collections of antibiotic resistance genes (Ammor et al., 2008a; Perreten et al., 2005; van Hoek et al., 2005). Finally, determinants have to be analysed by sequencing and comparison against the sequences deposited into the public databases. A crucial step in all these analyses is an unequivocal classification of the strains, because if related species are misidentified, confident results could never be obtained (Elliott and Facklam, 1996; Green et al., 1990).

3.2. Antibiotic Resistance-Susceptibility Profiles of LAB and Bifidobacteria

Based on the body of research published by European and EU member states monitoring programmes, the Scientific Committee on Animal Nutrition of the European Commission (SCAN) established microbiological breakpoints for certain antibiotics of human and veterinary importance. These were to be employed in the assessment of risks associated with the bacteria used in the production of animal feed, and were defined for most LAB species. The original SCAN breakpoints were recently updated by the Panel on Additives and Products or Substances used in Animal Feed (FEEDAP) (European Commission, 2005). However, breakpoints should be constantly updated as more results are available. Based on the results of most recent works, Table 1 summarizes the newest microbiological susceptibility-resistance cut-off values proposed for some of the principal LAB species and groups and representative antibiotics acting on Gram-positive microorganisms. The antibiotic resistance profiles of *Lactobacillus*, *Lactococcus*, *Streptococcus*, *Pediococcus*, *Leuconostoc*, *Bifidobacterium* and *Propionibacterium* species are rather different; thought clear-cut genus-specific resistance patterns have not been observed. MIC values varied widely between species and strains to most antimicrobials, although general guidelines could be highlighted.

LAB are resistant to metronidazole [MIC ≥ 32 μg/ml] (Flórez et al., 2005; Danielsen and Wind, 2003; Martel et al., 2003; Delgado et al., 2002) (an agent known for its anti-anaerobe activity) since they have no hydrogenase activity (the agent's target) (Church et al., 1996). Susceptibility to metronidazole, however, is variable in bifidobacterial strains (Masco et al., 2006; Delgado et al., 2005; Yazid et al., 2000; Lim et al., 1993). LAB are also usually resistant to sulphonamides (MIC ≥ 256 μg/ml) and trimethoprim (MICs ≥ 30 μg/ml) (Katla et al., 2001; Tenorio et al., 2001; Charteris et al., 1998a). Intrinsic resistance of lactobacilli and other LAB species to trimethoprim has been reported to be an experimental artefact. The use of appropriated testing media with low thymine concentration has proved these bacteria to be susceptible (Danielsen et al., 2004). In regard to sulphonamides, these organisms have limited biosynthetic capabilities and lack the folic acid synthesis pathway; and they are therefore regarded as intrinsically resistant to these agents (Katla et al., 2001). Strong resistance of LAB and bifidobacteria species to aminoglycosides such as kanamycin and streptomycin has repeatedly been reported (Masco et al., 2006; Danielsen and Wind, 2003; Katla et al., 2001), again suggesting the existence of inherent (intrinsic) resistance. A lack of cytochrome-mediated transport is thought to be responsible for the resistance of anaerobic and facultative bacteria to these antimicrobials (Bryan and Kwan, 1981). Similarly, polymyxin B is inactive against most Gram-positive bacteria (Kaye, 2004), suggesting that species and strains associated with high MICs (range from 4 to 1024 μg/ml) exhibit intrinsic resistance. Further, most *Lactobacillus*, *Pediococcus* and *Leuconostoc* species are resistant to high concentrations of vancomycin (MICs ≥ 256 μg/ml) (Flórez et al., 2006a; Danielsen and Wind, 2003; Zarazaga et al., 1999), whereas *Lactococcus* (Flórez et al., 2006a; Flórez et al., 2005; Delgado et al., 2002; Swenson et al., 1990) and *Bifidobacterium* (Masco et al., 2006; Delgado et al., 2005; Moubareck et al., 2005) species are very susceptible (MICs ≤ 2 μg/ml). The resistance of *Lactobacillus*, *Pediococcus* and *Leuconostoc* species to vancomycin is due to their possession of *D*-Alanine-*D*-Lactate in their peptidoglycan rather than the normal *D*-Alanine-*D*-Alanine dipeptide (Klein et al., 2000; Evers et al., 1996). Such resistance is thus intrinsic, as the antibiotic's target is absent. This type of resistance does not seem to pose a problem since its risk of transference is thought to be negligible (European Commission, 2005). Resistance to this antibiotic, however, is of major concern since it is the last remaining choice for combating clinical infections caused by multidrug-resistant Gram-positive pathogens (Woodford et al., 1995). Enterococcal species have already developed two independent routes of resistance (Reynolds and Courvalain, 2005), both of which have already been transmitted to *S. aureus* (Pfeltz and Wilkinson, 2004).

LAB are susceptible to ampicillin (MICs ≤ 1 μg/ml), piperacillin, piperacillin plus tazobactam (MICs ≤ 16 μg/ml), and other penicillins (Delgado et al., 2005; Flórez et al., 2005; Moubareck et al., 2005; Zhou et al., 2005; Delgado et al., 2002). *Lactobacillus*, *Lactococcus* and *Leuconostoc* species show resistance to high levels of cefoxitin (MICs ≥ 30 μg/ml) (Flórez et al., 2005; Delgado et al., 2002; Charteris et al., 1998b; Herrero et al., 1996), whereas *Bifidobacterium* species show variation in susceptibility to this antibiotic (Delgado et al., 2005; Moubareck et al., 2005; Zhou et al., 2005; Charteris et al., 1998a). Most LAB and bifidobacteria strains are very susceptible to rifampicin (MICs between 0.004 to 0.75 μg/ml), except for the *L. lactis* strains (MIC >32 μg/ml). Rifampicin interacts in a specific manner with the β-subunit of the bacterial RNA polymerase encoded by *rpoB* (Wehrli, 1983). Lactococci and enterococci are intrinsically resistant to this antibiotic (Eaton and Gasson, 2001; Orberg and Sandine, 1985). Uncharacterised resistance to high levels of rifampicin in *Lactobacillus* species has been recently reported (Kafili et al., 2009).

For other antibiotics, such as clindamycin, erythromycin, minocycline, and tetracycline most LAB proved to be very susceptible, presenting MICs <1 μg/ml. Although higher, the MIC of chloramphenicol to most LAB species and strains is also low (most MICs of 4 μg/ml). However, bimodal distributions of MICs for all these antibiotics have been detected in many works, suggesting the presence of truly resistant strains having acquired genes. Indeed, resistance determinants for all these antibiotics have been identify and characterized in LAB species (Ammor et al., 2007a).

In spite of many resistances being intrinsic, the cooperation of non-specific mechanisms, such as multi-drug transporters (Putman et al., 2001) and defective cell wall autolytic systems (Kim et al., 1982) may also account for MIC differences between species and strains in susceptible and resistant strains. In that sense, the secondary multidrug transporter (MDR) LmrP of *L. lactis* has been implicated in the resistance to a broad range of clinically important antibiotics (Putman et al., 2001). In addition, the *mdt*(A) gene from pK214 encoded an efflux system conferring increased resistance to macrolides, lincosamides, streptogramins and tetracyclines in *L. lactis* and *E. coli* (Perreten et al., 2001). Similarly, heterologous expression of a hypothetical MDR-encoding gene from *Bifidobacterium breve* conferred moderate resistance to macrolides in *L. lactis* (Margolles et al., 2005). Non-specific mechanisms could also enhance resistance to aminoglycosides, for which LAB species are thought to be naturally resistant (Bryan and Kwan, 1981).

3.3. Acquired Resistances in LAB and Characterization of Resistance Determinants

As seen in the previous section, a majority of the antibiotic resistances found in LAB species are thought to be intrinsic, although in some cases further studies are still required. Acquired resistances have also been characterized in detail at a molecular level in a few strains, and a majority has been found to be encoded by added genes.

Although it was not the first report on an antibiotic resistance determinant from a LAB strain, the presence of a multiple resistance plasmid in *L. lactis* K214 isolated from a raw-milk cheese shocked the scientific community (Perreten et al., 1997a). The 30-kbp plasmid pK214 encoded as much as four different acquired resistances, which had been collected, probably with the help of insertion-sequence (IS) elements, from different commensal and pathogenic microorganisms; namely, a *tet*(S) gene from *Listeria monocytogenes*, genes for a streptomycin-inactivating adenylase and a chloramphenicol acethyltransferase from *S. aureus*, and a macrolide efflux gene [*Mdt(A)*] of unknown origin conferring altogether multiple antibiotic resistance (Perreten et al., 1997a; Perreten et al., 2001).

Table 1. Microbiological susceptibility-resistance cut-offs for LAB and bifidobacteria species to representative antibiotics

Genus/Species (Group)	Antibiotic												
	Am	Vm	Gm*	Km*	Sm*	Neo*	Ery	Clin	Qui/ Dalf	Tc	Cm	Tri*	Lin
Bifidobacterium spp.	4	2	32	64	256	32	4	2	4	8	8	8	4
Lactobacillus (obligate homofermentative)	4	4	8	256	48	32	2	1	4	2	2	8	4
Lactobacillus (obligate heterofermentative)	4	IR	64	>256	256	32	2	1	4	32	2	8	4
Leuconostoc spp.	4	IR	32	256	64	32	4	1	4	8	8	8	4
Pediococcus spp.	4	IR	32	256	64	32	2	1	4	8	8	8	4
Lactobacillus plantarum	4	IR	16	256	64	32	4	16	4	64	8	8	4
Lactococcus lactis	4	4	8	16	64	8	2	1	4	2	16	IR	4
Streptococcus thermophilus	4	2	8	64	32	8	2	1	4	2	8	IR	4

Strains with MICs higher than the breakpoints are considered as resistant.

IR, intrinsic resistance.

*Interferences of antibiotics with the medium composition have been described.

Key of antibiotics: Ap, ampicillin; Vm, vancomycin; Gm, gentamicin; Km, kanamycin; Sm, streptomycin; Neo, neomycin; Ery, erythromycin; Clin, clindamycin; Qui/Dalf, quinupristin/dalfopristin; Tc, tetracycline; Cm, Chloramphenicol; Tri, trimethoprim; Lin, linezolid.

This clearly demonstrated that there is not barrier between pathogenic and commensal (or beneficial) bacteria regarding acquired antibiotic resistance.

Few other antimicrobial resistance determinants have been identified in *L. lactis* strains. Raha et al. (2002) identified an erythromycin resistance plasmid from a *L. lactis* strain isolated from the chicken cecum, although the resistance gene was not further characterized. Very recently, a *tet*(M) gene has been identified in two tetracycline resistant strains isolated from a raw milk traditional Spanish cheese (Flórez et al., 2008), and another resistant strain from a cat tonsil [A.B. Flórez and B. Mayo, unpublished results]. The partial sequences of all these genes proved to be 100% identical to each other and also to *tet*(M) from the transposon Tn*916* of *Enterococcus faecalis* (Flannagan et al., 1994). Further, two different tetracycline resistance genes were detected among tetracycline resistant *L. lactis* strains from Polish dairy products, a mosaic *tet*(L/S) gene along with *tet*(M) [J. Bardowski, personal communication].

Several antibiotic resistance determinants have been identified and characterized from *Lactobacillus* species; mostly from human and animal intestines. A non-conjugative plasmid (pCaT) mediating chloramphenicol resistance in *L. plantarum* was already reported in 1992 (Ahn et al., 1992). A similar plasmid-borne chloramphenicol-resistance gene highly related to the *cat* gene from the *S. aureus* plasmid pC194 was found in *Lactobacillus reuteri* G4 (Lin et al., 1996). Resistance genes encoding resistance to the macrolide-lincosamide-streptogramin (MLS) type B antibiotics have also been encountered in many lactobacilli species. Axelsson et al. (1988) identified a 5.5-kbp plasmid in *L. reuteri* encoding an erythromycin-resistance determinant related to the erythromycin resistance gene of the plasmid pIP501 from *Streptococcus agalactiae*. An *erm*(GT) gene encoded in a 9.8-kbp plasmid and related (ca. 81% nucleotide identity) to the *erm*(C) gene from *S. aureus* plasmid pE194 has been detected in *L. reuteri* 100-63 (Tannock et al., 1994), and a further gene encoding resistance to the macrolide tylosin [*erm*(T)] was also characterized on a *Lactobacillus* spp. strain from swine (Whitehead and Cotta, 2001). Our group has recently characterized an *erm*(B) gene in *Lactobacillus johnsonii* G41, a strain isolated from human faeces (Flórez et al., 2006c). Interestingly, the gene was found to be harboured in the chromosome of the strain. Within this region, a DNA segment of 1.5 kpb proved to be identical to a fragment of the erythromycin resistant plamids pRE25 from *E. faecalis* (Schwarz et al., 2001), which again stresses the exchange of genes between unrelated bacteria occupying the same environments. A series of tetracycline resistance genes –*tet* (K, M, S, W, 36) (Huys et al., 2006; Roberts, 2005; Rojo-Bezares et al., 2005; Gevers et al., 2003a; Chopra and Roberts, 2001), *tet*(O) (Villedieu et al., 2003; Chopra and Roberts, 2001) and *tet*(Q) (Chopra and Roberts, 2001)– have also been frequently detected on *Lactobacillus* species from different sources. An interesting organization for tetracycline resistance has been recently found in the Italian Sola cheese isolate *Lactobacillus sakei* Rits 9 (Ammor et al., 2007b). Two genes coding for a ribosomal protection protein [*tet*(M)] and a tetracycline efflux pump [*tet*(L)] were encountered. *tet*(M) was found to be chromosomally encoded and proved to be associated to a transposon-like structure. In contrast, *tet*(L) was located on a mobilizable 5.5-kbp Plasmid, pLS55 (Ammor et al., 2007b). Notably, the whole plasmid sequence, including the *tet*(L) region, proved to be identical (only 5 nucleotide changes) to that of the recently reported plasmid pMA67 from *Paenibacillus larvae*, a pathogenic species infecting honey bees (Murray et al., 2007).

The situation of *L. sakei* Rits 9 is uncommon, as in only a few cases single strains have been shown to harbour more than one antibiotic resistance determinant, which suggests that multiresistance is, at present, a rare event in LAB. Although this, a *Lactobacillus fermentum* strain resistant to novobiocin, tetracycline, erythromycin, and dalfopristin has been reported (Gfeller et al., 2003). A chromosomal *tet*(M) gene was identify as responsible for tetracycline resistance, while a macrolide-lincosamide-streptogramin B (MLS) methylase [*erm*(LF)] and a streptogramin A acetyltransferase [*vat*(E)] (identical to that of *Enterococcus faecium*) were seen to be located on a 19.3-kbp plasmid (Gfeller et al., 2003). Indeed, both *erm*(B) and *tet*(M) were detected in *Lactobacillus* species of dairy and human origin isolated in Turkey (Cataluk and Gogebakan, 2004). Similarly, *L. johnsonii* G41 has been shown to contain a mosaic tetracycline resistance gene, *tet*(O/W/32/O/W/O), in addition to the previously mentioned *erm*(B) gene (van Hoek et al., 2007). Hybrid *tet*(O)-*tet*(W), genes separated by 36% nucleotide sequence divergence, were firstly detected in isolates of the propionate-forming intestinal anaerobe *Megasphaera elsdenii* from swine (Stanton and Humphrey, 2003). Mosaic genes are thought to provide greater tetracycline resistance in this species (Stanton and McDowall, 2004); whether this is the biological significance of the hybrid gene detected in *L. johnsonii* remains to be determined.

The *tet*(W) gene encoding a ribosomal protection protein supporting tetracycline resistance has been reported in a variety of Gram positive and Gram negative, aerobic and anaerobic bacteria (Roberts, 2005; Chopra and Roberts, 2001). *tet*(W) is one of the most abundant tetracycline resistance genes found in bacteria from the mammalian gut and was first identified in the rumen anaerobe *Butyrivibrio fibrisolvens* 1.230 (Barbosa et al., 1999). Copies of *tet*(W) have also been found in other isolates of *B. fibrisolvens* and in isolates of *Selenomonas* spp., *Mitsuokella* spp., *Clostridium* spp., *Roseburia* spp., and *M. elsdenii* from bovine and sheep rumens, as well as from porcine and human feces (Roberts, 2005). Genes from the different bacteria display nearly identical nucleotide and protein sequences (Figure 2), suggesting a recent and rapid horizontal transfer from a common ancestor. The gene has been found to be spread among many intestinal *Bifidobacterium* species of human origin, including *Bifidobacterium longum, Bifidobacterium bifidum, Bifidobacterium catenulatum*, and *Bifidobacterium pseudocatenulatum* (Aires et al., 2007; Flórez et al., 2006d; Masco et al., 2006; Moubareck et al., 2005; Scott et al., 2000), but also in the type strains of *B. longum* biotype *suis, Bifidobacterium pseudolongum* subsp. *globosum*, *Bifidobacterium thermophilum* (Flórez et al., 2007a), and in the widely-used probiotic strain *Bifidobacterium animalis* subsp. *lactis* Bb-12 (Saarela et al., 2007; Masco et al., 2006). In *B. fibrisolvens* the gene is encoded on a highly mobile transposable chromosomal element, Tn*B1230* (Scott et al., 1997). However, transposons-related sequences have never been detected in *Bifidobacterium* species (Aires et al., 2007; Flórez et al., 2006d; Kazimierczak et al., 2006). The genetic organization found in the different species and strains proved to be highly variable (Ammor et al., 2008b; Kazimierczak et al., 2006). As discussed for *tet*(M), *tet*(W) genes seem to be associated to insertion sequences (IS) and transposon-related remnants, although the actual spreading mechanism remains unknown. One of the most streaking results of these works was the detection of silent genes in some isolates (Saarela et al., 2007; Flórez et al., 2006d). In at least one case, activation resulted from the insertion of two bases in the promoter region of the gene, generating an appropriate –35 region (A. Margolles, personal communication). The presence of such genes may suggest the necessity of analyzing for resistance determinants even susceptible strains, as, under certain circumstances, silent genes could be activated. Although less frequent, *tet*(O) and *tet*(M) (Aires et al., 2007; Mätto et al., 2007; McConnell et al., 1991), and hybrid genes *tet*(O/W), *tet*(W/32/O), and *tet*(O/32/O) have also been recently reported in *Bifidobacterium* species from animal (Patterson et al., 2007; van Hoek et al., 2007), and human fecal samples (Patterson et al., 2007).

A genetic determinant for MLS resistance [*erm*(*AM*)] from a *Pediococcus acidilactici* strain has been analysed and found to be localized on a 46-kbp non-transferable plasmid (Tankovic et al., 1993). A further macrolide resistance [*erm*(B)] gene in another *P. acidilactici* isolate was reported by Zarazaga et al. (1999). More recently, a further *erm*(E) gene located on a 10-kbp plasmid has been reported in several unrelated *P. acidilactici* strains (Danielsen et al., 2007). The encoding plasmid, pEOC01, has been fully sequenced (O'Connor et al., 2007) and found to contain a second antibictic resistance determinant [*add*(E)] encoding streptomycin resistance. Of particular note is the fact that the *aad*(E) gene holds 100% identity to a plasmid-encoded gene found in *Campylobacter jejuni*, although it has been suggested to originate from *Enterococcus* species (Nordnoy et al., 2005). This observation provides evidences for recent horizontal transfer of streptomycin resistance from a lactic acid bacterium to a Gram-negative intestinal pathogen. Tetracycline resistance determined by an efflux pump encoded by *tet*(L) has recently been reported in a *Pediococcus parvulus* strain from wine (Rojo-Bezares et al., 2006).

In spite of LAB and bifidobacteria species being intrinsically resistant to aminoglycosides, as stated in a previous paragraph, some aminoglycoside resistance genes, i.e., *aac*(6')-*aph*(2"), *ant*(6), *aph*(3')-IIIa, have been found in strains of *L. plantarum, Pediococcus pentosaceus,* and *P. parvulus* from animal origin and wine (Rojo-Bezares et al., 2006; Tenorio et al., 2001). Microarray analysis further detected *aph*(E) and *sat*(3) genes in one *B. bifidum,* and 2 *B. pseudocatenulatum* streptomycin-resistant strains, and in 3 streptomycin-susceptible isolates of *B. longum* and one *B. pseudocatenulatum* (Ammor et al., 2008a). Thus, the presence of these transferases has not been clearly associated with neither resistance nor higher streptomycin or kanamycin MICs (Ammor et al., 2008a; Rojo-Bezares et al., 2006). This suggests that the enzymes might play other biological roles.

Resistances can not only be acquired via gain of exogenous DNA, but also by chromosomal mutations by modifying the antibiotic-binding site, increasing the expression of intrinsic resistance determinants or altering the target molecule itself (European Commission, 2005). A human *Lactobacillus rhamnosus* strain isolated from faeces and resistant to macrolides, from which no resistance determinants have been detected by specific PCR and microarray screening, has been recently found to contain a heterozygous A-to-G transition mutation at position 2058 (*Escherichia coli* numbering) of its 23S rRNA genes (Flórez et al., 2007b). The same mutation has been recently observed in a series of vaginal *L. rhamnosus* and *L. fermentum* strains isolated in Serbia [L. Topisirovic, personal communication]. Rifaximin is a synthetic structural analogue of rifampicin an other rifamycin class antibiotics inhibiting the synthesis of RNA by blocking the bacterial RNA polymerase, which is used to treat gastrointestinal disorders. Recently, nucleotide mutations inside a core region of the *rpoB* gene (which encodes the β subunit of RNA polymerase) have been found in rifaximin-resistant strains of *Bifidobacterium* (Vitali et al., 2008). By their nature, acquired resistances by mutation are considered to pose a low risk of dissemination (European Commission, 2005).

3.4. Horizontal Transfer of Antibiotic Resistances to and from LAB Species

Although complete and active genes for late competence events have been described in the genome of *L. lactis* (Bolotin et al., 2001), *L. plantarum* (Kleerebezen et al., 2003) and other LAB species, natural competence has not been described in species of this group, except for a single report in *Leuconostoc carnosum* (Helmark et al., 2004). Transduction was the first gene transfer technique identified and used in LAB for the transfer of technologically-important traits between different strains (Fitzgerald and Gasson, 1988). However, its importance as a means of disseminating antibiotic resistance is questionable because of the high specificity of phages. Thus, as in other bacterial groups (Stanton et al., 2004; Normak and Normak, 2002), resistance to antibiotics by LAB might have mostly proceeded by horizontal transmission via conjugation. However, the molecular organization of some antibiotic resistance loci, such as those of *tet*(W) in bifidobacteria and *erm*(B) in *L. johnsonii* and others, suggests that DNA segments containing resistance genes might be internalised and integrated into the genomes by unknown mechanisms. Determinants are frequently flanked by IS elements, which may have a key role in the processes.

Conjugative plasmids are common in *Lactococcu, Leuconostoc, Pediococcus* and some species of *Lactobacillus* (Davidson et al., 1996). However, they are rare in other *Lactobacillus* strains, in *Bifidobacterium* and in *S. thermophilus* (Álvarez-Martín et al., 2007; Pouwels and Leer, 1994). However, although suspected in many cases, no antibiotic resistance determinants have ever been localized on conjugative plasmids in naturally occurring LAB. Conjugative transfer of resistance plasmids to LAB from enterococcal and staphylococcal species in laboratory conditions has repeatedly been achieved. For example, the well characterized broad-host-range plasmid pAMβ1, originally detected in *E. faecalis,* has been transferred to strains of *L. lactis* and *L. casei* (Tannock, 1987). Subsequently, pAMβ1 was transferred from *L. lactis* to a series of *Enterococcus* and *Lactobacillus* species both *in vitro* and *in vivo* (in the mouse GIT) (McConnell et al., 1991). Similar transfer ability has been found for the *S. agalactiae* plasmid pIP501 (Thompson and Collins, 1988) and the enterococcal resistance plasmid pRE25 (Schwarz et al., 2001). Seven out of 14 tetracycline-resistant *Lactobacillus* strains from meat were able to transfer *in vitro* at variable frequency this resistance to *E. faecalis* (Gevers et al., 2003b). However, the responsible element involved in the transference has yet to be identified.

Another category of plasmids includes those that do not confer conjugative ability but can be transferred between cells by conjugative plasmids or chromosomally-encoded sex-like factors (Gasson et al., 1995). For instance, pAMβ1 is able to mobilize the non-conjugative plasmid pCaT that mediates chloramphenicol resistance in *L. plantarum* caTC2R (Ahn et al., 1992). Among others, mobilizable plasmids may include the erythromycin and streptogramin A resistant plasmid pLME300 of *L. fermentum* ROT1 (Gfeller et al., 2003), and the tetracycline-resistance plasmid pMD5057 of *L. plantarum* 5057 (Danielsen, 2002). The high promiscuity of some plasmids illustrates the possibility of transfer of complete resistance units from at least related microorganisms; and this is though to be the case of pLS55 and pMA67 from *L. sakei* and *P. larvae*, respectively.

Conjugative transposons and integrative elements have also been described in nisin-producing lactococci (Rauch et al., 1994) and *S. thermophilus* (Burrus et al., 2000). Conjugative transposons appear to be an extremely efficient and important way of disseminating genes in some bacterial groups (Whittle et al., 2002); unlike plasmids, they neither replicate nor suffer surface exclusion or incompatibility problems. But there are no reports associating conjugative transposons with antibiotic resistance in LAB, except for those found in enterococcal species (Perreten et al., 1997b; Rice and Carias, 1998). Enterococcal transposons, especially those belonging to the Tn*916*-Tn*1545* family and encoding resistance to either tetracycline, erythromycin or chloramphenicol, have been broadly transferred by conjugation to Gram-positive bacteria such as *L. lactis* and *Leuconostoc mesenteroides* (Perreten et al., 1997b). *L. lactis* cells harbouring the transposon were able to act as donors in subsequent matings (Tomita et al., 2003; Perreten et al., 1997b). In a recent report (Flórez et al., 2007a), Tn*916* encoding tetracycline resistance has been detected in two *L. lactis* cheese isolates; the resistance could be transferred from the natural isolates to plasmid-free strains of *L. lactis* and *E. faecalis*. Transposon associated erythromycin and clindamycin resistance has recently been identified in *Bifidobacterium* (van Hoek et al., 2008) species and in *S. thermophilus* strains (M. Danielsen and L. Morelli, personal communication); whether these transposons and their encoding resistances are transmissible to other microorganisms remains to be determined. However, the presence of identical transposons in different bacterial species (van Hoek et al., 2008) argues for a certain spreading capability.

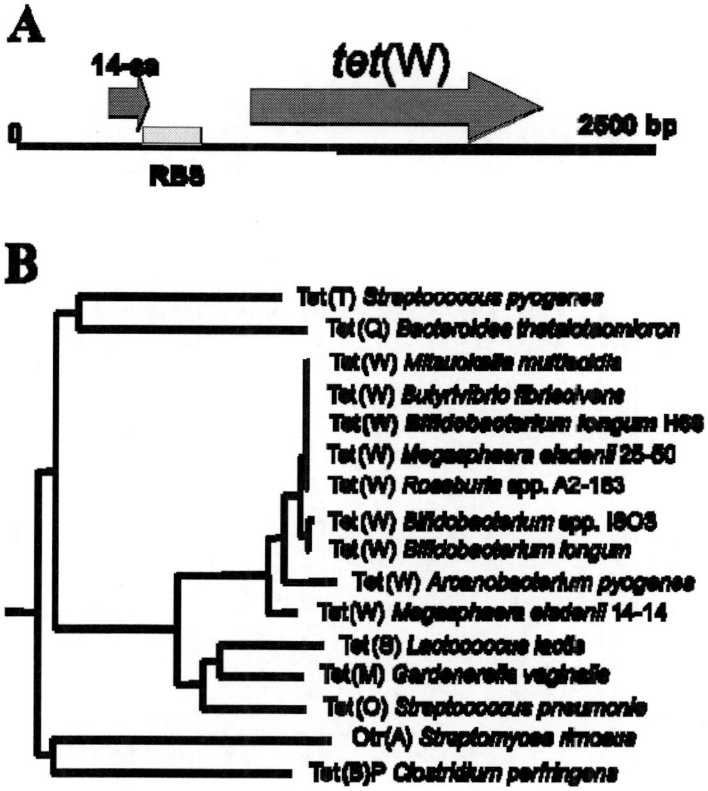

Figure 2. Panel A. Genetic organization of *tet*(W) in *Bifidobacterium longum* H66. The structural gene [*tet*(W)] is preceded by a small ORF in front of the ribosome binding site of *tet*(W) (RBS), which may encode a 14-amino acid peptide (14-αα) thought to be involved in regulation (Flórez et al., 2006d). Panel B. Phylogenetic relationships of the deduced Tet(W) from strain H66 with others and with different families of ribosome-protecting proteins involved in tetracycline resistance. Identical Tet(W) sequences are found in unrelated species and strains from the gastrointestinal environment.

Table 2. Antibiotic resistance genes identified and characterized from lactic acid bacteria and bifidobacteria species

Resistance gene(s)	Species	Origin	Transfer in culture	Reference(s)
Resistance to tetracycline				
tet(K)	*Lactobacillus* spp.	NR	-	Roberts, 2005
tet(L)	*Pediococcus parvulus*	Wine	-	Rojo-Bezares et al., 2006
tet(M)	*Bifidobacterium* spp.	Subgingival plaque	-	Lacroix and Walker, 1995
	Lactobacillus acidophilus	Human	-	Cataloluck and Gogebakan, 2004
	Lactobacillus alimentarius	Dry sausage	Yes	Gevers et al., 2003b
	Lactobacillus casei	Dairy products	-	Cataloluck and Gogebakan, 2004
	Lactobacillus crispatus	Human	-	Cataloluck and Gogebakan, 2004
	Lactobacillus curvatus	Dry sausage	Yes	Gevers et al., 2003b
	Lactobacillus fermentum	Raw milk	-	Gfeller et al., 2003b
	Lactobacillus gasseri	Human	-	Cataloluck and Gogebakan, 2004
	Lactobacillus jonhsonii	Human	-	Cataloluck and Gogebakan, 2004
	Lactobacillus paracasei	Foods	-	Huys et al., 2008
	Lactobacillus plantarum	Silage, dry sausage, wine	Yes	Gever et al., 2003a; Rojo-Bezares et al., 2006
	Lactobacillus rhamnosus	Human	-	Cataloluck and Gogebakan, 2004
	Lactococcus lactis subsp. *lactis*	Raw milk cheese	Yes	Flórez et al., 2008
tet(O)	*Bifidobacterium pseudolongum*	Pig faeces	-	Mayrhofer et al., 2007
	Lactobacillus spp.	Human	-	Chopra and Roberts, 2001; Villedieu et al., 2003
tet(Q)	*Lactobacillus* spp.	NR	-	Chopra and Roberts, 2001
tet(S),	*Lactobacillus* spp.	NR	-	Roberts, 2005
	Lactococcus lactis subsp. *lactis*	Raw milk cheese	-	Perreten et al., 2001
	Leuconostoc citreum	Sausage process line	-	Gevers et al., 2003a
	Streptococcus thermophilus	Cheese	-	Ge et al., 2007
tet(W)	*Bifidobacterium longum*	Human	Yes/No	Scott et al., 2000; Moubareck et al., 2005; Flórez et al., 2006d; Flórez et al., 2006d; Ammor et al., 2008a;
	Bifidobacterium thermophilum	Pig faeces	-	Mayrhofer et al., 2008
	Lactobacillus spp.	NR	-	Roberts, 2005; Huys et al., 2008
	Lactobacillus paracasei	Foods	-	

Resistance gene(s)	Species	Origin	Transfer in culture	Reference(s)
Resistance to tetracycline				
tet(36)	*Lactobacillus* spp.	NR	-	Roberts, 2005
Resistance to erythromycin				
erm(B)	*Lactobacillus acidophilus*	Human	-	Cataloluck and Gogebakan, 2004
	Lactobacillus animalis	Pig tonsil	-	Martel et al., 2003
	Lactobacillus casei	Dairy	-	Cataloluck and Gogebakan, 2004
	Lactobacillus crispatus	Human, pig	-	Strøman et al., 2003; Cataloluck and Gogebakan, 2004
	Lactobacillus fermentum	Human, pig	-	Cataloluck and Gogebakan, 2004
	Lactobacillus gasseri	Human	-	Cataloluck and Gogebakan, 2004
	Lactobacillus johnsonii	Human, pig	-	Martel et al., 2003; 2004; Flórez et al. 2006c
	Lactobacillus paracasei	Foods	-	Huys et al., 2008
	Lactobacillus plantarum	Dry sausage	Yes	Gevers et al., 2003b
	Lactobacillus reuteri	Chicken, pig	-	Axelsson et al., 1988; Lin et al., 1999; Martel et al., 2003
	Lactobacilus rhamnosus	Human	-	Cataloluck and Gogebakan, 2004
	Lactobacillus salivarius	Pig	-	Martel et al., 2003
	Pediococcus acidilactici	Stools, pig, wine	-	Tankovick et al., 1993; 1999; Rojo-Bezares et al., 2006
erm(LF)	*Lactobacillus fermentum*	Raw milk cheese	-	Gfeller et al., 2003
erm(T)	*Lactobacillus fermentum*	Poultry	-	Tannock et al., 1994
	Lactobacillus spp.	Swine faeces	-	Whitehead et al., 2001
	Lactococcus lactis subsp. *lactis*	Chicken caecum	-	Raha et al., 2002
erm(X)	*Bifidobacterium animalis*	Pig faeces	-	van Hoek et al., 2008
	Bifidobacterium thermophilum	Pig faeces	-	van Hoek et al., 2008
lnu(A)	*Lactobacillus reuteri*	Probiotic	-	Rosander et al., 2008

Table 2. (Continued)

Resistance gene(s)	Species	Origin	Transfer in culture	Reference(s)
Resistance to chloramphenicol				
cat-TC	*Lactobacillus plantarum*	Pork meat	-	Ahn et al., 1992
	Lactobacillus reuteri	Poultry	-	Lin et al., 1996
cat	*Lactococcus lactis* subsp. *lactis*	Raw milk cheese	-	Perreten et al., 2001
Resistance to aminoglycosides				
aac(6′)-*aph*(2″)	*Lactobacillus acidophilus*	Pig faeces	-	Tenorio et al., 2001
	Lactobacillus plantarum	Wine	-	Rojo-Bezares et al., 2006
	Lactobacillus salivarius	Pet faeces	-	Tenorio et al., 2001
	Pediococcus acidilactici	Pet faeces	-	Tenorio et al., 2001
	Pediococcus parvulus	Wine	-	Rojo-Bezares et al., 2006
	Pediococcus pentosaceus	Wine	-	Rojo-Bezares et al., 2006
ant(6)	*Lactobacillus plantarum*	Wine	-	Rojo-Bezares et al., 2006
	Pediococcus parvulus	Wine	-	Rojo-Bezares et al., 2006
vat(E-1)	*Lactobacillus fermentum*	Raw milk cheese	-	Gfeller et al., 2003
aph(3′)-IIIa	*Lactobacillus plantarum*	Wine	-	Rojo-Bezares et al., 2006
mdt(A)	*Lactococcus lactis* subsp. *lactis*	Raw milk cheese	-	Perreten et al., 2001
aad(E)	*Pediococcus acidilactici*	Dairy	-	O'Connor et al., 2007
str	*Lactococcus lactis* subsp. *lactis*	Raw milk cheese	-	Perreten et al., 2001

Key of resistance gene designation: aac, aminoglycoside acetyltransferase; ant, aminoglycoside adenylyltransferase; aph, aminoglycoside phosphotransferases; cat, chloramphenicol acetyltransferase; erm, erythromycin resistant gene; mdt, multiple drug transporter; lnu, lincosamide nucleotidyltransferase; str,streptomycin resistance gene; tet, tetracycline resistance gene; vat, streptogramin A acetyltransferase.

NR, not reported.

4. CONCLUSIONS

It is worth noting that bacteria from different genera and species, environments, and geographical locations appear to be exchanging a collection of circulating resistance genes to cope with an increasing antibiotic-containing environment. In spite of this, only a minority of LAB strains from dairy and intestinal sources show resistance levels to certain antibiotics compatible with acquired resistances. Acquired multi-resistance is a rare event in this group of bacteria, as only a few strains have been described. This relative small fraction, however, justifies monitoring antibiotic resistance programs for species intended to be used in food and feed systems, to avoid the inclusion of resistant strains in the formulation of starter and probiotic cultures. Although, resistant strains from other environments have been reported, most LAB having acquired antibiotic resistances come from the human and animal GIT, confirming the spread and transmission of resistance genes among the commensal organisms in this complex ecosystem. The analysis of the resistance determinants already identified and characterized in LAB and the comparison of their nucleotide sequences suggest that most antibiotic resistance genes emerged in microorganisms other than LAB; particularly in enterococci and other intestinal species. LAB species might be at the end of the resistance transmission chain, but they still can become a reservoir of acquired determinants. Identification of the molecular mechanisms underlying the transfer of resistance genes to and from LAB species would be essential for the control of their spread via the food chain.

5. ACKNOWLEDGMENTS

Work at the authors' laboratory was supported by projects from the VI EU Frame Program (ACE-ART, ref. CT-2003-506214) and from the Spanish Ministry of Science and Innovation (MICINN) (ref. AGL2007-61869-ALI). S. D. was awarded a postdoctoral contract from the MICINN under Juan de la Cierva Program (ref. JCI-2008-02391).

REFERENCES

Aarestrup, F. M. (2000). Characterization of glycopeptide-resistant-enterococci (GRE) from boilers and pigs in Denmark: genetic evidence that persistance of GRE in pig herds is associated with coselection by resistance to macrolides. *Journal of Clinical Microbiology*, *8,* 2774-2777.

Aires, J., Doucet-Populaire, F. & Butel, M. J. (2007). Tetracycline mediated by *tet*(W), *tet*(M), and *tet*(O) genes of *Bifidobacterium* isolates from humans. *Applied and Environmental Microbiology*, *73,* 2751-2754.

Ahn, C., Collins-Thompson, D., Duncan, C. & Stiles, M. E. (1992). Mobilization and location of the genetic determinant of chloramphenicol resistance from *Lactobacillus plantarum* caTC2R. *Plasmid*, *27,* 169-176.

Álvarez-Martín, P., Flórez, A. B. & Mayo, B. (2007). Screening for plasmids among human bifidobacteria species: sequencing and analysis of pBC1 from *Bifidobacterium catenulatum* L48. *Plasmid*, *57,* 165-174.

Ammor, M. S., Flórez, A. B. & Mayo, B. (2007a). Antibiotic resistance in non-enterococcal lactic acid bacteria and bifidobacteria. *Food Microbiology, 24,* 559-770.

Ammor, M. S., Gueimonde, M., Danielsen, M., Zagorec, M., de los Reyes-Gavilán, C. G., Mayo, B. & Margolles, A. (2007b). Two different tetracycline resistance mechanisms, a plasmid-carried *tet*(L) and a chromosomally-encoded transposon-associated *tet*(M), coexist in *Lactobacillus sakei* Rits 9. *Applied and Environmental Microbiology, 74,* 1394-1401.

Ammor, M. S., Flórez, A. B., van Hoek, A. H. A., de los Reyes-Gavilán, C. G., Aarts, H. J. M., Margolles, A. & Mayo, B. (2008a). Molecular characterization of intrinsic and acquired antibiotic resistance in lactic acid bacteria and bifidobacteria. *Journal of Molecular Microbiology and Biotechnology, 14,* 6-15.

Ammor, M. S., Flórez, A. B., Alvarez-Martín, P., Margolles, A. & Mayo, B. (2008b). Analysis of tetracycline resistance *tet*(W) genes and their flanking sequences in intestinal *Bifidobacterium* species. *Journal of Antimicrobials and Chemotherapy, 62,* 688-693.

Andersson, D. I. (2003). Persistence of antibiotic resistant bacteria. *Current Opinion in Microbiology, 6,* 452-456.

Andersson, D. I. & Levin, B. R. (1999). The biological cost of antibiotic resistance. *Current Opinion in Microbiology, 2,* 489-93.

Aquilanti, L., Garofalo, C., Osimani, A., Silvestri, G., Vignaroli, C. & Clementi, F. (2007). Isolation and molecular characterization of antibiotic-resistant lactic acid bacteria from poultry and swine meat products. *Journal of Food Protection, 70,* 557-565.

Axelsson, L.T., Ahrne, S. E., Andersson, M. C. & Stahl, S. R. (1988). Identification and cloning of a plasmid-encoded erythromycin resistance determinant from *Lactobacillus reuteri*. *Plasmid, 20,* 171-174.

Balsalobre, L., Ferrandiz, M. J., Linares, J., Tubau, F. & de la Campa, A. G. (2003). Viridans group streptococci are donors in horizontal transfer of topoisomerase IV genes to *Streptococcus pneumoniae*. *Antimicrobials Agents and Chemotherapy, 47,* 2072-2081.

Baquero, F., Baquero-Artigao, G., Canton, R. & García-Rey, C. (2002). Antibiotic consumption and resistance selection in *Streptococcus pneumoniae*. *Journal of Antimicrobials and Chemotherapy, 50,* 27-37.

Barbosa, T. M., Scott, K. P. & Flint, H. J. (1999). Evidence for recent intergeneric transfer of a new tetracycline resistance gene, *tet*(W), isolated from *Butyrivibrio fibrisolvens*, and the occurrence of *tet*(O) in ruminal bacterial. *Environmental Microbiology, 1,* 53-64.

Barbosa, T. M. & Levy, S. B. (2000). The impact of antibiotic use on resistance development and persistence. *Drug Resistance Updates, 3,* 303-311.

Bjorkman, J. & Andersson, D. I. (2000). The cost of antibiotic resistance from a bacterial perspective. *Drug Resistance Updates, 3,* 237-245.

Bolotin, A., Wincker, P., Mauger, S., Jaillon, O., Malarme, K., Weissenbach, J., Ehrlich, S.D. & Sorokin, A. (2001). The complete genome sequence of the lactic acid bacterium *Lactococcus lactis* ssp. *lactis* IL1403. *Genome Research, 11,* 731-753.

Borgen, K., Wasteson, Y., Kruse, H. & Willems, R. J. L. (2002). Vancomycin-resistant *Enterococcus faecium* (VREF) from Norwegian poultry cluster with VREF from poultry from the United Kingdom and the Netherlands in an amplified fragment length polymorphism genogroup. *Applied and Environmental Microbiology, 68,* 3133-3137.

Bronzwaer, S. L., Cars, O., Buchholz, U., Molstad, S., Goettsch, W., Veldhuijzen, I. K., Kool, J. L., Sprenger, M. J. & Degener, J. E. (2002). A European study on the relationship between antimicrobial use and antimicrobial resistance. *Emerging Infectious Diseases*, *8*, 278-282.

Bryan, L. E. & Kwan, S. (1981). Mechanisms of aminoglycoside resistance of anaerobic bacteria and facultative bacteria grown anaerobically. *Journal of Antimicrobials and Chemotherapy*, *8*, SD1-SD8.

Burrus, V., Roussel, Y., Decaris, B. & Guedon, G. (2000). Characterization of a novel integrative element, ICESt1, in the lactic acid bacterium *Streptococcus thermophilus*. *Applied and Environmental Microbiology*, *66*, 1749-1753.

Carr, F. J., Chill, D. & Maida, N. (2002). The lactic acid bacteria: a literature survey. *Critical Reviews in Microbiology*, *28*, 281-370.

Cataloluk, O. & Gogebakan, B. (2004). Presence of drug resistance in intestinal lactobacilli of dairy and human origin in Turkey. *FEMS Microbiology Letters*, *236*, 7-12.

Charteris, W. P., Kelly, P. M., Morelli, L. & Collins, J. K. (1998a). Antibiotic susceptibility of potentially probiotic *Bifidobacterium* isolates from the human gastrointestinal tract. *Letters in Applied Microbiology*, *26*, 333-337.

Charteris, W. P., Kelly, P. M., Morelli, L. & Collins, J. K. (1998b). Antibiotic susceptibility of potentially probiotic *Lactobacillus* species. *Journal of Food Protection*, *61*, 1636-1643.

Charteris, W. P., Kelly, P. M., Morelli, L. & Collins, J. K. (2001). Gradient diffusion antibiotic susceptibility testing of potentially probiotic lactobacilli. *Journal of Food Protection*, *64*, 2007-14.

Chopra, I. & Roberts, M. (2001). Tetracycline antibiotics: mode of action, applications, molecular biology, and epidemiology of bacterial resistance. *Microbiology and Molecular Biology Reviews*, *65*, 232-260.

Church, D. L., Bryant, R. D., Sim, V. & Laishley, E. J. (1996). Metronidazole susceptibility and the presence of hydrogenase in pathogenic bacteria. *Anaerobei*, *2*, 147-156.

CLSI (Clinical and Laboratory Standards Institute). (2004). *Methods for Antimicrobial Susceptibility Testing of Anaerobic Bacteria*: *Approved Standards*, (6th edition). M11-A6, *Vol. 24*, nº 2. Wayne, Pennsylvania.

Cogan, T. M. (1972). Susceptibility of cheese and yogurt starter bacteria to antibiotics. *Applied Microbiology*, *23*, 960-965.

Cohen, M. L., Wong, E. S. & Falkow, S. (1982). Common R-plasmids in *Staphylococcus aureus* and *Staphylococcus epidermidis* during a nosocomial *Staphylococcus aureus* outbreak. *Antimicrobial Agents and Chemotherapy*, *21*, 210-215.

Coppola, R., Succi, M., Tremonte, P., Reale, A., Salzano, G. & Sorrentino, E. (2005). Antibiotic susceptibility of *Lactobacillus rhamnosus* strains isolated from Parmigiano Reggiano cheese. *Lait*, *85*, 193-204.

Cosgrove, S. E. & Carmeli, Y. (2003). The impact of antimicrobial resistance *Clinical Infectious Diseases*, *36*, 1433-1437.

Danielsen, M. (2002). Characterization of the tetracycline resistance plasmid pMD5057 from *Lactobacillus plantarum* 5057 reveals a composite structure. *Plasmid*, *48*, 98-103.

Danielsen, M. & Wind, A. (2003). Susceptibility of *Lactobacillus* spp. to antimicrobial agents. *International Journal of Food Microbiology*, *82*, 1-11.

Danielsen, M., Andersen, H. S. & Wind, A. (2004). Use of folic acid casei medium reveals trimethoprim susceptibility of *Lactobacillus* species. *Letters in Applied Microbiology*, *38*, 206-210.

Danielsen, M., Mayrhofer, S., Domig, K. J., Amtmann, E., Mayer, H. K., Flórez, A. B., Mayo, B., Korhonen, J. & Tosi, L. (2007a). Assessment of the antimicrobial wild type minimum inhibitory concentration distributions of species of the *Lactobacillus delbrueckii* group. *Dairy Science and Technology, 88,* 183-191.

Danielsen, M., Simpson, P. J., O'Connor, E. B., Ross, R. P. & Stanton, C. (2007b). Susceptibility of *Pediococcus* spp. to antimicrobial agents. *Journal of Applied Microbiology, 102,* 384-389.

Davidson, B. E., Kordias, N., Dobos, M. & Hillier, A. J. (1996). Genomic organization of lactic acid bacteria. In, Venema, G., Huis in't Veld, J. H. J., Hugenholtz, J. (Eds.), *Lactic Acid Bacteria: Genetics, Metabolism and Applications,* (pp. 65-87). Dordrecht, The Netherlands: Kluwer Academic Publishers.

D'Acosta, V. M., McGrann, K., Huhges, D. W. & Wright, G. D. (2006). Sampling the antibiotic resistome. *Science, 311,* 374-377.

DeFlaun, M. F. & Levy, S. B. (1989). Genes and their varied hosts. In, Levy, S. B., Miller, R. V. (Eds.), *Gene Transfer in the Environment,* (pp. 1-32). New York: McGraw-Hill.

De Graaf, J., Elwell, L. P. & Falkow, S. (1976). Molecular nature of two β-lactamases-specifying plasmids isolated from *Haemophilus influenzae* type b. *Journal of Bacteriology, 126,* 439-446.

Delgado, S., Delgado, T. & Mayo, B. (2002). Technological performance of several *Lactococcus* and *Enterococcus* strains of dairy origin in milk. *Journal of Food Protection, 65,* 1590-1596.

Delgado, S., Flórez, A. B. & Mayo, B. (2005). Antibiotic susceptibility of *Lactobacillus* and *Bifidobacterium* species from the human gastrointestinal tract. *Current Microbiology, 50,* 202-207.

Donoghue, D. J. (2003). Antibiotic residues in poultry tissues and eggs: human health concerns? *Poultry Science, 82,* 618-621.

Eaton, T. J. & Gasson, M. J. (2001). Molecular screening of *Enterococcus* virulence determinants and potential for genetic exchange between food and medical isolates. *Applied and Environmental Microbiology, 67,* 1628-1635.

Egervärn, M., Danielsen, M., Roos, S., Lindmark, H. & Lindgren, S. (2007a). Antibiotic susceptibility profiles of *Lactobacillus reuteri* and *Lactobacillus fermentum. Journal of Food Protection, 70,* 412-418.

Egervärn, M., Lindmark, H., Roos, S., Huys, G. & Lindgren, S. (2007b). Effects of inoculum size and incubation time on broth microdilution susceptibility testing of lactic acid bacteria. *Antimicrobial Agents and Chemotherapy, 51,* 394-396.

Elliot, J. A. & Facklam, R. R. (1996). Antimicrobial susceptibilities of *Lactococcus lactis* and *Lactococcus garviae* and a proposed method to discriminate between them. *Journal of Clinical Microbiology, 34,* 1296-1298.

Elwell, L., Roberts, M., Mayer, L. & Falkow, S. (1977). Plasmid-mediated β-lactamase production in *Neisseria gonorrhoeae. Antimicrobial Agents and Chemotherapy, 11,* 528-533.

European Commission. (2005). Opinion of the scientific panel on additives and products in substances used in animal feed on the updating of the criteria used in the assessment of bacteria for resistance to antibiotics of human and veterinary importance. *The EFSA Journal, 223,* 1-12.

Evers, S., Casadewall, B., Charles, M., Dutka-Malen, S., Galimand, M. & Courvalin, P. (1996). Evolution of structure and substrate specificity in D-alanine:D-alanine ligases and related enzymes. *Journal of Molecular Evolution, 42,* 706-712.

FDA. (1997). *Substances Generally Recognized As Safe. Federal Register.* Food and Drug Administration, *62*, 18937-64.

Felten, A., Barreau, C., Bizet, C., Lagrange, P. H., & Philippon, A. (1999). *Lactobacillus* species identification, H_2O_2 production, and antibiotic resistance and correlation with human clinical status. *Journal of Clinical Microbiology*, *37,* 729-733.

Fitzgerald, G. F. & Gasson, M. J. (1988). In vivo gene transfer systems and transposons. *Biochimie*, *70,* 489-502.

Flannagan, S. E., Zitzow, L. A., Su, Y. A. & Clewell, D. B. (1994). Nucleotide sequence of the 18-kb conjugative transposon Tn*916* from *Enterococcus faecalis*. *Plasmid, 32,* 350-354.

Flórez, A. B., Delgado, S. & Mayo, B. (2005). Antimicrobial susceptibility of lactic acid bacteria isolated from a cheese environment. *Canadian Journal of Microbiology*, *51,* 51-58.

Flórez, A. B., Danielsen, M., Korhonen, J., Zycka, J., von Wright, A., Bardowski, J. & Mayo, B. (2006a). Antibiotic survey of *Lactococcus lactis* strains to six antibiotics by Etest, and establishment of new susceptibility-resistance cut-off values. *Journal of Dairy Research*, *74,* 262-268.

Flórez, A. B., Egervarn, M., Danielsen, M., Tosi, L., Morelli, L., Lindgren, S. & Mayo, B. (2006b). Susceptibility of *Lactobacillus plantarum* strains to six antibiotics and definition of new susceptibility-resistance cutoff values. *Microbial Drug Resistance*, *12,* 252-256.

Flórez, A. B., Ammor, M. S., Delgado, S. & Mayo, B. (2006c). Molecular analysis of a chromosome-carried *erm*(B) gene and its flanking insertion points in *Lactobacillus johnsonii* G41. *Antimicrobial Agents and Chemotherapy*, *50,* 4189-4190.

Flórez, A. B., Ammor, M. S., Álvarez-Martín, P., Margolles. A. & Mayo, B. (2006d). Molecular analysis of *tet*(W) gene-mediated tetracycline resistance in dominant intestinal *Bifidobacterium* species from healthy humans. *Applied and Environmental Microbiology*, *72,* 7377-9737.

Flórez, A. B., Ammor, M. S., van Hoeck, A. H. A. M., Aarts, H. J. M., Huys, G. & Mayo, B. (2007a). Antimicrobial susceptibility profiles of 32 type strains of *Lactobacillus*, *Bifidobacterium*, *Lactococcus* and *Streptococcus* spp. *International Journal of Antimicrobial Agents*, *31,* 484-6.

Flórez, A. B., Ladero, V., Álvarez-Martín, P., Ammor, M. S., Álvarez, M. A. & Mayo, B. (2007b). Acquired macrolide resistance in the human intestinal strain *Lactobacillus rhamnosus* E41 associated with a transition mutation in 23S rRNA genes. *International Journal of Antimicrobial Agents*, *30,* 341-4.

Flórez, A. B., Ammor, M. S. & Mayo, B. (2008). Identification of *tet*(M) in two *Lactococcus lactis* strains isolated from a Spanish traditional starter-free cheese made of raw milk and conjugative transfer of tetracycline resistance to lactococci and enterococci. *International Journal of Food Microbiology*, *121,* 189-194.

Foulquié Moreno, M. R., Sarantinopoulos, P., Tsakalidou, E. & De Vuyst, L. (2006). The role and application of enterococci in food and health. *International Journal of Food Microbiology*, *106,* 1-24.

Gasson, M. J., Godon, J. J., Pillidge, C. J., Eaton, T. J., Jury, K. & Shearman, C. A. (1995). Characterization and exploitation of conjugation in *Lactococcus lactis*. *International Dairy Journal, 5,* 757-762.

Ge, B., Jiang, P., Han, F., Saleh, N. K., Dhiman, N., Fedorko, D. P., Nelson, N. A. & Meng, J. (2008). Identification and antimicrobial susceptibility of lactic acid bacteria from retail fermented foods. *Journal of Food Protection, 70,* 2606-2612.

Gevers, D., Masco, L., Baert, L., Huys, G., Debevere, J. & Swings, J. (2003a). Prevalence and diversity of tetracycline resistant lactic acid bacteria and their *tet* genes along the process line of fermented dry sausages. *Systematic and Applied Microbiology, 26,* 277-283.

Gevers, D., Danielsen, M., Huys, G. & Swings, J. (2003b). Molecular characterization of *tet*(M) genes in *Lactobacillus* isolates from different types of fermented dry sausage. *Applied and Environmental Microbiology, 69,* 1270-1275.

Gfeller, K. Y., Roth, M., Meile, L. & Teuber, M. (2003). Sequence and genetic organization of the 19.3-kb erythromycin- and dalfopristin-resistance plasmid pLME300 from *Lactobacillus fermentum* ROT1. *Plasmid, 50,* 190-201.

Green, M., Wadowsky, R. M. & Barbadora, K. (1990). Recovery of vancomycin-resistant Gram-positive cocci from children. *Journal of Clinical Microbiology, 28,* 484-488.

Hall, R. M., Collis, C. M., Kim, M. J., Partridge, S. R., Recchia, G. D. & Stokes, H. W. (1999). Mobile gene cassettes and integrons in evolution. *Annals of the New York Academy of Sciences, 870,* 68-80.

Helmark, S., Hansen, M. E., Jelle, B., Sorensen, K. I. & Jensen, P. R. (2004). Transformation of *Leuconostoc carnosum* 4010 and evidence for natural competence of the organism. *Applied and Environmental Microbiology, 70,* 3695-3699.

Herrero, M., Mayo, B., González, B. & Suárez, J. E. (1996). Evaluation of technologically important traits in lactic acid bacteria isolated from spontaneous fermentations. *Journal of Applied Bacteriology, 81,* 565-570.

Herreros, M. A., Sandoval, H., González, L., Castro, J. M., Fresno, J. M. & Tornadijo, M. E. (2005). Antimicrobial activity and antibiotic resistance of lactic acid bacteria isolated from Armada cheese (a Spanish goats' milk cheese). *Food Microbiology, 22,* 455-459.

Hummel, A. S., Hertel, C., Holzapfel, W. H. & Franz, C. M. A. P. (2007). Antibiotic resistances of starter and probiotic strains of lactic acid bacteria. *Applied and Environmental Microbiology, 73,* 730-739.

Huys, G., D'Haene, K. & Swings, J. (2002). Influence of the culture medium on antibiotic susceptibility testing of food-associated lactic acid bacteria with the agar overlay disc diffusion method. *Letters in Applied Microbiology, 34,* 402-406.

Huys, G., D'Haene K. & Swings, J. (2006). Genetic basis of tetracycline and minocycline resistance in potentially probiotic *Lactobacillus plantarum* strain CCUG 43738. *Antimicrobial Agents and Chemotherapy, 50,* 1550-1551.

Huys, G., D'Haene, K., Danielsen, M., Mättö, J., Egervärn, M. & Vandamme, P. (2008). Phenotypic and molecular assessment of antimicrobial resistance in *Lactobacillus paracasei* strains of food origin. *Journal of Food Protection, 71,* 339-344.

Kafili, T., Razavi, S. H., Djomeh, Z. E., Salehi, G. R., Álvarez-Martín, P. & Mayo, B. (2009). Antibiotic resistance-susceptibility profiles of *Lactobacillus* strains from Lighvan, a traditional Iranian raw milk cheese. *Milchwissenschaft, (in press).*

Katla, A. K., Kruse, H., Johnsen, G. & Herikstad, H. (2001). Antimicrobial susceptibility of starter culture bacteria used in Norwegian dairy products. *International Journal of Food Microbiology*, *67,* 147-152.

Kaye, D. (2004). Current use for old antibacterial agents: polymyxins, rifampin, and aminoglycosides. *Infectious Diseases and Clinics of North America, 18,* 669-689.

Kazimierczak, K. A., Flint, H. J. & Scott, K. P. (2006). Comparative analysis of sequences flanking *tet*(W) resistance genes in multiple species of gut bacteria. *Antimicrobial Agents and Chemotherapy, 50,* 2632-2639.

Kim, K. S., Morrison, J. O. & Bayer, A .S. (1982). Deficient autolytic enzyme activity in antibiotic-tolerant lactobacilli. *Infection and Immunity, 36,* 582-585.

Klare, I., Konstabel, C., Müller-Bertling, S., Reissbrodt, R., Huys, G., Vancanneyt, M., Swings, J., Goossens, H. & Witte, W. (2005). Evaluation of new broth media for microdilution antibiotic susceptibility testing of lactobacilli, lactococci, pediococci, and bifidobacteria. *Applied and Environmental Microbiology, 71,* 8982-8986.

Klein, G. (2003). Taxonomy, ecology and antibiotic resistance of enterococci from food and the gastro-intestinal tract. *International Journal of Food Microbiology, 88,* 123-131.

Klein, G., Hallmann, C., Casas, I. A., Abad, J., Louwers, J. & Reuter, G. (2000). Exclusion of *vanA, vanB* and *vanC* type glycopeptide resistance in strains of *Lactobacillus reuteri* and *Lactobacillus rhamnosus* used as probiotics by polymerase chain reaction and hybridization methods. *Journal of Applied Microbiology, 89,* 815-824.

Kleerebezem, M., Boekhorst, J., van Kranenburg, R., Molenaar, D., Kuipers, O. P., Leer, R., Tarchini, R., Peters, S. A., Sandbrink, H. M., Fiers, M. W., Stiekema, W., Lankhorst, R. M., Bron, P. A., Hoffer, S. M., Groot, M. N., Kerkhoven, R., de Vries, M., Ursing, B., de Vos, W. M. & Siezen, R. J. (2003). Complete genome sequence of *Lactobacillus plantarum* WCFS1. *Proceeding of the National Academy of Sciences USA, 100,* 1990-1995.

Lacroix, J. M. & Walker, C. B. (1995). Detection and incidence of the tetracycline resistance determinant *tet*(M) in the microflora associated with adult periodontitis. *Journal of Periodontology, 66,* 102-108.

Leroy, F. & de Vuyst, L. (2004). Functional lactic acid bacteria starter cultures for the food fermentation industry. *Trends in Food Science and Technology, 15,* 67-78.

Levy, S. B. (1985). Ecology of plasmids and unique DNA sequences. In, Haalvorson, H.C., Pramer, D., Rogul, M. (Eds.). *Engineered Organisms in the Environment: Scientific Issues. Washington DC,* ASM Press.

Levy, S. B. (1997). Antibiotic resistance: an ecological imbalance. In, *Antibiotic Resistance: Origins, Evolution, Selection and Spread. Ciba Foundation Symposium, S207,* 1-14. Chichester: Wiley.

Levy, S. B. & Marshall, B. (2004). Antibacterial resistance worldwide: causes, challenges and responses. *Nature Medicine, 10,* 122-129.

Lim, K. S., Huh, C. S. & Baek, Y. J. (1993). Antimicrobial susceptibility of bifidobacteria. *Journal of Dairy Science, 76,* 2168-2174.

Lin, C. F., Fung, Z. F., Wu, C. L. & Chung, T. C. (1996). Molecular characterization of a plasmid-borne (pTC82) chloramphenicol resistance determinant (*cat-TC*) from *Lactobacillus reuteri* G4. *Plasmid, 36,* 116-124.

Margolles, A., Moreno, J. A., van Sinderen, D. & de los Reyes-Gavilán, C. G. (2005). Macrolide resistance mediated by a *Bifidobacterium breve* membrane protein. *Antimicrobials Agents and Chemotherapy*, *49,* 4379-4381.

Martel, A., Meulenaere, V., Devriese, L. A., Decostere, A. & Haesebrouck, F. (2003). Macrolide and lincosamide resistance in the gram-positive nasal and tonsillar flora of pigs. *Microbial Drug Resistance*, *9,* 293-297.

Masco, L., Van Hoorde, K., De Brandt, E., Swings, J. & Huys, G. (2006). Antimicrobial susceptibility of *Bifidobacterium* strains from humans, animals and probiotic products. *Journal of Antimicrobial Chemotherapy*, *58,* 85-94.

Mathur, S. & Singh, R. (2005). Antibiotic resistance in food lactic acid bacteria. A review. *International Journal of Food Microbiology*, *105,* 281-295.

Matteuzzi, D., Crociani, F. & Brigidi, P. (1983). Antimicrobial susceptibility of *Bifidobacterium*. *Annals of Microbiology (Paris)*, *134,* 339-249.

Mättö, J., Suihko, M. L. & Saarela, M. (2006). Comparison of three test media for antimicrobial susceptibility testing of bifidobacteria using the Etest method. *International Journal of Antimicrobial Agents*, *28,* 42-48.

Mättö, J., van Hoek, A. H. A. M., Domig, K. J., Saarela, M., Flórez, A. B., Brockmann, E., Amtmann, E., Mayo, B., Aarts, H. J. M. & Danielsen, M. (2007). Susceptibility of human and probiotic *Bifidobacterium* spp. to selected antibiotics as determined by the Etest method. *International Dairy Journal*, *9,* 1123-1131.

Mayrhofer, S., Domig, K. J., Amtmann, E., van Hoek, A. H., Petersson, A., Mair, C., Mayer, H. K. & Kneifel, W. (2007). Antibiotic susceptibility of *Bifidobacterium thermophilum* and *Bifidobacterium pseudolongum* isolates from animal sources. *Journal of Food Protection*, *70,* 119-124.

Mayrhofer, S., Domig, K. J., Mair, C., Zitz, U., Huys, G. & Kneifel, W. (2008). Comparison of broth microdilution, Etest, and agar disk diffusion methods for antimicrobial susceptibility testing of *Lactobacillus acidophilus* group members. *Applied and Environmental Microbiology*, *74,* 3745-3748.

McConnell, M. A., Mercer, A. A. & Tannock, G. W. (1991). Transfer of plasmid pAMß1 between members of the normal microflora inhabiting the murine digestive tract and modification of the plasmid in a Lactobacillus *reuteri* host. *Microbial Ecology in Health and Disease*, *4,* 343-355.

McManus, P. S., Stockwell, V. O., Sundin, G. W. & Jones, A. L. (2002). Antibiotic use in plant agriculture. *Annual Reviews in Phytopathology*, *40,* 443-465.

Miller, J. H. (1996). Spontaneous mutators in bacteria: insights into pathways of mutagenesis and repair. *Annual Reviews in Microbiology*, *50,* 625-643.

Moubareck, C., Gavini, F., Vaugien, L., Butel, M. J. & Doucet-Populaire, F. (2005). Antimicrobial susceptibility of bifidobacteria. *Journal of Antimicrobial Chemotherapy*, *55,* 38-44.

Murray, K. D., Aronstein, K. A. & de Leon, J. H. (2007). Analysis of pMA67, a predicted rolling-circle replicating, mobilizable, tetracycline-resistance plasmid from the honey bee pathogen, *Paenibacillus larvae*. *Plasmid*, *58,* 89-100.

Nandi, S., Maurer, J. J., Hofacre, C. & Summers, A. O. (2004). Gram-positive bacteria are a major reservoir of Class 1 antibiotic resistance integrons in poultry litter. *Proceeding of the National Academy of Sciences USA*, 101, 7118-7122.

Netherwood, T., Bowden, R., Harrosin, P., O'Donnel, A. G., Parker, D. S. & Gilbert, H. J. (1999). Gene transfer in the gastrointestinal tract. Appl Environ Microbiol 65:5139-5141

Normark BH, Normark S (2002) Evolution and spread of antibiotic resistance. *Journal of International Medicine*, *252,* 91-106.

Nirdnoy, W., Mason, C. J. & Guerry, P. (2005). Mosaic structure of a multiple-drug-resistant, conjugative plasmid from *Campylobacter jejuni. Antimicrobial Agents and Chemotherapy*, *49,* 2454-2459.

O'connor, E. B., O'sullivan, O., Stanton, C., Danielsen, M., Simpson, P. J., Callanan, M. J., Ross, R. P. & Hill, C. (2007). pEOC01: A plasmid from *Pediococcus acidilactici* which encodes an identical streptomycin resistance (*aadE*) gene to that found in *Campylobacter jejuni. Plasmid*, *58,* 115-126.

Orberg, P. K. & Sandine, W. E. (1985). Survey of antimicrobial resistance in lactic streptococci. *Applied and Environmental Microbiology*, *49,* 538-542.

Ouwehand, A. C., Salminen, S. & Isolauri, E. (2002). Probiotics: an overview of beneficial effects. *Antonie van Leeuwenhoek*, *82,* 279-289.

Patterson, A. J., Rincon, M. T., Flint, H. J. & Scott, K. P. (2007). Mosaic tetracycline resistance genes are widespread in human and animal fecal samples. *Antimicrobial Agents and Chemotherapy*, *51,* 1115-1118.

Perreten, V., Schwarz, F., Cresta, L., Boeglin, M., Dasen, G. & Teuber, M. (1997a). Antibiotic resistance spread in food. *Nature*, *389,* 801-802.

Perreten, V., Kolloffel, B. & Teuber, M. (1997b). Conjugal transfer of the Tn*916*-like transposon TnFO1 from *Enterococcus faecalis* isolated from cheese to other Gram-positive bacteria. *Systematic and Applied Microbiology*, *20,* 27-38.

Perreten, V., Schwarz, F. V., Teuber, M. & Levy, S. B. (2001). Mdt(A), a new efflux protein conferring multiple antibiotic resistance in *Lactococcus lactis* and *Escherichia coli*. *Antimicrobial Agents and Chemotherapy*, *45,* 1109-1114.

Perreten, V., Vorlet-Fawer, L., Slickers, P., Ehricht, R., Kuhnert, P. & Frey, J. (2005). Microarray-based detection of 90 antibiotic resistance genes of Gram-positive bacteria. *Journal of Clinical Microbiology*, *43,* 2291-2302.

Pfeltz, R. F. & Wilkinson, B. J. (2004). The escalating challenge of vancomycin resistance in *Staphylococcus aureus*. *Current Drug Targets and Infectious Disorders*, *4,* 273-294.

Pouwels, P. H. & Leer, R. J. (1994). Genetics of lactobacilli: plasmids and gene expression. *Antonie van Leeuwenhoek*, *64,* 85-107.

Putman, M., van Veen, H. W., Degener, J. E. & Konings, W. N. (2001). The lactococcal secondary multidrug transporter LmrP confers resistance to lincosamides, macrolides, streptogramins and tetracyclines. *Microbiology*, *147,* 2873-2880.

Raha, A. R., Ross, E., Yusoff, K., Manap, M. Y. & Ideris, A. (2002). Characterisation and molecular cloning of an erythromycin resistance plasmid of *Lactococcus lactis* isolated from chicken cecum. *Journal of Biochemistry, Molecular Biology and Biophysics*, *6,* 7-11.

Rauch, P. J., Beerthuyzen, M. M. & de Vos, W. M. (1994). Distribution and evolution of nisin-sucrose elements in *Lactococcus lactis*. *Applied and Environmental Microbiology*, *60,* 1798-1804.

Reynolds, P. E. & Courvalin, P. (2005). Vancomycin resistance in enterococci due to synthesis of precursors terminating in D-alanyl-D-serine. *Antimicrobial Agents and Chemotherapy*, *49,* 21-25.

Rice, L. B. & Carias, L. L. (1998). Transfer of Tn5385, a composite, multiresistance chromosomal element from *Enterococcus faecalis*. *Journal of Bacteriology, 180,* 714-721.

Roberts, M. C. (2005). Update on acquired tetracycline resistance genes. *FEMS Microbiology Letters, 245,* 195-203.

Rojo-Bezares, B., Saénz, Y., Poeta, P., Zarazaga, M., Ruiz-Larrea, F. & Torres, C. (2006). Assessment of antibiotic susceptibility within lactic acid bacteria strains isolated from wine. *International Journal of Food Microbiology, 111,* 234-240.

Rosander, A., Connolly, E. & Roos, S. (2008). Removal of antibiotic resistance gene-carrying plasmids from *Lactobacillus reuteri* ATCC 55730 and characterization of the resulting daughter strain, *L. reuteri* DSM 17938. *Applied and Environmental Microbiology, 74,* 6032-6040.

Saarela, M., Mogensen, G., Fonden, R., Mättö, J. & Mattila-Sandholm, T. (2002). Probiotic bacteria: safety, functional and technological properties. *Journal of Biotechnology, 84,* 197-215.

Saarela, M., Maukonen, J., von Wright, A., Vilpponen-Salmela, T., Patterson, A. J., Scott, K. P., Hamynen, H. & Mättö, J. (2007). Tetracycline susceptibility of the ingested *Lactobacillus acidophilus* LaCH-5 and *Bifidobacterium animalis* subsp. *lactis* Bb-12 strains during antibiotic/probiotic intervention. *International Journal of Antimicrobial Agents, 29,* 271-280.

Salyers, A. A., Shoemaker, N. B. & Bonheyo, G. T. (2002). The ecology of antibiotic resistance genes. In, Lewis, K., Salyers, A.A., Taber, H.W., Wax, R.G. (Eds.). *Bacterial Resistance to Antimicrobials*, pp. 1-17. New York: Marcel Dekker Inc.

Salyers, A. A., Gupta, A. & Wang, Y. (2004). Human intestinal bacteria as reservoirs for antibiotic resistance genes. *Trends in Microbiology, 12,* 412-416.

Schwarz, F. V., Perreten, V. & Teuber, M. (2001). Sequence of the 50-kb conjugative multiresistance plasmid pRE25 from *Enterococcus faecalis* RE25. *Plasmid, 46,* 170-187.

Scott, K. P., Barbosa, T. M., Forbes, K. J. & Flint, H. J. (1997). High-frequency transfer of a naturally occurring chromosomal tetracycline resistance element in the ruminal anaerobe *Butyrivibrio fibrisolvens*. *Applied and Environmental Microbiology, 63,* 3405-3411.

Scott, K. P., Melville, C. M., Barbosa, T. M. & Flint, H. J. (2000). Occurrence of the new tetracycline resistance gene *tet*(W) in bacteria from the human gut. *Antimicrobial Agents and Chemotherapy, 44,* 775-777.

Shales, D. M., Projan, S. J. & Edwards, J. E. (2004). Antibiotic discovery: state of the state. *ASM News, 70,* 275-281.

Stanton, T. B. & Humphrey, S. B. (2003). Isolation of tetracycline-resistant *Megaspaera elsdenii* strains with novel mosaic gene combinations of *tet*(O) and *tet*(W) from swine. *Applied and Environmental Microbiology, 69,* 3874-3882.

Stanton, T. B., McDowall, J. S. & Rasmussen, M. A. (2004). Diverse tetracycline resistance genotypes of *Megasphaera elsdenii* strains selectively cultured from swine feces. *Applied and Environmental Microbiology, 70,* 3754-3757.

Strøman, P., Müller, C. C. & Sørensen, K. I. (2003). Heat shock treatment increases the frequency of loss of an erythromycin resistance-encoding transposable element from the chromosome of *Lactobacillus crispatus* CHCC3692. *Applied and Environmental Microbiology, 69,* 7173-7180.

Strommenger, B., Kettlitz, C., Werner, G. & Witte, W. (2003). Multiplex PCR assay for simultaneous detection of nine clinically relevant antibiotic resistance genes in *Staphylococcus aureus*. *Journal of Clinical Microbiology*, *41*, 4089-4094.

Swenson, J. M., Facklam, R. R. & Thornsberry, C. (1990). Antimicrobial susceptibility of vancomycin-resistant *Leuconostoc*, *Pediococcus*, and *Lactobacillus* species. *Antimicrobial Agents and Chemotherapy*, *34*, 543-549.

Swenson, J. M., Ferraro, M. J., Sahm, D. F., Charache, P. & Tenover, F. C. (1992). New vancomycin disk diffusion breakpoints for enterococci. The National Committee for Clinical Laboratory Standards Working Group on Enterococci. *Journal of Clinical Microbiology*, *30*, 2525-2528.

Tankovic, J., Leclercq, R. & Duval, J. (1993). Antimicrobial susceptibility of *Pediococcus* spp. and genetic basis of macrolide resistance in *Pediococcus acidilactici* HM3020. *Antimicrobial Agents and Chemotherapy*, *37*, 789-792.

Tannock, G. W. (1987). Conjugal transfer of plasmid pAMß1 in *Lactobacillus reuteri* and between lactobacilli and *Enterococcus faecalis*. *Applied and Environmental Microbiology*, *53*, 2693-2695.

Tannock, G. W., Luchansky, J. B., Miller, L., Connell, H., Thode-Andersen, S., Mercer, A. A. & Klaenhammer, T. R. (1994). Molecular characterization of a plasmid-borne (pGT633) erythromycin resistance determinant (*ermGT*) from *Lactobacillus reuteri* 100-63. *Plasmid*, *31*, 60-71.

Tenorio, C., Zarazaga, M., Martínez, C. & Torres, C. (2001). Bifunctional enzyme 6'-N-aminoglycoside acetyltransferase-2"-O-aminoglycoside phosphotransferase in *Lactobacillus* and *Pediococcus* isolates of animal origin. *Journal of Clinical Microbiology*, *39*, 824-825.

Tenover, F. C. (2006). Mechanisms of antimicrobial resistance in bacteria. *American Journal of Medicine*, *119*, S3-S10.

Teuber, M.,. Meile, L. & Schwarz, F. (1999). Acquired antibiotic resistance in lactic acid bacteria from food. *Antonie van Leeuwenhoek*, *76*, 115-137.

Thompson, J. K. & Collins, M. A. (1988). Evidence for the conjugal transfer of the broad host range plasmid pIP501 into strains of *Lactobacillus helveticus*. *Journal of Applied Bacteriology*, *65*, 309-319.

Tomita, H., Tanimoto, K., Hayakawa, S., Morinaga, K., Ezaki, K. & Oshima, H. I. (2003). Highly conjugative pMG1-Like plasmids carrying Tn*1546*-Like transposons that encode vancomycin resistance in *Enterococcus faecium*. *Journal of Bacteriology*, *185*, 7024-7028.

Tosi, L., Berruti, G., Danielsen, M., Wind, A., Huys, G. & Morelli, L. (2007). Susceptibility of *Streptococcus thermophilus* to antibiotics. *Antonie van Leeuwenhoek*, *92*, 21-28.

van Hoek, A. H., Scholtens, I. M., Cloeckaert, A. & Aarts, H. J. (2005). Detection of antibiotic resistance genes in different *Salmonella* serovars by oligonucleotide microarray analysis. *Journal of Microbiological Methods*, *62*, 13-23.

van Hoek, A. H. A. M, Mayrhofer, S., Domig, K. J., Flórez, A. B., Ammor, M. S., Mayo, B. & Aarts, H. J. M. (2007). Mosaic tetracycline resistance genes and their flanking regions in *Bifidobacterium thermophylum* and *Lactobacillus johhsonii*. *Antimicrobial Agents and Chemotherapy*, *52*, 248-252.

van Hoek, A. H. A. M., Mayrhofer, S., Domig, K. J. & Aarts, H. J. M. (2008). Resistance determinant *erm*(X) is borne by transposon Tn*5432* in *Bifidobacterium thermophilum* and

Bifidobacterium animalis subsp. *lactis*. *International Journal of Antimicrobial Agents, 31,* 544-548.

Vitali, B., Turroni, S., Dal Piaz, F., Candela, M., Wasinger, V. & Brigidi, P. (2008). Genetic and proteomic characterization of rifaximin resistance in *Bifidobacterium infantis* BI07. *Research in Microbiology, 158,* 355-362.

Villedieu, A., Díaz-Torres, M. L., Hunt, N., McNab, R., Spratt, D. A., Wilson, M. & Mullany, P. (2003). Prevalence of tetracycline resistance genes in oral bacteria. *Antimicrobial Agents and Chemotherapy, 47,* 878-882.

Volkmann, H., Schwartz, T., Bischoff, P., Kirchen, S. & Obst, U. (2004). Detection of clinically relevant antibiotic-resistance genes in municipal wastewater using real-time PCR (TaqMan). *Journal of Microbiological Methods, 56,* 277-286.

Wegener, H. C. (2003). Antibiotics in animal feed and their role in resistance development. *Current Opinion in Microbiology, 6,* 439-445.

Wehrli, W. (1983). Rifampin: mechanisms of action and resistance. *Reviews in Infectious Diseases, 5,* S407-S411.

Whitehead, T. R. & Cotta, M. A. (2001). Sequence analyses of a broad host-range plasmid containing *ermT* from a tylosin-resistant *Lactobacillus* sp. isolated from swine feces. *Current Microbiology, 43,* 17-20.

Whittle, G., Shoemaker, N. B. & Salyers, A. A. (2002). The role of Bacteroides conjugative transposons in the dissemination of antibiotic resistance genes. *Cellular and Molecular Life Sciences, 59,* 2044-2054.

Witte, W. (2000). Selective pressure by antibiotic use in livestock. *International Journal of Antimicrobial Agents, 16,* S19-S24.

Wood, B. J. B. & Warner, P. J. (2003). *Genetics of Lactic Acid Bacteria.* New York: Kluwer Academic/Plenum Publishers.

Woodford, N., Johnson, A. P., Morrison, D. & Speller, D. C. (1995). Current perspectives on glycopeptide resistance. *Clinical Microbiology Reviews, 8,* 585-615.

Wright, G. D. (2003). Mechanisms of resistance to antibiotics. *Current Opinion in Chemical Biology, 7,* 563-569.

Yazid, A. M., Ali, A. M., Shuhaimi, M., Kalaivaani, V., Rokiah, M. Y. & Reezal, A. (2000). Antimicrobial susceptibility of bifidobacteria. *Letters in Applied Microbiology, 31,* 57-62.

Zarazaga, M., Sáenz, Y., Portillo, A., Tenorio, C., Ruiz-Larrea, F., Del Campo, R., Baquero, F. & Torres, C. (1999). In vitro activities of ketolide HMR3647, macrolides, and other antibiotics against *Lactobacillus, Leuconostoc,* and *Pediococcus* isolates. *Antimicrobial Agents and Chemotherapy, 43,* 3039-3041.

Zhou, J. S., Pillidge, C. J., Gopal, P. K. & Gill, H. S. (2005). Antibiotic susceptibility profiles of new probiotic *Lactobacillus* and *Bifidobacterium* strains. *International Journal of Food Microbiology, 98,* 211-217.

In: Encyclopedia of DNA Research (3 Volume Set)
Editors: S. J. Duncan and P. H. Wiley

ISBN 978-1-61324-305-3

Chapter 28

ANTIBODY VARIABLE-REGION SEQUENCING FOR HYBRIDOMA AUTHENTICATION

Miha Kosmač[1,2,*], Simon Koren[1,2,*], Nives Škrlj[2], Marko Dolinar[2] and Vladka Čurin Šerbec[1,2]

[1] Department for Production of Diagnostic Reagents and Research, Blood Transfusion Centre of Slovenia, Šlajmerjeva 6, SI-1000 Ljubljana, Slovenia

[2] Faculty of Chemistry and Chemical Technology, University of Ljubljana, Aškerčeva 5, SI-1000 Ljubljana, Slovenia

ABSTRACT

Miss-identification and cross-contamination of cell lines used in research and biotechnology represent widespread problems. Appropriate cell authentication methods should thus be used regularly to avoid invalidating scientific conclusions and to assure the quality of biotechnological products. DNA fingerprinting is often used to establish and verify the identities of various cell lines, but this approach is unsuitable for hybridoma cell lines that are derived from syngeneic animals. Here, we describe the sequencing of antibody variable regions, the only highly diverse region of the hybridoma genome, as a robust and accessible method for hybridoma cell line authentication. The protocol involves RNA isolation, reverse transcription of immunoglobulin variable regions, and amplification of the resulting cDNA using highly degenerate primers, chosen to amplify the majority of the possible variable regions. Depending on the quality of PCR amplification, the PCR products can then be sequenced either directly or after cloning into a plasmid vector. Additionally, we provide an alternative protocol based on rapid amplification of cDNA ends, which can be used to obtain variable region sequences where successful amplification with the degenerate primers is not achieved.

INTRODUCTION

Cell lines owe much of their fame to the involuntary contribution of a young woman

* These authors contributed equally to this work.

named Henrietta Lacks, whose cervical cancer cells were harvested and grown in 1951. Since the discovery and consequent mass distribution of what became known as HeLa cells [1], cultured cell lines have become indispensible tools in medicine, biochemistry and biotechnology. Their constant behaviour, ease of manipulation, and immortality have secured them a place in most academic and industrial biotechnological laboratories around the world. Many very complicated experiments have been skilfully performed using various cell lines, yet so few researchers have performed the simple tests to authenticate the cells they believe to be working with, to be sure that they are truly the right ones.

It was not long after the use of cultured cell lines became worldwide that the first reports of cell-line cross-contamination surfaced. It was again HeLa cells that proved to be the culprit, reappearing in the United States after having been sent from the Soviet Union as perfectly unique cell lines [2]. More than four decades later, and armed with a wealth of knowledge about cell-line cross-contamination, the level of incorrectly identified cell lines that continue to be used in research worldwide is shockingly high. This is estimated to occur in 16% to 35% of studies [3-7]. Although research papers reporting the use of confirmed cross-contaminated cell lines still fill scientific journals, small but steady steps in the right direction are being made by the scientific community. A lot has been done to inform researchers around the world about the severity of this problem, and user-friendly protocols, such as those described here, are being developed to enable laboratories to authenticate their cell lines. Perhaps the biggest hurdle that remains is to persuade those that still turn a blind eye to this problem that contamination of cell lines can happen to anyone, and that it is in their best interests to embrace rather than disregard cell-line authentication practices in their own laboratory.

Appropriate methods for cell-line authentication have been developed and should be applied on a regular basis to avoid spurious scientific conclusions and false results. Early methods for cell-line authentication relied on karyotyping [8,9] and isoenzyme profiling [10]. While cytogenetics remains indispensable for identifying tumour cell lines with published karyotypes, the advent of DNA fingerprinting has enabled the discrimination between different cell lines with unparalleled resolution and accuracy [11]. Today, the most commonly used method for human cell-line identification is short tandem repeat (STR) profiling. It provides fast, easy and reproducible identification of cells, analogous to its application in the field of forensic science [12].

Most of the cell-line authentication protocols now in use were developed for human cancer cell lines and have little or no use for cell lines of different origins [13,14]. Much like things began with their human counterparts, the problem of cell-line cross-contamination in the case of hybridomas has so far gone relatively unnoticed. This is surprising, especially when considering (a) the enormous impact that hybridoma technology and the applications to which the resulting monoclonal antibodies (mAbs) have been applied in modern life sciences, and (b) the commercial viability of many hybridoma cell lines. However outdated hybridoma technology may appear at times, the final product, the mAb, is still a very important tool in various biomedical fields, such as immunology, histopathology and biochemistry, notwithstanding diagnostics and therapy.

Therefore, a fast, reproducible, robust and easy-to-perform method for hybridoma cell-line authentication is warranted. STR profiling might at first appear the logical choice also for hybridomas, however most hybridoma cell lines stem from inbred animals, and this makes the DNA fingerprinting approach much more difficult. Moreover, as is the case with most hybridoma cell lines, both fusion partners, i.e. the B cell isolated from the immunised animal and the myeloma cell, originate from the same inbred animal strain (e.g. BALB/c mice). This results in a high genetic uniformity between the various unique hybridoma cell lines, thus rendering the STR profiling approach inapplicable. To the best of our knowledge, there has only been one report of successful discrimination of hybridoma cell lines by DNA fingerprinting, however this involved human-mouse hybridomas [15].

In the case of mAbs, biochemical characterisation is by far the most widely applied approach to address this issue, although characterisation of binding and stability are labour intensive, insensitive, and especially problematic in cases where the antigen is difficult to obtain, or if it is toxic or infectious. Laboratories that develop and produce mAbs usually deal with large numbers of hybridoma cell lines. Therefore, several cell lines that produce different mAbs directed against the same antigen molecule are normally obtained from one cell fusion. The resulting mAbs can differ only slightly in their binding properties, making identification of the cell lines by this approach even more difficult. Clearly, introducing hybridoma authentication protocols that are quicker and simpler, require less material, and offer higher discriminatory powers would be of great benefit to laboratories and companies that handle these cell lines. Effective hybridoma authentication procedures could also be used for improved protection of intellectual property.

An elegant way of applying both approaches, i.e. using the only feature of hybridoma cell lines that is unique – the mAb they produce, while at the same time exploiting the robust, reproducible and easy to perform nature of DNA typing methods is to sequence the mAb variable regions and apply the results to the identification of the cell lines that produced the mAb. In our laboratory, we have assembled a constantly growing list of sequences that provide a molecular fingerprint of the mAbs we use on a daily basis, and we have thus eliminated the possibility of undetected cross-contamination of our hybridoma cell lines with cells of a similar nature. Due to its lower genetic variability, the mAb light chain is especially useful for cell-line authentication, meaning that only a small number of oligonucleotide primers are needed to cover the majority of possible sequences, while still providing enough discriminatory power to differentiate between unique cell lines.

In this chapter, we will introduce two protocols that we have implemented for sequencing variable regions of murine mAbs produced by hybridoma technology [16]. The basic protocol describes the use of highly degenerated primers for reverse transcription, followed by PCR amplification of most of the variable region sequences (heavy and light chains) that we have come across. Due to the vast number of murine Ab genes[1], it is practically impossible to amplify them all with a limited number of primers. Therefore, we have optimised an alternative, although more tedious, protocol, using rapid amplification of 5'-cDNA ends (5'-RACE). This we have applied in instances where we have not been able to obtain a variable region sequence using the basic protocol.

[1] In this chapter, the term gene will refer to the rearranged DNA encoding a single Ab chain that is unique to a specific B-cell clone. This is in contrast to the variable region gene segments, which are also sometimes referred to as variable or V genes, and which are present in their germ-line order in all nucleated cells.

Basic Protocol

Three different germ-line loci code for three possible Ab chains: the heavy chain, the kappa light chain, and the lambda light chain. These are located on mouse chromosomes 12, 6 and 16, respectively [17]. During B-cell development, the germ-line sequences become rearranged so that one of the possible V gene segments is juxtaposed to one of the J gene segments in the case of light chains, or to a similarly rearranged D-J segment in the case of heavy chains [18]. It is only after transcription and RNA processing that all of the non-coding sequences are discarded and the entire translatable region, combined with the appropriate constant regions, lies in one contiguous sequence. There are 119-124 functional Ig*h* gene segments in the murine genome, and 99-101 functional Igκ and 8-9 Igλ gene segments, depending on the strain of mice [19]. Even though not all of the combinations give rise to functional immunoglobulins (Igs), the actual number of unique Ig rearrangements is still too vast to manage easily. To amplify all of the possible mouse Ig rearrangements, conserved stretches within the Ig coding sequence must be found. At the 3'-end of mature Ig (heavy or light chain) mRNA, there is the sequence that codes for the constant domains, the sequences of which are invariable in all Abs of one isotype within one strain of inbred mice (there are, however, allotypic differences between mice of different strains [20]). The 5'-end, however, is much more diverse. There is a considerable number of different sequences coding for the variable domains, and therefore sets of primers must be used that will anneal to all of the possible different sequences. There have been a number of reports claiming the design of "universal" primers that can amplify all encountered Ig variable region sequences [21-23]; however, in our experience, none are truly "universal". We have therefore opted to use the small number of highly degenerate primers reported by Wang et al. [23], which can anneal to a large number of different sequences and resort to other alternatives only if we are unsuccessful with these.

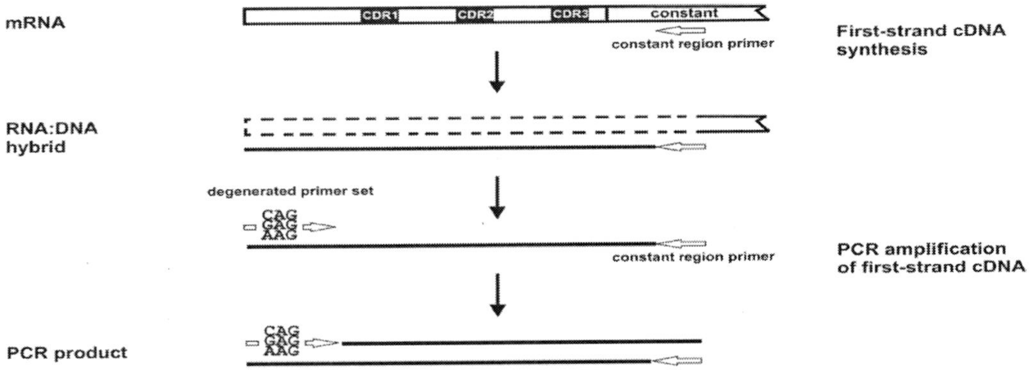

Figure 1. Schematic representation of the basic protocol for amplification of immunoglobulin variable regions using degenerate primers.

Materials

We list here the materials that we use for the experiments described below. Of note, many different reagents and suppliers can be chosen. When using reagents or kits other than those listed here, please refer to the manufacturer instructions regarding the materials that should be supplied by the user. It is assumed that the materials and equipment for the proper handling of cell cultures are available.

RNA extraction

RNeasy Mini kit (QIAGEN GmbH, Hilden, Germany);
2-mercaptoethanol (Sigma-Aldrich Chemie GmbH, Steinheim, Germany);
Syringes and needles;
Table-top centrifuge with rotor for 2-ml tubes.

cDNA synthesis

M-MLV Reverse Transcriptase (Promega, Madison, WI, USA);
Nuclease-free water (Promega, Madison, WI, USA);
dNTP Mix (Promega, Madison, WI, USA);
RNasin® Ribonuclease Inhibitor (Promega, Madison, WI, USA);
Isotype-specific constant region primers (Invitrogen, Paisley, UK).

PCR amplification of Ig variable regions

GoTaq Polymerase (Promega, Madison, WI, USA);
dNTP Mix (Promega, Madison, WI, USA);
Primers for amplification of Ig variable regions (Invitrogen, Paisley, UK);
Thermal cycler with block for 200-µl tubes.

Note: Thermostable polymerases with 3'-5'-exonuclease activity (e.g. *Pfu* or *Vent*) cannot be used here, as the use of degenerate primers generates primer:template mismatches, which are not amplified by the so-called "proof-reading" polymerases.

Electrophoretic separation and purification of PCR products

Agarose (Sigma-Aldrich Chemie GmbH, Steinheim, Germany);
Wizard® SV Gel and PCR Clean-Up System (Promega, Madison, WI, USA);
DNA gel electrophoresis chamber and power supply.

DNA sequencing

BigDye Terminator v3.1 Cycle Sequencing Kit (Applied Biosystems, Foster City, CA, USA);
DyeEx 2.0 Spin Kit (QIAGEN GmbH, Hilden, Germany);
Thermal cycler with block for 200-µl tubes;
ABI PRISM® 3100 Genetic Analyser (Applied Biosystems, Foster City, CA, USA);
Formamide (Sigma-Aldrich Chemie GmbH, Steinheim, Germany).

Cloning of PCR products into cloning vector

CloneJET™ PCR Cloning Kit (Fermentas, Vilnius, Lithuania);
GeneJET™ Plasmid Miniprep Kit (Fermentas, Vilnius, Lithuania);
T4 DNA Ligase (Fermentas, Vilnius, Lithuania);
LB Broth powder (Sigma-Aldrich Chemie GmbH, Steinheim, Germany);
Bacteriological agar (Sigma-Aldrich Chemie GmbH, Steinheim, Germany);

Competent *Escherichia coli* DH5α cells for transformation;

Autoclave.

The primers used for cDNA synthesis and PCR amplification of the hybridoma immunoglobulin variable regions are essentially the same as those described by Wang et al. [23]. Although the occurrence of lambda light chains is rare in mice [24], which led Wang et al. [23] not to include primers for PCR amplification of lambda light chain variable regions, we have nevertheless come across a cell line producing a mAb with a lambda light chain. We have therefore constructed our own variable and constant region primers based on the sequences published by Krebber et al. [25] and those in the IMGT database [19].

Methods

1. Preparation of hybridoma cells for RNA extraction

Although mRNA represents only about 1% to 5 % of the total RNA in most cells, hybridoma cells are usually rich in Ig-coding mRNA. Purification of RNA from different cell lines will yield very different amounts of target mRNA, and therefore a good number of cells to start with is 1×10^6 to 5×10^6 cells of good viability. This will ensure enough RNA for all of the downstream applications. Pour the cell suspension into a 50-ml centrifuge tube and centrifuge the cells for 5 min at 300× g to remove the culture medium from the cells. Decant the medium. Wash the cells with 20 ml PBS and repeat the centrifugation to remove residual traces of the medium.

2. Purification of RNA from hybridoma cells

"In-house methods *versus* commercial kits?" If your budget does not allow you to buy a commercial RNA extraction kit, a good protocol for purification of RNA from scratch is described by Sambrook and Russel [26]. There are a number of commercial kits from various suppliers that you can use to extract either total RNA or only the mRNA fraction, although, bearing in mind that hybridoma cell cultures are very good samples to work with, perhaps the simplest kit you can find is the best. Also, there is no need to isolate only the mRNA fraction, because the subsequent downstream reactions specifically amplify regions from Ig-coding mRNA only. Another useful feature of commercial kits is that it is possible to avoid toxic organic solvents completely, such as phenol and chloroform, both of which are commonly used in "in-house" protocols. Whatever the preferred method, follow the manufacturer instructions to ensure maximal yield and purity of the resultant RNA.

The cells should first be lysed using lysis buffer containing guanidinium salts that deactivate RNases. Be sure to loosen the cell pellet completely and mix the cells with the lysis buffer thoroughly, otherwise incomplete lysis will result in much lower RNA yields. After lysis, genomic DNA (gDNA) should be fragmented (e.g. using a syringe and needle, and passing the sample through the needle many times), otherwise the high viscosity of the sample will prevent further purification. The gDNA can be eliminated by passing the sample through an affinity column or by adding RNase-free DNase to the sample. After cell disruption and elimination of genomic DNA, the sample is applied to a silica-based affinity column where RNA is adsorbed, and multiple washes are performed.

Note: Most wash buffers contain ethanol, so be sure to eliminate all of the residual traces of ethanol after the final washing step, otherwise the residual ethanol can inhibit downstream reactions.

The RNA is eluted from the column using RNase-free water or 10 mM Tris-HCl, pH 7.5. The concentration of the eluted RNA should be as high as achievable, and preferably exceeding 50 μg/ml. The expected yield from 1×10^6 cells ranges from 5 μg to 50 μg total RNA.

Once the RNA is extracted, its purity and concentration can be assessed spectrophotometrically. Some spectrophotometers already have a built-in conversion algorithm, but as a rule of thumb, the concentration can be calculated using the following equation:

$$\text{RNA concentration } [\mu g/ ml] = A_{260} \times 40$$

where 40 is the average extinction coefficient of RNA [27].

A more precise estimation of RNA concentrations can be obtained by measuring the absorbance of serially diluted samples of standard preparations of RNA and comparing the absorbance values obtained for the purified RNA to a standard curve. A common method to measure RNA purity is to measure the ratio of absorbance of A_{260}/A_{280}. A value close to 2.0 is preferred, but the ratio can vary widely, especially due to fluctuations in pH [28]. Values ranging from 1.7 to 2.3 are normally obtained, and do not interfere with downstream reactions.

The integrity of the RNA obtained can also be assessed on a denaturing agarose gel. Two prominent bands should be seen, corresponding to the 28S and 18S ribosomal RNA, respectively. The intensity of the 28S rRNA band should be roughly twice that of the 18S rRNA band. Much less prominent smears above, between and below both bands correspond to other RNA species of various lengths, including mRNA [27].

Contamination with DNA fragments

Although most protocols, and especially those using commercial kits, involve a step for elimination of genomic DNA, small amounts of DNA will inevitably be co-purified with the RNA. Performing RT-PCR with the gDNA contaminants should yield considerably larger products (if they are amplified at all) than from Ig-coding mRNA. Nevertheless, there is a possibility that certain gDNA fragments will interfere with the downstream reactions, so it is important to check the PCR product size prior to purification and sequencing. A database search should always be performed with the sequences obtained, to verify that the amplified products are in fact rearranged Ig variable regions. A negative control sample can also be added, where genomic DNA isolated from the same hybridoma cell culture is substituted for cDNA in the PCR reaction (see next section). This way, the sizes of the potential contaminating PCR products can be seen on an agarose gel.

Note: RNA is more prone to degradation than DNA, especially at higher temperatures and higher pH. Also, the RNA-degrading enzymes, the RNases, are very stable and are virtually everywhere (e.g. hands, glassware). The combined effects of this mean that the extraction and storage of RNA must be carried out with the utmost care, to avoid contamination with RNases. Use sterile plastic-ware instead of glassware, wear gloves, and keep samples on ice as much as possible (note, however, that RNases are still active at 4 °C). Most commercial

RNA extraction protocols use highly denaturing guanidinium-salt-containing buffers that deactivate RNases as soon as cell lysis is carried out. A variety of RNase inhibitors can also be added.

3. First-strand cDNA synthesis

RNA is reverse transcribed into cDNA using a viral enzyme called reverse transcriptase, which catalyses the addition of dNTPs to a growing DNA chain on the basis of an RNA template. For the enzyme to begin polymerisation, a priming site has to be present on the RNA chain.

For first-strand cDNA synthesis, use the following reaction mixture in a total reaction volume of 25 μl:

2 μg purified total RNA;
1 μg of appropriate constant region primer;
5 μl 5× reaction buffer;
25 U RNasin Ribonuclease Inhibitor;
200 U M-MLV reverse transcriptase;
0.5 mM dNTP;
Nuclease-free water.

Note: AMV reverse transcriptase can also be used, but because of its comparatively weaker RNase H activity, M-MLV is the preferred enzyme for cDNA synthesis.

Before adding the rest of the reaction mixture, prepare the RNA and corresponding constant region primer in a separate tube, and heat to 70 °C for 10 min, to disrupt any RNA secondary structure that might have formed. After heating, place the tube immediately on ice to prevent secondary structure reforming. RNA can form very stable secondary structures that can markedly reduce the efficacy of reverse transcription. Add the rest of the reverse transcriptase mix to the tube, mix gently, spin down, and incubate at 42 °C for 60 min. Although heat inactivation of reverse transcriptase can be performed afterwards, in our experience it is not necessary.

4. PCR amplification of Ig variable region sequences

For PCR amplification of the resultant cDNA, both constant and variable region primers must be used. Prepare the following PCR reaction mix in a total reaction volume of 50 μl:

10 μl 5× Green GoTaq reaction buffer;
0.2 μM variable region primer;
0.2 μM constant region primer;
0.2 μM dNTP mix;
1.25 U GoTaq polymerase;
1 μl cDNA reaction mix from the previous step;
Nuclease-free water.

Mix gently, spin down, and place the tube in a thermal cycler pre-heated to 95 °C. Incubate in a thermal cycler using the following cycling conditions:

For heavy chains

Initial melting step at 95 °C for 3 min, followed by 30 cycles of:
95 °C for 45 s;
45 °C for 45 s;
72 °C for 1 min.
Hold the reactions at 72 °C for 7 min, then cool to 4 °C and store at -20 °C.

For light chains

Initial melting step at 95 °C for 3 min, followed by 10 cycles of:
95 °C for 45 s;
45 °C for 45 s:
72 °C for 1 min.

Then 10 cycles of:
95 °C for 45 s;
47 °C for 45 s;
72 °C for 1 min.

Then 10 cycles of:
95 °C for 45 s;
50 °C for 45 s;
72 °C for 1 min.

Finally, hold the reactions at 72 °C for 7 min, then cool to 4 °C, and store at -20 °C.

5. Agarose gel electrophoresis

After the reaction, separate the PCR-obtained products by gel electrophoresis on 2% agarose gels stained with either ethidium bromide or another appropriate DNA-staining dye. The gel electrophoresis should be run at 8 V/cm in TAE buffer for approximately 30-40 min. Estimate the sizes and quantities of the products by comparing them to molecular-weight markers. The predicted sizes of the amplified variable regions are between 350 bp and 400 bp. Be sure to inspect the gel for any unexpected products. An additional band that we frequently observe is the amplified chain from the heavy chain mRNA of the parent myeloma NS1 cells that were used in the fusion experiment (see Figure 2, lane 1). This band can be seen at a slightly lower molecular weight than most amplified heavy chain variable regions due to a 50-bp deletion, which is characteristic of the NS1 heavy chain [29]. Even though a functional polypeptide chain is not secreted, the mRNA is still present in the cell and it is amplified in the PCR reaction.

After electrophoretic separation, excise the products of the correct sizes and proceed to DNA extraction.

6. DNA extraction from agarose gels

There are numerous commercial kits and in-house protocols for DNA extraction from agarose gels. It is important to note that most protocols can only extract DNA from gels run with TAE buffer, or they require the addition of specific reagents for extraction from TBE-containing gels.

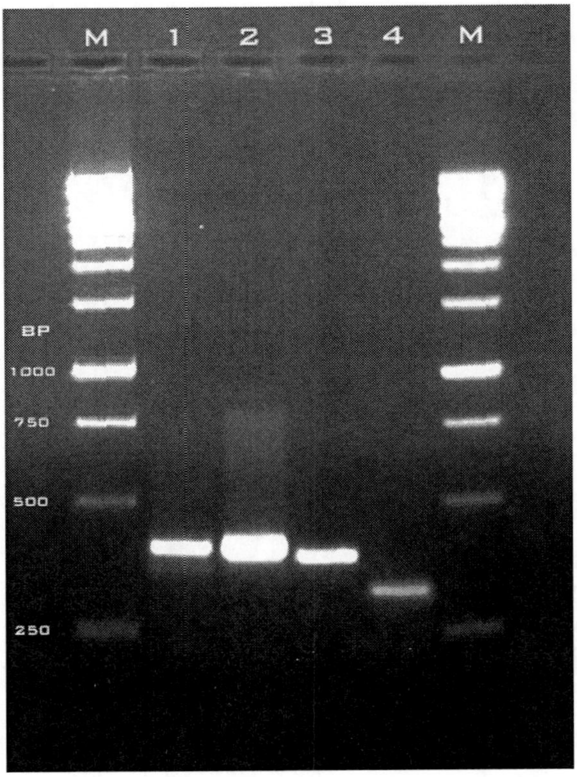

Figure 2. Electrophoretic separation of the PCR products. M, molecular-weight markers; lane 1, mAb A - amplified γ1 heavy chain variable region; lane 2, mAb B - amplified μ heavy chain variable region; lane 3, mAb A - amplified κ light chain variable region; lane 4, mAb B - amplified λ light chain variable region.

The reason for this is that borate ions form complexes with agarose, thus hindering extraction. There will also be decreased yields if the agarose concentration is above 2%. After purification of the appropriate bands, run an aliquot of the purified DNA again on a 2% agarose gel, and estimate its concentration compared to the molecular-weight markers.

Note: Although DNA concentrations can be assessed spectrophotometrically, there are a number of contaminants that absorb ultraviolet light, thereby falsely increasing the estimated concentrations. Therefore, a much more precise estimate is obtained by simply running the sample on another agarose gel, and comparing the signal obtained with that of the molecular-weight markers.

7. Sequencing

a) Direct sequencing of the PCR products

The purified fragments excised from the gel can be directly sequenced if only one sharp band corresponding to the correct size is seen on the gel. In cases where the hybridoma cell line was not monoclonal or where the amplified product cannot be separated from the parent myeloma chain, cloning of the PCR products into a cloning vector must be carried out (see next section).

Most of the time, however, there will only be one PCR product, which can be sequenced directly. The same primers as those used for PCR can be used for the sequencing reaction. As the first 50 bp or so cannot be read, sequencing must be carried out in both directions, to provide overlap at both ends.

b) Cloning of the PCR products

When you are certain that the hybridoma cell line you wish to study is not monoclonal, or as is most often the case, when the sequencing results obtained show multiple overlapping sequences, the PCR product must be cloned into a cloning vector, so that single variable region sequences can be obtained. PCR products obtained with *Taq* polymerase can easily be cloned into an appropriate vector (e.g. pGEM-T; Promega, Madison, WI, USA) by taking advantage of the 3'-adenosine overhang that is left by the polymerase; alternatively, the ends can be polished using a blunting enzyme, and cloned into a blunt-ended cloning vector (e.g. pJET1.2; Fermentas, Vilnius, Lithuania). After ligation of the PCR products and transformation of the vector into *E. coli* bacterial cells, around 20-50 random insert-positive bacterial colonies should be assayed by colony PCR, and the colonies bearing the insert of the right size grown in 5 ml liquid medium. The plasmids should be purified and used for sequencing with vector-specific primers.

DNA sequencing

Regardless of which of the aforementioned options you choose, the products still need to be sequenced using the Sanger dideoxy-termination method. If an automated genetic analyser and a skilled technician are available, sequencing can be performed "in-house" by first incorporating fluorescently labelled dideoxynucleotides in a cycle sequencing reaction, followed by capillary separation of labelled single-stranded DNA fragments.

For the cycle sequencing reaction, in a total volume of 10 µl, add:

4 µl Terminator Ready Reaction Mix;

2 µl purified PCR product (approximately 20-40 ng) or purified plasmid DNA (150-300 ng);

5-10 pmol primer;

Nuclease-free water.

Mix gently, spin down, and incubate in a thermal cycler pre-heated to 95 °C. Use the following thermal cycling conditions:

30 cycles of:

95 °C for 30 s;

45 °C for 5 s;

60 °C for 2 min.

Cool to 4°C, and store at -20 °C.

After the sequencing reaction, the products should be purified to remove excess primers and unincorporated dye terminators. Although ethanol or isopropanol precipitation protocols are much cheaper, we recommend using commercially available spin columns. These are fast, easy to use, and provide much higher yields and purities of the fragments to be sequenced.

After the sequencing reaction products have been purified, dry them using a vacuum manifold or thermal cycler. Be careful to avoid overdrying. Products of the sequencing reaction should be dissolved in 10 μl formamide prior to analysis. When performing the sequencing reaction using BigDye sequencing chemistry, the sequencing fragments can be analysed on one of the ABI PRISM Genetic Analysers (Applied Biosystems; see manufacturer protocol for all possible makes and models).

Not all laboratories dealing with hybridomas, however, can afford such an analyser, therefore a commercial sequencing service can be used. These days there are many commercial sequencing providers that use robotic sequencers, and that offer fast and good quality results at very competitive prices. Some providers even offer bacterial colony picking and DNA purification along with their sequencing service, thus greatly reducing the amount of work hours that are needed to sequence your samples. Both of the approaches described can be performed, i.e. direct sequencing of the PCR products or sequencing of inserts in a cloning vector, omitting the steps involving the sequencing reaction, purification of sequencing fragments, and analysis on a genetic analyser.

8. Data analysis

Whichever method of sequencing is used, the signal obtained must be converted into a readable sequence. When using the dye-terminator sequencing method, the readout is in the form of a chromatogram that corresponds to the signal intensities of the different dye terminators incorporated. Most computer software packages that come with sequencing apparatuses can automatically transform the signal intensities into a sequence file. However, there are a number of common sequencing problems (e.g. dye blobs or base dropouts); therefore, it is essential that you also inspect the chromatograms manually to check for difficult-to-read sections or miscalled bases. When sequencing PCR products directly, there will be a stretch of about 40 bp to 50 bp missing at the start of the sequence, corresponding to the first 30 or so bp of the incorporated unlabelled primer, followed by a stretch of about 10 bp to 20 bp of unreadable sequence. Sequencing the same PCR product in both directions enables the entire variable region from one end to the other to be read reliably. Superimposed signal peaks should be seen at positions corresponding to the degenerate sites in the variable-region primers, but nowhere else in the sequence. When cloning the PCR products into a cloning vector, vector-specific primers can be used and there is no need to sequence in both directions, because the start of the insert will usually be well within the stretch of the readable sequence. Due to the incorporation of a single insert into one vector molecule, there will only be one base at the degenerate sites. After extracting the correct sequence from the readout, the entire sequence should be checked against databases of other immunoglobulin sequences (e.g. IMGT and BLAST [19,30]), and the results should show a rearranged murine Ig variable region sequence of either the heavy chain or the light chain.

Anticipated results

The protocol described should yield the sequences of most mouse heavy and light chain Ig variable regions. There are several points at which the results can be assessed and the steps that follow can be modified accordingly. The integrity of the purified total RNA can be assessed on a denaturing agarose gel. Two prominent bands should be seen, corresponding to 28S and 18S rRNA, respectively, and a slight smear of variously sized other RNA molecules, including mRNA, in the background. If the rRNA species show a high degree of degradation, there is a good chance that the target mRNA is also degraded. Agarose gel electrophoresis of PCR products is also a useful point, where the effectiveness of the cDNA amplification can be examined. Due to the high fold of degeneracy of the variable region primers, two unwanted effects can occur:

1. Non-specific priming. Due to the relatively low annealing temperature and high degeneracy of primers, non-specific annealing is possible, especially in cases where the target sequence is primed sub-optimally. This is seen on gels as bands migrating at unexpected sizes, or multiple bands in one lane. Bands corresponding to the amplified target sequence should be sharp and intense (see Figure 2).
2. No priming at all. Some variable-region sequences simply cannot be amplified because of the degree of non-complementarity between the template and the primers. This is seen as a smear on gels, with no prominent band present.

After DNA sequencing, the results are usually obtained in the form of a chromatogram that corresponds to the signal intensities and colour of the incorporated dye terminators (i.e. if dye-terminator sequencing is performed). When sequencing PCR products directly, a good way to ensure that no bases are called incorrectly is to align both the forward and the reverse direction sequences, and to inspect the overlapping region for mismatches. As mentioned above, there will inevitably be stretches of 40 bp to 50 bp at both ends where there will be no overlap. After obtaining the entire end-to-end sequence, this can be aligned with other immunoglobulin sequences using common internet databases. A very good place to start is the ImMunoGeneTics (IMGT) Database at http://imgt.cines.fr/. [19]. There is a wealth of information on various types of immune-system-related proteins, and their web resources can directly predict Ig CDR regions from the nucleotide sequences provided. A BLAST [30] sequence search for similar sequences should also show that the sequence obtained is a rearranged Ig variable region from either a heavy or a light chain.

Troubleshooting

Low RNA yield.
Possible causes:

a. cell-culture medium not completely removed;
b. cell pellet not homogenised and solubilised completely;
c. genomic DNA not removed;
d. ethanol carry-over;
e. too much starting material;
f. cells of poor quality/ viability;
g. RNA degradation.

No PCR product seen on agarose gels.
Possible causes:

a. poor/no priming of template;
b. primer degradation;
c. RNA degradation.

Multiple PCR products seen on agarose gels.

Possible causes:

a. non-specific priming;
b. priming of myeloma cell variable-region sequence;
c. contamination with gDNA.

No readable sequence – low or no signal.
Possible causes:

a. not enough template in sequencing reaction;
b. contamination of sequencing reaction (e.g. other nucleic acids, proteins, salts, detergents);
c. poor/no priming of template (see previous section).

No readable sequence – "noisy" chromatogram.
Possible causes:

a. too much template DNA in sequencing reaction;
b. DNA template not pure;
c. incorrect cycle sequencing conditions;
d. primer self annealing (primer dimer);
e. leftover dye (dye blobs);
f. high GC content – difficult stretches;
g. secondary structure formation;
h. polymerase slippage;
i. primer degradation.

Decline of signal strength.
Possible causes:

a. too much template DNA in sequencing reaction;
b. secondary structure formation;
c. high GC content – difficult stretches;
d. polymerase slippage.

Alternative Protocol

Rapid amplification of 5'-cDNA ends (5'-RACE) is a technique for obtaining complete DNA sequences from mRNAs with unknown 5'-end sequences [31]. The general strategy involves three sequential enzymatic steps: (1) reverse transcription of specific mRNA; (2) homopolymeric tailing of first-strand cDNA; and (3) amplification by PCR (see Figure 3). Two gene-specific primers are required, one for reverse transcription of the mRNA into the first strand cDNA, and the second one for PCR amplification, binding upstream of the first primer.

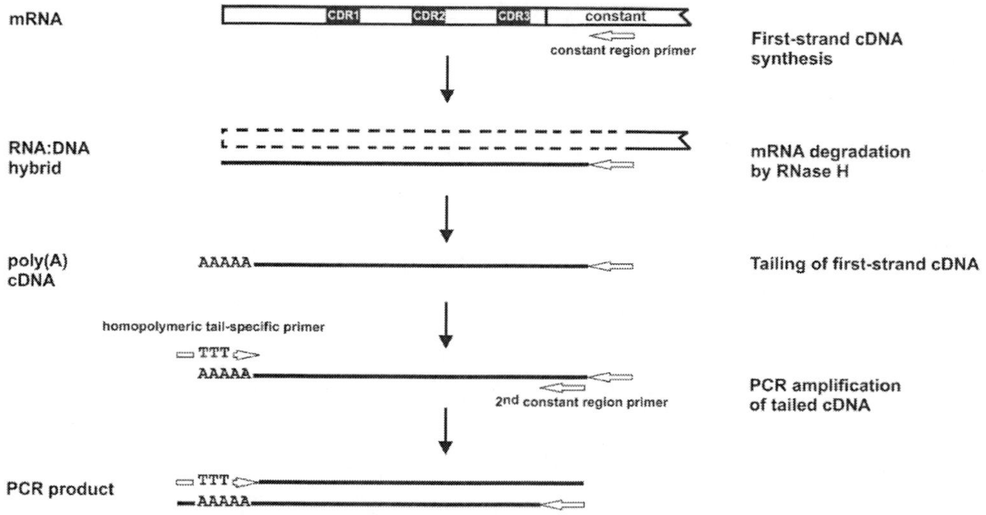

Figure 3. Schematic representation of the 5'-RACE method.

A primer-binding site upstream of the unknown 5'-sequence of the target mRNA is provided by a synthetic tail added by terminal transferase (see Figure 3, for review, see [32]). The modified PCR technique described here can be used to obtain DNA sequences from Ig variable heavy and light chain genes that fail to be amplified with degenerate primers. A very important additional feature of this method, and one that is required when recombinant Abs or fragments of Abs are desired, is that the exact 5'-end sequences that include signal sequences can be obtained, since no degenerate primers are used.

Materials

The materials used for amplification of Ig-variable regions by 5'-RACE are as given below.

RNA extraction and purification

See Basic Protocol.

cDNA synthesis

ImProm-II Reverse Transcriptase (Promega, Madison, WI, USA);
dNTP Mix (Fermentas, Vilnius, Lithuania);
RiboLock RNase Inhibitor (Fermentas, Vilnius, Lithuania);
Isotype-specific constant-region primer 1 (e.g. RO1_IgG1, RO1_Igk; see Table 1; Invitrogen, Paisley, UK);
RNase H (Applied Biosystems/Ambion, Austin, TX, USA).

Tail addition

Terminal Deoxyribonucleotide Transferase (Promega, Madison, WI, USA);
dNTP (e.g. dATP) (Fermentas, Vilnius, Lithuania).

PCR amplification

Pfu polymerase or *Taq* polymerase (Fermentas, Vilnius, Lithuania);

dNTP Mix (Fermentas, Vilnius, Lithuania);

Isotype-specific constant region primer 2 (e.g. constant region primer used in the Basic Protocol; Invitrogen, Paisley, UK);

Homopolymeric tail-specific primer (e.g. d(t)17);

Thermal cycler.

Table 1. Primers used for 5'-RACE

Primer name	Region	Primer sequence
RO1_Igk	constant	TGTTAACTGCTCACTGGATGGTGG
RO1_IgG1	constant	GTTTGATTGGGCCGCAGATCCAGG
RO1_IgG2a	constant	GGTCAAGGTGACTGGCTCAGGGAA
d(t)17	poly(A) tail	GACTCGAGTCGACATCGATTTTTTTTTTTTTTTTT

PCR product cloning and analysis

See Basic Protocol.

Primers used for 5'-RACE are listed in Table 1. A commercial DNA extraction kit is recommended for the purification of the DNA. All incubations can be performed in a thermal cycler.

The primers for the first-strand cDNA synthesis of the Ig variable region are designed on the basis of the sequences of the light and heavy constant regions (RO1_Igk for kappa light chain, RO1_IgG1 and RO1_IgG2a for IgG1 and IgG2a heavy chains, respectively). For the synthesis of the second-strand cDNA and for PCR amplification, a second constant region primer (e.g. the constant region primer as used in the Basic Protocol) and a homopolymeric tail complement primer (d(t)17) are used. Primer d(t)17 consists of a 17-nt oligo-(dT) sequence that primes the poly(A) tail, and a non-homologous sequence encoding for restriction sites at the 5'-end of the primer, which facilitates cloning. The primers for other Ab isotypes can be constructed based on sequences from the IMGT database [19].

Methods

1. Extraction of RNA from hybridoma cells

See Basic Protocol.

2. First-strand cDNA synthesis

Reverse transcriptase is used for transcription of specific mRNA into the first-strand cDNA using an appropriate constant region primer (e.g. RO1_IgG1). In general, 0.5 µg to 1.0 µg total RNA is sufficient for the amplification of specific Ig variable regions by 5'-RACE. See Basic Protocol for reaction mixture details.

The result of the mRNA reverse transcription is an RNA:DNA heteroduplex. When M-MLV reverse transcriptase is used, its RNase H activity degrades RNA in the RNA:DNA hybrid. If the reverse transcriptase reaction mixture is used directly for PCR amplification (as described in the Basic Protocol), the heteroduplex is further denaturated by heating. In RACE, the first-strand cDNA is used in a reaction with terminal transferase, which exhibits a preference for single-stranded substrates. Therefore, 5 U RNase H is added to the reaction mixture after thermal inactivation of reverse transcriptase (70 °C, 10 min) and incubated at 37 °C for 30 min for optimal results. Afterwards, a DNA purification kit should be used for purification of the first-strand cDNA from the reaction mixture, which can be used directly for poly(A) tailing. It is important to eliminate unused primer and free dNTPs from the first strand product, to avoid tailing of the residual primer and the consequent multiple PCR products.

3. Addition of tails

A homopolymeric tail is added to the 3'-end of the first-strand cDNA in a reaction that is catalysed by terminal deoxynucleotidyl transferase (TdT) using only one of the possible four dNTPs. Usually, 20-100 nucleotides are added to the DNA fragment in the tailing reaction, and these represent a primer binding site for second-strand synthesis.

For addition of the poly(A) tail, prepare the following reaction mixture in a sterile microcentrifuge tube on ice, in a total reaction volume of 20 µl:

5-10 µl purified first-strand cDNA (the optimal amount depends on the amount of RNA used in the first-strand reaction and on the purification efficiency);

4 µl 5× reaction buffer;

150 pmol dATP (or any other dNTP);

30-40 U TdT;

Nuclease-free water.

Incubate for 1 h at 37 °C, and then for 10 min at 70 °C, to inactivate the enzyme. Place the tube on ice. Isolate the poly(A)-tailed cDNA with a DNA Purification kit. TdT buffer is not compatible with the PCR buffer system, therefore the tailed cDNA has to be purified before PCR.

Note: Any nucleotide can be used for the homopolymeric tailing, along with the appropriate tail-specific primer. Since A:T base pairs are less stable than C:G base pairs, longer stretches are required for priming and fewer internal primings occur during amplification.

4. Synthesis of second-strand cDNA and PCR amplification of the Ig variable-region sequence

Finally, the Ig-coding variable region is amplified in a PCR reaction using two primers: one complementary to the 5'-end (constant region) and the other complementary to the homopolymeric tail at the 3'-end of first-strand cDNA. In contrast to the Basic Protocol, where *Taq* polymerase is required for amplification with degenerate primers, *Pfu* polymerase is preferred for RACE, due to its lower error rate.

Prepare the following PCR reaction mixture, in a total reaction volume of 50 µl:

5 µl 10× reaction buffer;

1.5 mM $MgSO_4$;

0.2 µM constant region primer (e.g. constant region primer as used in the Basic Protocol);

0.2 µM homopolymeric tail primer (d(t)17);

0.2 µM dNTP mix;

2.5 U *Pfu* polymerase;

1-5 μl purified tailed first-strand cDNA;
Nuclease-free water.
Incubate in a thermal cycler preheated to 95 °C, using the following conditions:
Initial denaturation at 95 °C for 5 min, followed by 33 cycles of:
95 °C for 1 min;
55 °C for 1 min;
68 °C for 1 min.
Incubate at 68 °C for 10 min, then cool to 4 °C and store at -20 °C.

Note: Set up a series of PCR reactions to determine the optimal amounts of tailed first-strand cDNA and to obtain a sufficient amount of PCR product.

5. Agarose gel electrophoresis

The PCR products should be analysed by gel electrophoresis, as described in the Basic Protocol. Usually, only a single distinct band at 600 bp or 500 bp is seen for amplified cDNA of Ig heavy and light variable regions, respectively.

6. DNA extraction, sequencing, data analysis, and cloning of PCR products

The PCR products are purified from the agarose gel and cloned into an appropriate cloning vector for sequencing, as described in the Basic Protocol.

In summary, 5'-RACE is an efficient and straightforward method (see Figure 3) in which total RNA is first reverse transcribed by reverse transcriptase, using a constant-region primer. First-strand cDNA is then tailed by terminal transferase, and finally amplified by PCR using a second gene-specific primer and a homopolymeric tail-specific primer.

Troubleshooting

No PCR product seen on agarose gel.
Possible causes:

a) RNA degradation or secondary structure;
b) error in first-strand cDNA synthesis;
c) low purity of cDNA;
d) inefficient tailing of first-strand cDNA;
e) polymerase inhibition by TdT buffer;
f) inappropriate primers.

Multiple PCR products or smear seen on agarose gels.
Possible causes:

a) RNA degradation;
b) contamination by gDNA;
c) partial reverse transcriptase products;
d) non-specific priming (annealing temperature too low).

CONCLUSION

Cell-line authentication protocols need to be used in all laboratories that deal with cell culture material if the integrity of the cell lines used is to be ensured. Experience shows that morphological observations and cell culture assays show too much inter-assay variability to be reliable sources of discrimination. Short tandem repeat profiling is the method of choice when dealing with human cancer tissue cultures. When dealing with cell lines of different origins, a trait inherent to the cells in question has to be found that can be used to differentiate among the myriad cell lines that most laboratories deal with every day. For hybridoma cells, that trait is the variable region sequences from the Ab they produce. We have described two protocols here that can be used to obtain sequences of Ig variable regions of both heavy and light chains. We would suggest that a list is made (and kept updated) of all of the cell cultures that are in use in your laboratory, and that sequencing of their Ig variable regions is carried out, to see which are unique and which are identical. Who knows, you might be in for a surprise, should you see that cell lines A and B, which were thought to be different, are actually identical.

REFERENCES

[1] Gey, GO; Coffman, WD; Kubicek, MT. Tissue culture studies of the proliferative capacity of cervical carcinoma and normal epithelium. *Cancer Res.*, 1952, 1952, 12, 264-5.

[2] Gold, MA. Conspiracy of Cells: One Woman's Immortal Legacy and the Medical Scandal It Caused. Albany: *State University of New York Press*, 1986.

[3] Lacroix, M. Persistent use of "false" cell lines. *Int. J. Cancer*, 2008 Jan, 1, 122(1), 1-4

[4] MacLeod, RA; Dirks, WG; Matsuo, Y; Kaufmann, M; Milch, H; Drexler, HG. Widespread intraspecies cross-contamination of human tumor cell lines arising at source. *Int. J. Cancer*, 1999 Nov, 12, 83(4), 555-63.

[5] Drexler, HG; Dirks, WG; Matsuo, Y; MacLeod, RA. False leukemia-lymphoma cell lines: an update on over 500 cell lines. *Leukemia*, 2003 Feb, 17(2), 416-26.

[6] Dirks, WG; Drexler, HG. Authentication of scientific human cell lines: easy-to-use DNA fingerprinting. *Methods Mol. Biol.*, 2005, 290, 35-50.

[7] Nardone, RM. Eradication of cross-contaminated cell lines: a call for action. *Cell Biol. Toxicol.*, 2007 Nov, 23(6), 367-72.

[8] Nelson-Rees, WA; Flandermeyer, RR; Hawthorne, PK. Banded marker chromosomes as indicators of intraspecies cellular contamination. *Science*, 1974 Jun, 7, 184(4141), 1093-6.

[9] Lee, JY; Lee, CH; Shim, SH; Seo, HK; Kyhm, JH; Cho, S; et al. Molecular cytogenetic analysis of the monoblastic cell line U937. karyotype clarification by G-banding, whole chromosome painting, microdissection and reverse painting, and comparative genomic hybridization. *Cancer Genet. Cytogenet*, 2002 Sep, 137(2), 124-32.

[10] O'Brien, SU; Kleiner, G; Olson, R; Shannon, JE. Enzyme polymorphisms as genetic signatures in human cell cultures. *Science*, 1977 Mar, 25, 195(4284), 1345-8.

[11] Gilbert, DA; Reid, YA; Gail, MH; Pee, D; White, C; Hay, RJ; et al. Application of DNA fingerprints for cell-line individualization. *Am. J. Hum. Genet.*, 1990 Sep, 47(3), 499-514.

[12] Masters, JR; Thomson, JA; Daly-Burns, B; Reid, YA; Dirks, WG; Packer, P; et al. Short tandem repeat profiling provides an international reference standard for human cell lines. *Proc. Natl. Acad. Sci. U S A*, 2001 Jul, 3, 98(14), 8012-7.

[13] Freshney, RI. Cell line provenance. *Cytotechnology*, 2002 Jul, 39(2), 55-67.

[14] Lery, X; LaRue, B; Cossette, J; Charpentier, G. Characterization and authentication of insect cell lines using RAPD markers. *Insect Biochem. Mol. Biol.*, 2003 Oct, 33(10), 1035-41.

[15] Hampe, J; Nurnberg, P; Epplen, C; Jahn, S; Grunow, R; Epplen, JT. Oligonucleotide fingerprinting as a means to identify and survey long-term cultured B cell hybridomas and T cell lines. *Hum. Antibodies Hybridomas*, 1992 Oct, 3(4), 186-90.

[16] Koren, S; Kosmac, M; Colja Venturini, A; Montanic, S; Curin Serbec, V. Antibody variable-region sequencing as a method for hybridoma cell-line authentication. *Appl Microbiol Biotechnol.*, 2008 Apr, 78(6), 1071-8.

[17] D'Eustachio, P; Bothwell, AL; Takaro, TK; Baltimore, D; Ruddle, FH. Chromosomal location of structural genes encoding murine immunoglobulin lambda light chains. Genetics of murine lambda light chains. *J Exp Med.*, 1981 Apr, 1, 153(4), 793-800.

[18] Tonegawa, S. Somatic generation of antibody diversity. *Nature*, 1983 Apr, 14, 302(5909), 575-81.

[19] Giudicelli, V; Chaume, D; Lefranc, MP. IMGT/GENE-DB: a comprehensive database for human and mouse immunoglobulin and T cell receptor genes. *Nucleic Acids Res.*, 2005 Jan, 1, 33(Database issue), D256-61.

[20] Parsons, M; Oi, VT; Huang, CM; Herzenberg, LA. Structural characterization of mouse immunoglobulin allotypic determinants (allotopes) defined by monoclonal antibodies. *Immunogenetics*, 1983, 18(4), 323-34.

[21] Essono, S; Frobert, Y; Grassi, J; Creminon, C; Boquet, D. A general method allowing the design of oligonucleotide primers to amplify the variable regions from immunoglobulin cDNA. *J. Immunol. Methods*, 2003 Aug, 279(1-2), 251-66.

[22] Seijen, AM; Seijen, HG; Bos, NA. Systematic design of mouse Vh gene family-specific oligonucleotides. *J. Immunol. Methods*, 2001 Aug, 1, 254(1-2), 161-8.

[23] Wang, Z; Raifu, M; Howard, M; Smith, L; Hansen, D; Goldsby, R, et al. Universal PCR amplification of mouse immunoglobulin gene variable regions: the design of degenerate primers and an assessment of the effect of DNA polymerase 3' to 5' exonuclease activity. *J. Immunol. Methods*, 2000 Jan, 13, 233(1-2), 167-77.

[24] Honjo, T; Alt F, W. Immunoglobulin genes, 2nd ed. London: *Academic Press*, 1995.

[25] Krebber, A; Bornhauser, S; Burmester, J; Honegger, A; Willuda, J; Bosshard, HR; et al. Reliable cloning of functional antibody variable domains from hybridomas and spleen cell repertoires employing a reengineered phage display system. *J. Immunol. Methods*, 1997 Feb, 14, 201(1), 35-55.

[26] Sambrook, J; Russell, DW. Molecular Cloning: A Laboratory Manual, 3rd ed.: *Cold Spring Harbor Laboratory Press*, 2001.

[27] Farrell, RE; Farrell, RE; Jr. RNA Methodologies: A Laboratory Guide for Isolation and Characterization, 3 ed. *Amsterdam: Elsevier/Academic Press*, 2005.

[28] Wilfinger WW, Mackey K, Chomczynski P. Effect of pH and ionic strength on the spectrophotometric assessment of nucleic acid purity. *BioTechniques*. 1997 Mar, 22(3), 474-481.

[29] Thammana P. Characterization of a deletion in the P3 myeloma heavy chain V region sequence in the non-secreting cell line NS1. *Mol. Immunol.* 1994 Jan, 31(1),77-78.

[30] Altschul, SF; Gish, W; Miller, W; Myers, EW; Lipman DJ. Basic local alignment search tool. *J. Mol. Biol.*, 1990 Oct, 5, 215(3), 403-10.

[31] Frohman, MA; Dush, MK; Martin, GR. Rapid production of full-length cDNAs from rare transcripts: amplification using a single gene-specific oligonucleotide primer. *Proc Natl. Acad. Sci. U S A.*, 1988 Dec, 85(23), 8998-9002.

[32] Schaefer, BC. Revolutions in rapid amplification of cDNA ends: new strategies for polymerase chain reaction cloning of full-length cDNA ends. *Anal. Biochem.*, 1995 May, 20, 227(2), 255-73.

In: Encyclopedia of DNA Research (3 Volume Set)
Editors: S. J. Duncan and P. H. Wiley
ISBN 978-1-61324-305-3

Chapter 29

DNA SEQUENCING: METHODS, STRATEGIES AND PROTOCOLS

***M. Naum*[1], *R. Bell*[1], *M. Mammel*[2], *J. Zheng*,[1] *and E. W. Brown*[1]**
[1]Food and Drug Administration, Division of Microbiology, 5100 Paint Branch Parkway, HFS-712, College Park, Maryland U. S.
[2] Food and Drug Administration, Division of Molecular Biology, 8301 Muirkirk Road, HFS-025, Laurel, Maryland U. S.

ABSTRACT

DNA sequencing is a common and requisite practice for molecular biologists today working in all areas of biology, including microbiology. A number of nucleotide and amino acid-based sequence typing methods are now widely recognized and have transformed the way in which complex evolutionary and epidemiological relationships are elucidated among bacterial strains. Recent advances in DNA sequencing underpin a vast array of bacterial subtyping methods and have become an integral component to traceback investigations of foodborne outbreaks caused by specific bacterial pathogenic strains. Direct DNA sequence-based approaches include multi-locus sequence analysis (MLSA), which simultaneously analyzes the nucleotide substitutions found within concatenated sequence alignments from multiple housekeeping genes. This collective and simultaneous analysis of nascent single nucleotide polymorphisms (SNPs) is effective in differentiating highly homogeneous strain clusters including the closely related strains and serovars of *Salmonella* enterica. In addition to the extraordinary power this approach offers in molecular evolution/epidemiology, the study of concatenated gene sequences has been instrumental in defining a new paradigm for horizontal gene transfer in the genetic structure of many bacterial species. Facile variations of MLSA are now being developed that rely on the culling of SNP-based targets from relatively uncharted and under-exploited regions of the bacterial genome. As an example, SNPs that reside in the selectively neutral intergenic non-transcribed spacer regions of the genome hold great promise in resolving bacterial strains down to a level of epidemiological relevance. In parallel with these advances in nucleotide sequence analysis, the development of rapid and highly accurate SNP discovery and detection tools is and includes methods such as Pyrosequencing, Luminex probe recognition technology, and single-basepair primer extension assays to name but a few. Together, these successes clearly position DNA sequence-based strategies at the helm for future discoveries in the genetics, evolution, and epidemiology of bacterial strains and species.

INTRODUCTION

1995 marked the publication of the first completed genome. At that time, a total of 1,830,137 DNA base pairs of *Haemophilus influenzae* were sequenced successfully, albeit over many months. Today, as was the case over ten years ago, molecular biology is on the cusp of a new era, and scientists are about to take giant leaps in developing faster, more efficient and cost effective nucleotide sequencing methods. The rapid progress attained in this area resounds a decade later in the publication of the *Homo sapien* genome, encompassing an astounding 3.3 billion bases of human DNA. Current and continued success in the development of sequencing methodologies is evident in facile sequence-based methods and strategies such as multi-locus sequence typing (MLST), and pyrosequencing (now used widely in the generation of whole-genome sequence data), which have become common words in the vocabularies of most molecular biologists.

DNA sequences, an invaluable component of modern biology, contain the encrypted instructions detailing the assembly and operation of all living tissues and, for evolutionary biologists, are accepted as an invaluable and unimpeachable record of the history of life on earth (Xiong et al.., 2000). Indeed, many molecular biologists and in more recent years microbiologists depend heavily on phylogenetics to make biological sense of the natural history of prokaryotes as well as resolve complex taxonomic relationships among bacteria using new sequence mining strategies such as SNP harvesting. Only through molecular phylogenetic studies can we understand fully and exploit completely this new form of rapidly accumulating genomic data which is being used now on a daily basis, particularly by researchers studying rapidly evolving organisms such as bacteria (Welsh and McClelland, 1990).

Arguably, the most accurate means for strain identification and phylogenetic analysis stems from the direct examination of DNA sequences. Random nucleotide mutation and horizontal gene transfer events often give rise to substitutions in the genomes of bacterial strains (Weissman et al., 2003). These sequence changes provide potential targets for strain identification and are built upon variation that is highly amenable to evolutionary reconstruction and can be readily analyzed in a phylogenetic and population genetic context to (i) assign unknown strains to known strain clusters, (ii) reveal sibling relationships among strains, and (iii) examine the distribution of specific alleles in a bacterial population.

DNA sequences offer several advantages over other fingerprinting methods.

Because each individual nucleotide is a useful genetic character, the cumulative differences in two or more sequences provide a larger number of discriminators, thus allowing a more determinant, stable, and evolutionarily reliable bacterial identification. Results based on nucleotide variation are unambiguous and belie secondary measurements such as amplicon size variations of DNA fragments.

These characteristics of DNA sequence data make this approach valuable in legal arenas as well. The technique and the data are well-established in human forensic casework and largely incontrovertible as a molecular determinant for attribution (Parsons et al., 1998; Sokurenko et al., 2001). The utility of DNA sequences was demonstrated recently with the characterization of *B. anthracis* which was associated with the anthrax postal attacks of 2001 (Read et al., 2002). For both *E. coli*, *S. enterica* and other enteric bacterial species, where substantial sequence variation exists between individuals, nucleotide substitutions are informative and can be used to establish a robust sequence-based strain identification system. Such techniques have already brought us a more refined understanding of the molecular evolutionary and molecular epidemiological relationships of many strains that are of great significance to public health. Here, we discuss several of the most commonly applied strategies in applying DNA sequencing to understanding the genetic and phylogenetic relationships that exist between closely related bacterial strains.

MLST vs MLSA: The Application of Raw SNP Data to the Delineation of Strains

Multi Locus Sequence Typing (MLST)

Multi locus sequence typing (MLST) was first proposed by Maiden et al. (1998) as an unambiguous and portable alternative to multi locus enzyme electrophoresis (MLEE). In MLEE, several (25 or more) metabolically important enzymes are separated on a starch gel according to size. For each enzyme, an allele number is assigned based on its position on the gel. Since the size of each enzyme is determined by its amino acid content and the amino acid content is determined by the organism's DNA sequence, inferences can be drawn about how different strains or species of bacteria are related to one another. Similarly, the concept behind MLST is much like MLEE, except MLST is a sequence based typing method where portions of several housekeeping genes, usually 7, are sequenced and compared. Alleles are assigned based on nucleotide sequence changes, and any single change in the nucleotide sequence constitutes a new allele.

The allele profiles are concatenated and a sequence type (ST) is assigned to the strain under investigation. Phylogenetic relationships between bacteria are inferred by clustering isolates with identical or similar STs. These clusters however, may or may not provide any useful evolutionary information depending on the organism being studied. For highly clonal bacteria, where it is assumed that most genetic variation arose via mutation, the clusters will reveal recent divergence events. In weakly clonal species, however, use of MLST in identification of clones and clonal complexes should only be viewed in the short term recent outbreak situation, as long term relationships are quickly distorted by recombination (Maiden et al., 1998; Feil et al., 2001).

Overall, multi locus methods are advantageous over single locus methodologies because the combination of multiple data sets enhances underlying phylogenetic signal, leading to more robust hypotheses. Additionally, higher discriminatory power is attainable when working with closely related organisms. For example, the species of the genus *Mycobacterium* share 94% to 100% sequence similarity in their 16S rRNA gene, making differentiation of some species impossible using only this marker. By using a multi locus approach however, all *Mycobacterium* species can be clearly differentiated from one another (Devulder et al., 2005). Furthermore, multi locus approaches counter the effects of gene conversion and recombination, which can mask the true evolutionary history between a bacterium and its nearest neighbors, as was the case in *Neisseria* (Hanage et al., 2005). By using more than one locus, recombinational misassignments are often diluted and the proper position of the individual species may be successfully delineated.

Multi Locus Sequence Analysis (MLSA) vs. MLST

Multilocus sequence analysis (MLSA) is an extension of MLST, although in MLSA concatenated nucleotide sequences are used instead of the allele profiles and sequence types. Use of concatenated sequences of the gene datasets and appropriate phylogenetic analyses may result in better supported deeper branches on an evolutionary tree (Spratt, 1999). Additionally, even though both MLSA and MLST proffer advantages over MLEE, MLST has some shortcomings that MLSA is able to surmount. While the sequence data is retained in MLST, clustering is based on the differences in the allele profiles and not the DNA sequences. As a result, when assigning the allele profiles to the data, essential information about similarities or specific changes in the DNA is lost. Thus the only phylogenetic information that MLST can provide is on recent clonal diversification of the strains being studied. MLSA on the other hand, is capable of revealing additional evolutionary information because inferences are based on the raw sequence information. This is clearly evident in the increase of bootstrap support values, especially at deeper nodes in the phylogenies (Thompson et al., 2005; Devulder et al., 2005; Kotetishvili et al. 2005). By retaining all nucleotide information and not reducing it to a mere sequence type, more robust assessments of the relatedness of two bacterial species or strains can be made. One of the best examples identifying the above mentioned difference between MLST and MLSA stem from an investigation of *Burkholderia pseudomallei*, *B. mallei* and *B. thailandensis* (Godoy et al., 2003). MLST allele profiles were unable to discriminate the deeper evolutionary relationships between the *Burkholderia* species, and clustering was unsuccessful at assessing the closest relative of one unique sequence type (ST73). MLSA concatenated sequences however, discerned these relationships with very high statistical support, and phylogenetic analysis of the sequences unmistakably related ST73 to *B. thailandensis*. This relationship was based on the similarity of the DNA sequences even though ST73 shared no common alleles with any of the other STs.

However, probably the most striking difference between MLST and MLSA is the identification and use of single nucleotide polymorphisms (SNPs). MLSA is a great tool for identifying and assigning SNPs to specific clades of strains because all the raw sequence data is retained and not reduced down to a ST. It is the discovery of these diagnostic SNPs that enable us to discriminate a specific strain of a bacterium, and it is these SNPs that will be useful in the development of rapid identification systems for use in outbreak situations.

MLSA call informative genetic signatures

While multi locus methods are better than their single locus counterparts, the choice of genes to be used in a given analysis is critical to the questions looking to be answered. If the goal is to identify the species an unknown isolate belongs to, then highly conserved genes under stablilizing selection are necessary elements of the concatenated sequence datasets

(Stackebrandt et al., 2002; Spratt, 1999; Hanage et al., 2006; Gevers, 2005). Typically, these are the genes which encode proteins essential for the general growth and metabolism of an organism (*i.e.* housekeeping genes), they are usually more diverse than 16S rRNA with a selectively neutral variation, and they are present in at least all the species of a given genus due to their essential nature. These characteristics permit the differentiation of closely related species while preserving the phylogenetic signal of the relationships between the organisms being studied.

However, for very closely related species, such as the members of the *Mycobacterium tuberculosis* complex, the use of housekeeping genes was unable to differentiate them from one another (Devulder, 2005). Subsequently, for very closely related species or to differentiate strains within a species, other more variable gene targets need to be explored. For example, in *Salmonella enterica* subsp. *enterica* there is very little variation in housekeeping genes especially at the serovar level (Torpdahl et al., 2005; Fakhr et al., 2005; Harbottle et al., 2006). Therefore, in order to achieve better discrimination of strains within a serovar, virulence gene sequences are also included with some housekeeping gene sequences (Foley et al., 2006; Sukhnanand et al., 2005; Tankouo-Sandjong et al. 2007). The combination of both housekeeping and virulence gene sequences did reveal differences within serovars, however one must be cautious when drawing inferences about the evolution of strains when using virulence gene sequences because virulence genes are highly recombinogenic and subject to positive selection by the host immune response. Finally, MLSA may also be used to answer questions about hyper-virulent lineages or host specificity of pathogenic bacteria (Lemee et al., 2005; Alcaine et al., 2006). Once again however, care must be taken when interpreting these results because the genes may not reflect the true 'vertical' evolutionary history of the pathogen.

Consequently, tools such as MLST and MLSA are useful to molecular biologists as long as caution is used when selecting which 7 genes to sequence and use in the phylogenetic analysis. Mildly variable genes are going to be extremely useful in detecting speciation events, while genes subject to rampant recombination may hide within them SNP targets ideal for subtyping (Figure 1). However, in any given level the same mantra applies; genes with enough variability that allow for the differentiation of the organisms of interest must be chosen.

SINGLE NUCLEOTIDE POLYMORPHISM DETECTION AND POPULATION PREVALENCE BY COMBINING CONVENTIONAL AND PYRO-SEQUENCING TECHNOLOGIES

SNP discovery

Single nucleotide polymorphisms (SNPs) can be useful genetic markers and can be quickly determined and evaluated. A series of SNPs which are evenly split between two bases in any given population are useful in binning strains into groups. On the other hand, a SNP which shows a rare variant in the population may not be as useful for grouping the collection of strains, but may be a useful marker for identifying individual strains.

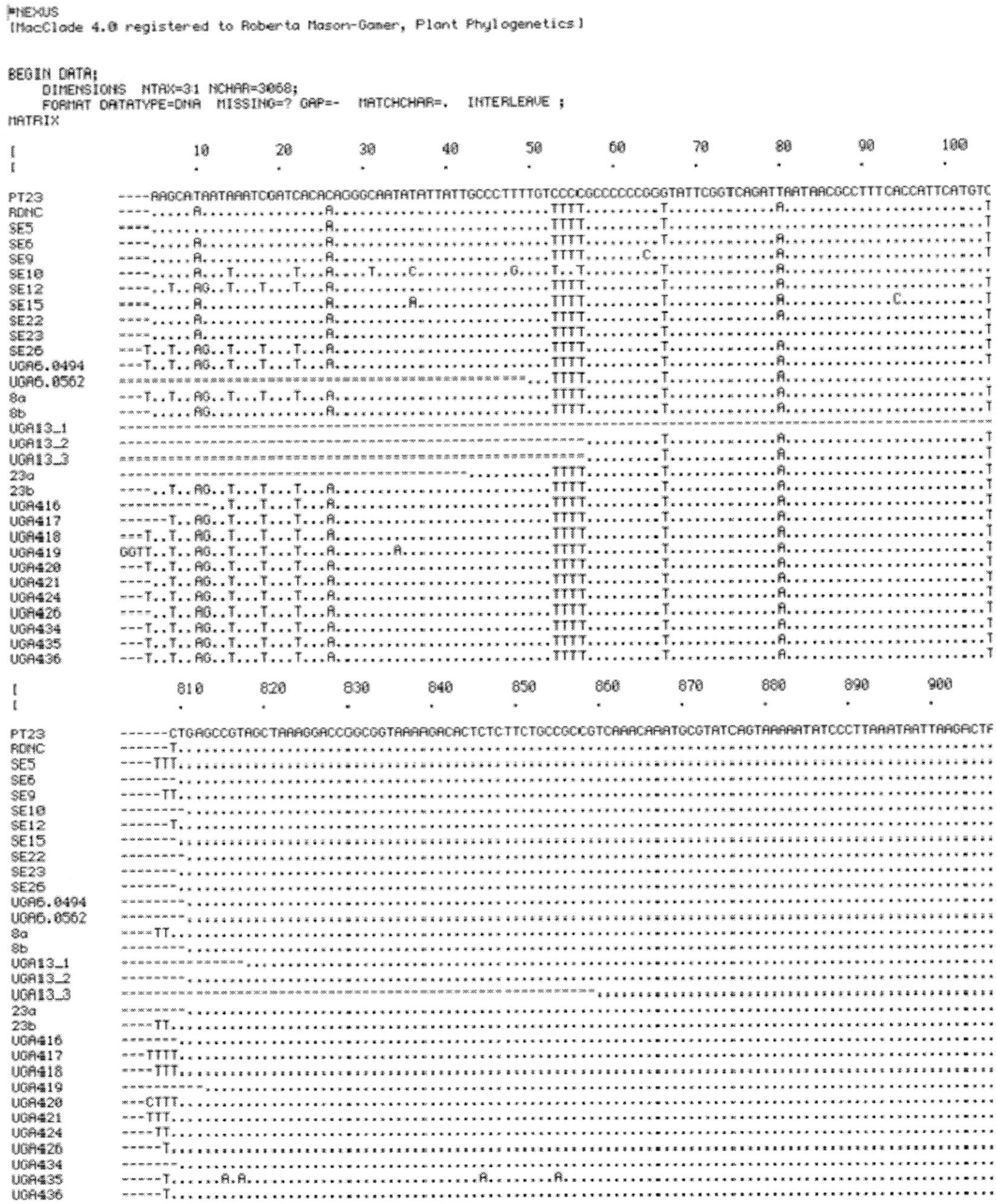

Figure 1. A portion of the MLSA 7 gene dataset depicting the variable and conserved nucleotides between the Salmonella enterica subsp. enterica serovar Enteritidis strains included in the phylogenetic analysis. Outputs such as this are used as the primary source of single nucleotide polymorphisms identification because they very clearly portray the variable regions between the taxa.

To find SNPs which will differentiate strains within a group of microorganisms, DNA sequence comparisons are made, either over a whole genome or over individually sequenced genes. Conventional Sanger sequencing, or now sequencing by 454 technology, yields a whole genome sequence in a number of assembled contigs. From these sequences, or sequenced PCR products, alignments of genes matched by BLAST or other comparison methods can be made, allowing the detection of base changes. Initially, as we have previously stated, genes used in MLST or better yet MLSA studies can be used as a source of SNPs as other less conventional targets are explored, particularly in recently diverged organisms such as *Salmonella*. In *Salmonella* subtyping many un-conventional targets have been explored because of the clonal nature of the serovars. Of particular use have been intergenic regions which are subject to neutral selection and evolve at a steady rate. By examining a combination of sequenced genetic markers and the space between them, researchers have been able to collect numerous SNPs which are successfully used in the subtyping of *Salmonella enterica* serovars.

Principles of pyrosequencing

Pyrosequencing is a rapid sequencing technology which uses the principle of "sequencing by synthesis". A sequencing primer is annealed to single stranded PCR template and individual deoxynucleotide triphosphates (dNTPs) are added to the reaction. If the dispensed dNTP base is complementary to the base in the template strand, it hydrolyzes to generate dNMP and is pyrophosphated by DNA polymerase. The released pyrophosphate is then converted to ATP by ATP sulfurylase. Luciferase, uses this newly formed ATP molecule to convert luciferin, generating a bioluminescent signal. When the pyrosequencing technology is used for genotyping, a few positive control bases are included in the sequence of dNTPs dispensed in addition to the SNP site, in order to ensure that the sequencing primer is in the right location. After the run, a program is generated, and the DNA sequence is read from the displayed peaks associated with each base added (Figure 2). The software included with each pyrosequencer makes the base calls and displays the quality of each call, as each light signal produced is directly proportional to the number of nucleotides incorporated.

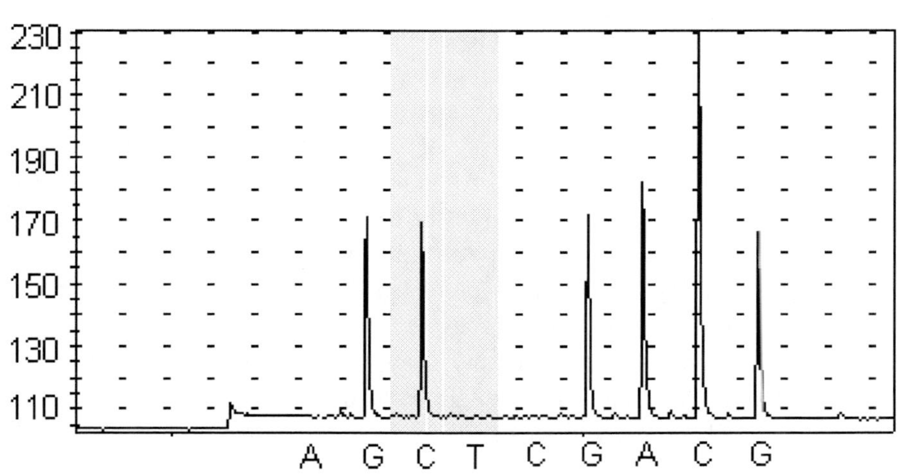

Figure 2. The pyrogram shown here is indicative of pyrosequencing results for a SNP assay of Escherichia coli O157:H7 EDL933. In this particular example the strain tested has a C rather than a T at the SNP position shown in yellow, resulting in the following sequence read: GCGACCG.

Validating and using SNP sites to characterize microbial strains

Once SNPs are identified by conventional sequencing or other methods, they must be verified. This is accomplished by the design of pyrosequencing assays intended to test for the SNPs. Using assays to test reference and predicted variant strains enables us to determine whether the SNP is truly present in variant strains or whether the sequencing data was in error. The next step is to determine the frequency of the nucleotide polymorphism in a population by testing a pool of DNA from representative with the assay. The SNP frequency in the pool of sample strains will give an estimation of how useful the SNP can be at differentiating the population of strains. A single pool of DNA containing 10-20 samples can be used to determine the frequency of a SNP in the population without testing individual strains greatly increasing the high-throughput capabilities of any laboratory. Once a series of validated SNPs is assembled, testing individual strains with a panel of SNPs will result in a profile for each strain. This profile or "barcode" can later be used to cluster strains into groups or possibly identify individual strains depending on the level of discrimination attained from each SNP. Running the panel on a set of strains that represent the diversity of the group being studied helps establish a reference database, and as new strains are encountered, such as strains isolated from outbreaks, profiles of the unknown strain can be compared to the reference database in order to identify its nearest neighbors.

Comparative Genome Sequencing

With the ever increasing number of complete reference genome sequences publically available, investigations have been focused on analyzing sequence variation in closely related organisms. As opposed to direct, exhaustive sequencing of entire genomes, hybridization-based re-sequencing strategies provides an alternative to Sanger dideoxy sequencing allowing a more rapid identification of genome-wide sequence polymorphisms. In general, test and reference genomic DNA samples are independently labeled with fluorescent dyes either by one-color or two-color protocols. For one-color comparative genome sequencing (CGS), the 2 pools are hybridized to two separate whole genome re-sequencing arrays. For two-color CGS, samples are mixed and hybridized to a single array, and the ratios of test DNA to reference DNA are plotted against their genomic position. Alignment between the test and reference genomes enables investigators to identify the regions or bases that are different between the two DNA copies. Re-sequencing arrays typically contain over 10^5 oligonucleotides (~25-mers) to interrogate a genomic DNA with size over 10 Kb (Wong et al., 2004) and provide investigators with actual sequence data. Techniques such as photolithography and ink-jet deposition are used to manufacture such high-density arrays (Chee et al., 1996; Hughes et al., 2001; Pease et al., 1994). Different tiling strategies are being used for low, medium, and high resolution resequencing, in which DNA probes are chosen from contiguous stretches of the genome (Mockler et al., 2005). In the highest resolution tiling array design, probes are typically tiled at 1 base resolution (Frazer et al., 2001; Hacia, 1999), with eight unique 25-mer oligonucleotides being incorporated per base pair, that is, one for each potential base pair at each position on forward and reverse strands of the genome being investigated. The false positive rate for SNPs using this CGS approach was one per 244 Kb of genome sequence after confirmation (Herring and Palsson, 2007).

Recently, laboratories from the Center for Food Safety and Applied Nutrition at the FDA have used a novel tiling strategy that employs 40-fold fewer probes than the current re-sequencing approaches (Mukherjee et al., 2006). This tiling method uses 29-mers which overlap by 24 base-pairs each. Each 29-mer is complementary to a region of a reference bacterial genome, and single base pair mismatches (insertions-deletions and inversions) within a test genome are detected by unique hybridization signals relative to the reference genome. Since the sequence of each probe in the array is partially redundant with its neighboring probe by 24 nucleotides, the confidence of true SNP calling is dramatically increased due to the unique hybridization signature which is signaled by multiple oligonucleotide probes. *In silico* comparison between *E. coli* K-12 strain MG 1655 and *E. coli* O157:H7 strain EDL933 using the O157:H7 tiling array indicated that over 99% of the polymorphisms that occur between the two strains in the 60kb genomic range could be accurately detected (Cebula et al., 2005).

CGS methods are fast and inexpensive due to the high-throughput capabilities however, the tilling arrays identify only the position of the variable region in the genome but not the base composition of the nucleotide polymorphism or insertion/deletion. Therefore, in order to identify the composition of the SNPs, this type of array approaches have to be coupled with other analytical methods such as pyrosequencing.

Rapid Sequencing Methodologies Used in SNP Detection

Bioplex Targets

The invention of high-density arrays has greatly enhanced and simplified the process of biological research and development (Fodor et al., 1991), however, the different technologies used to fabricate microarrays suffer from high manufacturing costs which make this spatially addressable platform difficult to approach for most laboratories. An alternative approach to the spatially addressable method is the concept of 'liquid array' or 'suspension array' by unfixing array elements (Nolan and Sklar, 2002). Flow cytometry which has been used for multi-parametric cellular analyses for over 20 years has the capability to detect any particles in the lower micron range, including inert microspheres of different sizes, dyed with various fluorochromes. Flow cytometry, in combination with fluorescent dye-incorporated polymeric beads is ideal for constructing biological multiplexed arrays.

Luminex Corporation developed multiplex analytes profiling the xMAP system around the "suspension array" concept (Kettman et al., 1998). Instead of the traditional approach of using flat surfaces in microarray making, Luminex used 5.6-micron polystyrene microspheres, which are internally dyed with two spectrally distinct fluorochromes. Using different intensities of one of these dyes, an 'array' of up to 100 fluorescent microsphere sets is offered by several companies. Each set has a unique spectral address determined by its dye mixture, allowing different sets of microspheres to be pooled together and measured simultaneously in a single reaction. Sample throughput is extremely rapid, requiring only a couple of seconds per sample.

Suspension array, such as the one used by Luminex, differs from flat array technology (microarray) in various ways even though both are based on similar concepts. In suspension arrays, differentiation of each element is based on intrinsic optical parameters instead of the physical location on a surface as in the flat arrays. Further, in terms of preparation, suspension arrays are conducive to mass preparation, as opposed to flat arrays which require each array element to be prepared individually using a spotting robot or photolithography. Once coupled with the appropriate receptor, a suspension of microspheres can be used to prepare thousands of microsphere arrays. The small size and three-dimensional exposure of the microspheres allow for nearly liquid phase kinetics, which renders suspension array faster reaction times and better reproducibility. Since flow cytometry has the ability to distinguish free from particle-bound probes over a wide range of free-probe concentrations, suspension array can process samples without a wash step (Nolan and Sklar, 2002), and although suspension arrays do not have the comparable high density content of flat arrays, for many applications the ability to conduct a few dozen analyses simultaneously is entirely adequate.

The application of this platform as a tool to probe both genomic and proteomic functions has clearly been demonstrated, and multiple assays have been described using this technology for gene expression profiling, HLA DNA typing, microbial detection, and SNP genotyping (Armstrong et al., 2000; Diaz and Fell, 2005; Dunbar et al., 2003; Fulton et al., 1997; Lee et al., 2004; Wilson et al., 2005; Yang et al., 2001; Ye et al., 2001). SNPs, one of the most abundant forms of genomic variation, provide important biomarkers which can greatly facilitate molecular subtyping of fairly clonal microorganisms such as *Salmonella*. With SNP data obtained from sequencing strategies such as MLSA or intergenic regions as we discussed in previous sections, one can use the xMAP platform to design multiplexed SNP genotyping assays. A common approach for single nucleotide discrimination is to use a direct hybridization format, that is, the hybridization of a labeled complementary sequence amplified by PCR with microsphere sets coupling oligonucleotide capture probes. However, as multiple SNPs with different levels of specificity may be used in genotyping strains in a single reaction, the design of capture probes and PCR primers for direct hybridization assays is challenging. In addition to direct hybridization assays, other approaches including allele-specific primer extension (ASPE) (Taylor et al., 2001), oligonucleotide ligation assay (OLA) (Iannone et al., 2000) and single base chain extension (SBCE) (Chen et al., 2000) have also been described for the xMAP system. These approach utilize tagged or cZipCode microsphere sets and solution-based chemistries (Dunbar, 2006) to detect SNPs, allowing multiple sets of SNPs to be analyzed by a limited set of tagged microspheres. Nevertheless, direct hybridization is more flexible in SNP detection and is not restricted by established patents.

Ibis Targets

The Ibis T5000 biosensor, marketed by Ibis Biosciences, is a high throughput, PCR/mass spectrometric based identification system capable of identifying bacterial, fungal, viral and protozoal pathogens as well as mutations in human mitochondrial DNA associated with disease. The principle behind this instrument is quite simple (Hofstadler et al., 2005; Ecker DJ et al., 2006). PCR is used to amplify small (80-140 bp) regions of DNA or cDNA that contain unique changes specific to a bacterial, fungal or viral species, whose masses are then measured using electrospray ionization time of flight mass spectrometry. Since the mass accuracy of the spectrometer is to the exact base count, the total number of A, G, T and C, nucleotides is calculated. These base counts are then compared to a library of known organisms in order to determine the identification of the unknown organisms. The basic workflow of the Ibis T5000 from sample collection to data analysis is shown in Figure 3. Additionally, this instrument is capable of identifying several organisms in a mixed sample without a pre-enrichment step which is frequently used to increase the bacterial load. Quantification is also possible through the addition of an internal PCR calibrant, allowing for the determination of the relative amount of each organism in an unknown sample.

Further, depending on the questions being asked, there are several assay kits available which address many common clinical questions, with new assays being developed for use in the areas of foodborne pathogen detection, plant pathogen detection and human forensics (Hall et al., 2005; Postnikova et al., 2008). The broad bacterial surveillance kit currently available can distinguish all major lineages of eubacteria to the genus level and some to the species level (Hofstadler et al., 2005), even in mixed culture samples. More specific genotyping can be achieved using kits designed for specific organisms, such as those for *Acinetobacter*, *Bacillus anthracis*, group A *Streptococcus,* and *Staphylococcus aureus* (Ecker DJ et al., 2006; Ecker JA et al., 2006). The power of this instrument however, stems from the bioinformatics that support its database. Most of the DNA changes that the Ibis T5000 is able to detect are single nucleotide polymorphisms, and many of the assays currently available have been developed based on previously completed MLST/MLSA analyses. Once unique SNPs that delineate a species or strain are identified, primers are designed that flank that site. Although the primers must be positioned in a way which allows base count differences to be seen, since the mass spectrometer does not generate sequences, but rather base counts.

SUMMARY

DNA sequences provide a versatile and widely applicable opportunity to study the evolution, epidemiology, and genetic structure of pathogenic bacteria. Enteric organisms have accrued numerous nucleotide differences over millions of years of evolution. The DNA sequence-based methodologies described here now make this trove of genetic diversity available to those who require a detailed understanding of the genetic relatedness of a variety of bacterial strains. Moreover, the power of these methods should allow for us to discern the unique personalities of numerous bacterial strains of clinical relevance. Moreover, analysis of these data in a phylogenetic context will greatly aid in elucidating the unique genetic signatures of pathogenic bacterial strains making these approaches helpful for public health laboratories in their continual surveillance to ensure a safe food supply.

The examination of sequence variation among genes scattered around the chromosome often yields a snapshot of the general taxonomic or evolutionary relationships of the strains or species in question. These types of binned sequence or allele data when wed to thoughtful phylogenetic clustering can readily identify epidemiological clones of pathogens while pinpointing the ancestry of important groups of bacteria. The impact of this discriminatory power cannot be overstated.

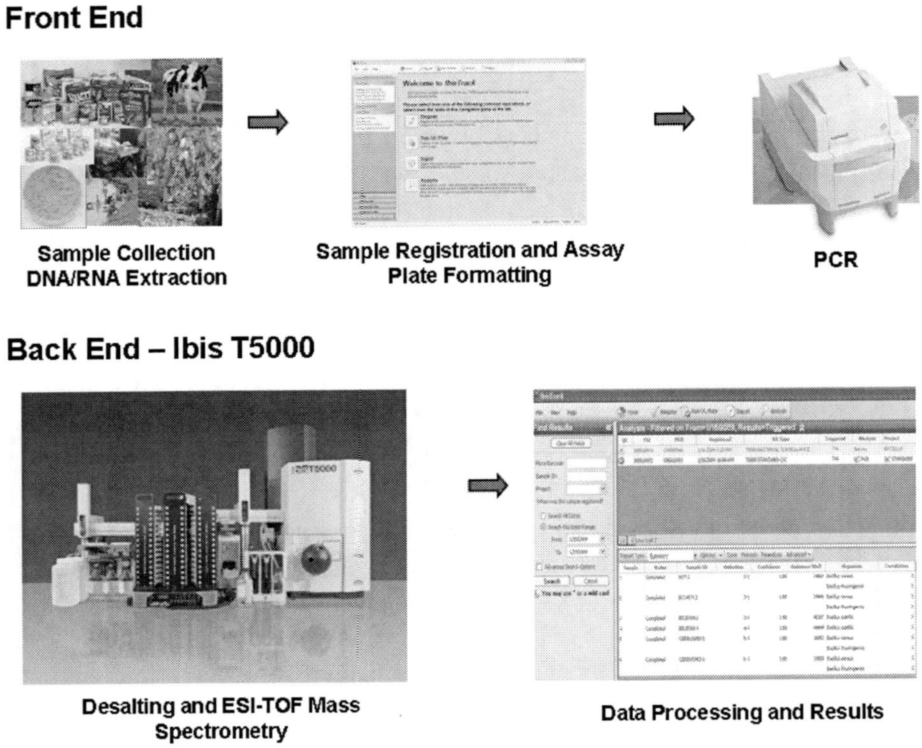

Figure 3. A simplified schematic representation of the multiple steps samples undergoe prior to being sprayed onto the mass spectrometer. Briefly, samples from multiple food matrices are collected and microbial DNA is extracted. Sample are then electronically registered and the appropriate identification plates are determined. After sample amplification, the plates are loaded onto the Ibis-T5000 platform where the amplified products are cleaned by magnetic beads and prepared for the mass spectrometer. Once all samples have been processed, the algorithm associated with the Ibis-T5000 generates the results in the form of a table as shown on the last pannel of the figure.

Often times, these data can provide clues to the reservoir or source of a particular pathogenic cluster of strains including complex veterinary, environmental or clinical points of origin (Maiden et al., 1998). Taken together, the advantages of these sequence-based strategies have made them highly effective for subtyping as many of these have been used widely to subtype populations of many bacterial species including *Salmonella enterica, Escherichia coli, Vibrio cholerae, Listeria monocytogenes*, and *Mycobacterium* spp. (Sukhnanand et al., 2005).

An ever expanding arsenal of automated subtyping technologies continues to flow from the science of DNA sequencing. Techniques such as sequence tiling on DNA microarray, mass spectroscopy-based SNP detection, optical mapping, and whole-genome pyrosequencing offer higher throughput approaches for characterizing the nucleotide diversity between strains. With multiple whole-genome sequences readily available in the public database for many notable enteric pathogens, it has become possible now more than ever to pinpoint those sequences that are under more subtle selective constraints allowing them to evolve at more rapid rates between strains. This provides a promising strategy for harvesting a greater number of nucleotide differences between two strains. Numerous recent publications have demonstrated the effectiveness of SNP-based typing schemes for a number of pathogenic clones and lineages of bacteria (Hu et al., 2006).

In his Nobel Prize acceptance address in 1987, the sagacious organic chemist, Donald James Cram, quipped that "few scientists acquainted with the chemistry of biological systems at the molecular level can avoid being inspired..." Arguably, there exists no place where this has rung more true than in microbiology over the last decade where a torrent of discoveries in evolution, taxonomy, and genetic science-all stemming from the availability of bacterial DNA sequences-continues to flow. As more and more bacterial genomes continue to surrender their evolutionary secrets, microbiologists will continue to develop more effective genotyping schemes capable of culling the subtle genomic differences that distinguish all strains and particularly those that threaten human life.

REFERENCES

Alcaine SD, Y. Soyer, L.D. Warnick, W-L. Su, S. Sukhnanand, J. Richards, E.D. Fortes, P. McDonough, T.P. Root, N.B. Dumas, Y. Gröhn, M. Weidmann, (2006). Multilocus sequence typing supports the hypothesis that cow- and human-associated *Salmonella* isolates represent distinct and overlapping populations. *Appl. Environ. Microbiol.* 72: 7575-7585.

Armstrong B, M. Stewart, and A. Mazumder, (2000). Suspension arrays for high throughput, multiplexed single nucleotide polymorphism genotyping. *Cytometry* 40: 102-8.

Cebula TA, E.W. Brown, S.A. Jackson, M.K. Mammel, A. Mukherjee, J.E. LeClerc, (2005). Molecular applications for identifying microbial pathogens in the post-9/11 era. *Expert Rev. Mol. Diagn.* 5(3): 431-445.

Chee M, R. Yang, E. Hubbell, A. Berno, X. C. Huang, D. Stern, J. Winkler, D. J. Lockhart M. S. Morris, and S. P. Fodor, (1996). Accessing genetic information with high-density DNA arrays. *Science* 274: 610-4.

Chen J, M. A. Iannone, M. S. Li, J. D. Taylor, P. Rivers, A. J. Nelsen, K. A. Slentz-Kesler, A. Roses, and M. P. Weiner, (2000). A microsphere-based assay for multiplexed single nucleotide polymorphism analysis using single base chain extension. *Genome Res* 10: 549-57.

Devulder G, M. Pérouse de Montclos, J.P. Flandrois, (2005). A multigene approach to phylogenetic analysis using the genus *Mycobacterium* as a model. *Inter. J. Syst. Evol. Microbiol.* 55: 293-302.

Diaz MR, and J. W. Fell, (2005). Use of a suspension array for rapid identification of the varieties and genotypes of the Cryptococcus neoformans species complex. *J Clin Microbiol* 43: 3662-72.

Dunbar SA, C. A. Vander Zee, K. G. Oliver, K. L. Karem, and J. W. Jacobson, (2003). Quantitative, multiplexed detection of bacterial pathogens: DNA and protein applications of the Luminex LabMAP system. *J Microbiol Methods* 53: 245-52.

Dunbar SA (2006). Applications of Luminex xMAP technology for rapid, high-throughput multiplexed nucleic acid detection. *Clin. Chim. Acta* 363: 71-82.

Ecker DJ, J.J. Drader, J. Gutierrez, A. Gutierrez, J.C. Hannis, A. Schink, R. Sampath, L.B. Blyn, M.W. Eshoo, T.A. Hall, M. Tobarmosquera, Y. Jiang, K.A. Sannes-Lowery, L.L.

Cummins, B. Libby, D.J. Walcott, C. Massire, R.Ranken,S. Manalili, C. Ivy, R. Melton, H. Levene, V. Harpin, F. Li, N. White, M. Pear, J.A. Ecker, V. Samant, D. Knize, D. Robbins, K. Rudnick, F. Hajjar, S.A. Hofstadler, (2006). The Ibis T5000 universal biosensor: an automated platform for pathogen identification and strain typing. *JALA*. 11: 341-351.

Ecker JA, C. Massire, T.A. Hall, R. Ranken, T-T.D. Pennella, C.A. Ivy, L.B. Blyn, S. A Hofstadler, T. P. Endy, P.T. Scott, L. Lindler, T. Hamilton, C. Gaddy, K. snow, M. Pe, J. Fishbain, D. Craft, G. Deye, S. Riddell, E. Milstrey, B. Petruccelli, S. Brizze, V. Harpin, A. Schink, D.J. Ecker, R. Sampath, M.W. Eshoo, (2006). Identification of *Acinetobacter* species and genotyping of *Acinetobacter baumannii* by multilocus PCR and mass spectrometry. *J. Clin. Microbiol*. 44: 2921-2932.

Fakhr MK, L.K. Nolan, C.M. Logue, (2005). Multilocus sequence typing lacks discriminatory ability of pulse-field gel electrophoresis for typing *Salmonella enterica* serovar Typhimurium. *J. Clin. Microbiol*. 43: 2215-2219.

Feil EJ, E.C. Holmes, D.E. Bessen, M-S. Chan, N.P.J. Day, M.C. Enright, R. Goldstein, D.W. Hood, A. Kalia, C.E. Moore, J. Zhou, B.G. Spratt, (2001). Recombination within natural populations of pathogenic bacteria: short-term empirical estimates and long-term phylogenetic consequences. *PNAS* 98: 182-187.

Fodor SP, J. L. Read, M. C. Pirrung, L. Stryer, A. T. Lu, and D. Solas, (1991). Light-directed, spatially addressable parallel chemical synthesis. *Science* 251: 767-73.

Foley SL, D.G. White, P.F. McDermott, R.D. Walker, B. Rhodes, P.J. Fedorka-Cray, S. Simjee, S. Zhao, (2006). Comparison of subtyping methods for differentiating *Salmonella enterica* serovar Typhimurium isolates obtained from food animal sources. *J. Clin. Microbiol.* 44: 2569-3577.

Frazer KA, J. B. Sheehan, R. P. Stokowski, X. Chen, R. Hosseini, J. F. Cheng, S. P. Fodor, D. R. Cox, and N. Patil, (2001). Evolutionarily conserved sequences on human chromosome 21. *Genome Res* 11: 1651-9.

Fulton RJ, R. L. McDade, P. L. Smith, L. J. Kienker, and J. R. Kettman, Jr., (1997). Advanced multiplexed analysis with the FlowMetrix system. *Clin. Chem.* 43: 1749-56.

Gevers D, F.M. Cohan, J.G. Lawrence, B.G. Spratt, T. Coenye, E.J. Feil, E. Stackebrandt, Y. Van de Peer, P. Vandamme, F.L. Thompson, J. Swings, (2005). Re-evaluating prokaryotic species. *Nature Rev.* 3: 733-739.

Godoy D, G. Randle, A.J. Simpson, D.M. Aanensen, T.L. Pitt, R. Kinoshita, B.G. Spratt, (2003). Multilocus sequence typing and evolutionary relationships among the causative agents of melioidosis and glanders, *Burkholderia pseudomallei* and *Burkholderia mallei*. *J. Clin. Micobiol*. 41: 2068-2079.

Hacia JG, (1999). Resequencing and mutational analysis using oligonucleotide microarrays. *Nat Genet* 21: 42-7.

Hall TA, B. Budowle, Y. Jiang, L. Blyn, M. Eshoo, K. A. Sannes-Lowery, R. Sampath, J. J Drader, J. C. Hannis, P. Harrell, V. Samant, N. White, D. J. Ecker, S.A. Hofstadler, (2005). Base composition analysis of human mitochondrial DNA using electrospray ionization mass spectrometry: A novel tool for the identification and differentiation of humans. *Anal. Biochem*. 244: 53-69.

Hanage WP, C. Fraser, B.G. Spratt, (2006). Fuzzy species among recombinogenic bacteria. *BMC Biol*. 3: 6.

Harbottle H, D.G. White, P.F. McDermott, R.D. Walker, S. Zhao, (2006). Comparison of multilocus sequence typing, pulse-field gel electrophoresis, and antimicrobial susceptibility typing for characterization of *Salmonella enterica* serotype Newport isolates. *J. Clin. Microbiol.* 44: 2449-2457.

Herring CD, and BO Palsson. 2007. An evaluation of Comparative Genome Sequencing (CGS) by comparing two previously-sequenced bacterial genomes. *BMC Genomics* 8.

Hofstadler SA, R. Sampath, L.B. Blyn, M.W. Eshoo, T.A Hall, Y. Jiang, J.J Drader, J.C. Hannis, K. A. Sannes-Lowery, L.L. Cummins, B. Libby, D.J. Walcott, A. Schink, C. Massire, R. Ranken, J. Gutierrez, S. Manalili, C. Ivy, R. Melton, H. Leveene, G. Barrett_Wilt, F. Li, V. Zapp, N. White, V. Samant, J.A. McNeil, D. Knize, D. Robbins, K. Rudnick, A. Desai, E. Moradi, D.J. Ecker, (2005). TIGER: the universal biosensor. *Mass Spectrom*. 242: 23-41.

Hu H., R. Lan, P.R. Reeves, (2006). Adaptation of multilocus sequencing for studying variation within a major clone: evolutionary relationships of *Salmonella enterica* serovar Typhimurium. *Genetics* 172: 743-50.

Hughes TR, M. Mao, A. R. Jones, J. Burchard, M. J. Marton, K. W. Shannon, S. M. Lefkowitz, M. Ziman, J. M. Schelter, M. R. Meyer, S. Kobayashi, C. Davis, H. Dai, Y. D. He, S. B. Stephaniants, G. Cavet, W. L. Walker, A. West, E. Coffey, D. D. Shoemaker, R. Stoughton, A. P. Blanchard, S. H. Friend, and P. S. Linsley, (2001). Expression profiling using microarrays fabricated by an ink-jet oligonucleotide synthesizer. *Nat Biotechnol* 19: 342-7.

Iannone MA, J. D. Taylor, J. Chen, M. S. Li, P. Rivers, K. A. Slentz-Kesler, and M. P. Weiner, (2000). Multiplexed single nucleotide polymorphism genotyping by oligonucleotide ligation and flow cytometry. *Cytometry* 39:131-40.

Kettman JR, T. Davies, D. Chandler, K. G. Oliver, and R. J. Fulton, (1998). Classification and properties of 64 multiplexed microsphere sets. *Cytometry* 33: 234-43.

Korlach J, P.J. Marks, R.L. Cicero, J.J. Gray, D.L. Murphy, D.B. Roitman, T.T. Pham, G.A. otto, M. Foquet, S.W. Turner, (2008). Selective aluminum passivation for targeted immobilization of single DNA polymerase molecules in zero-mode waveguide nanostructures. *PNAS* 105: 1176-1181.

Kotetishvili M, A. Kreger, G. Wauters, J.G. Morris, Jr., A. Sulakvelidze, O.C. Stine, (2005). Multilocus sequence typing for studying genetic relationships among *Yersinia* species. *J. Clin. Microbiol.* 43: 2674-2684.

Lee SH, D. R. Walker, P. B. Cregan, and H. R. Boerma. 2004. Comparison of four flow cytometric SNP detection assays and their use in plant improvement. *Theor. Appl. Genet.* 110: 167-74.

Lemée L, I. Bourgeois, E. Ruffin, A. Collignon, J-F. Lemeland, J-L. Pons, (2005). Multilocus sequence analysis and comparative evolution of virulence-associated genes and housekeeping genes of *Clostridium difficile*. *Microbiol.*151: 3171-3180.

Levene MJ, J. Korlach, S.W. Turner, M. Foquet, H.G. Craighead, (2003). W.W. Webb, Zero-Mode Waveguides for Single-Molecule Analysis at high concentrations. *Science* 299: 682-686.

Maiden MCJ, J.A. Bygraves, D. Feil, G. Morelli, J.E. Russell, R. Urwin, Q. Zhang, J. Zhou, K. Zurth, D.A, Caugant, I.M. Feavers, M. Achtman, B.G. Spratt, (1998). Multilocus sequence typing: a portable approach to the identification of clones within populations of pathogenic microorganisms. *PNAS* 95: 3140-3145.

Mockler TC, S. Chan, A. Sundaresan, H. Chen, S. E. Jacobsen, and J. R. Ecker, (2005). Applications of DNA tiling arrays for whole-genome analysis. *Genomics* 85: 1-15.

Mukherjee A, S. A. Jackson, J. E. LeClerc, and T. A. Cebula, (2006). Exploring genotypic and phenotypic diversity of microbes using microarray approaches. *Toxicol. Mechan. Methods* 16: 121-128.

Nolan JP, and L. A. Sklar. 2002. Suspension array technology: evolution of the flat-array paradigm. *Trends Biotechnol* 20: 9-12.

Parsons W, T. J. Parsons, R. Scheithauer, and M. M. Holland, (1998). Population data for 101 Austrian Caucasian mitochondrial DNA D-loop sequences: amplification of mtDNA sequence analysis for a forensic case. *Int. J. Legal Med.* 111: 124–132.

Pease AC, D. Solas, E. J. Sullivan, M. T. Cronin, C. P. Holmes, and S. P. Fodor, (1994). Light-generated oligonucleotide arrays for rapid DNA sequence analysis. *PNAS.* 91: 5022-6.

Read TD, S. L. Salzberg, M. Pop, M. Shumway, L. Umayam, L. Jiang, E. Holtzapple, J. D. Busch, K. L. Smith, J. M. Schupp, D. Solomon, P. Keim, and C. M. Fraser, (2002). Comparative genome sequencing for discovery of novel polymorphisms in *Bacillus anthracis. Science* 296: 2028–2033.

Sokurenko EV, V. Tchesnokova, A. T. Yeung, C. A. Oleykowski, E. Trintchina, K. T. Hughes, R. A. Rashid, J. M. Brint, S. L. Mose-ley, and S. Lory, (2001). Detection of simple mutations and polymorphisms in large genomic regions. *Nucl. Acids Research* 29(22): 15.

Spratt BG, (1999). Multilocus sequence typing: molecular typing of bacterial pathogens in and era of rapid DNA sequencing and the internet. *Curr. Opin. Microbiol.* 2: 312-316.

Stackebrandt E, W. Frederiksen, G.M. Garrity, P.A.D. Grimont, P. Kämpfer, M.C.J. Maiden, X. Nesme, R. Rosselló-Mora, J. Swings, H.G. Trüper, L. Vauterin, A.C. Ward, W.B. Whitman, (2002). Report of the ad hoc committee for the re-evalutation of the species definition in bacteriology. *Inter. J. Syst. Evol. Micobiol.* 52: 1043-1047.

Sukhnanand S, S. Alcaine, L.D. Warnick, W-L. Su, J. Hof, M.P.J. Craver, P. McDonough, K.J. Boor, M. Weidmann, (2005). DNA sequence-based subtyping and evolutionary analysis of selected *Salmonella enterica* serotypes. *J. Clin. Microbiol.* 43: 3688-3698.

Tankouo-Sandjong B, A. Sessitsch, E. Liebana, C. Kornschober, F. Allerberger, H. Hächler, L. Bodrossy, (2007). MLST-v, multilocus sequence typing based on virulence genes, for molecular typing of *Salmonella enterica* subsp. *enterica* serovars. *J. Microbiol. Methods.* 69: 23-36.

Taylor JD, D. Briley, Q. Nguyen, K. Long, M. A. Iannone, M. S. Li, F. Ye, A. Afshari, E. Lai, M. Wagner, J. Chen, and M. P. Weiner, (2001). Flow cytometric platform for high-throughput single nucleotide polymorphism analysis. *Biotechniques* 30: 661-6, 668-9.

Thompson FL, D. Gevers, D.D. Thompson, P. Dawyndt, S. Naser, B. Hoste, C.B. Munn, J. Swings, (2005). Phylogeny and molecular identification of vibrios on the basis of mulitlocus sequence analysis. *Appl. Environ. Micribiol.* 71: 5107-5115.

Torpdahl M, M.N. Skov, D. Sandvang, D.L. Baggesen, (2005). Genotypic characterization of *Salmonella* by multilocus sequence typing, pulse-field gel electrophoresis and amplified fragment length polymorphism. *J. Microbiol. Methods.* 63: 173-184.

Weissman SJ, S. L. Moseley, D. E. Dykhuizen, and E. V. Sok-urenko, (2003). Enterobacterial adhesins and the case for studying SNPs in bacteria. *Trends Microbiol.* 11: 115–117.

Welsh J, and M. McClelland, (1990). Fingerprinting genomes using isolates. *J. Clin. Microbiol.* 31: 1927–1931.

Wilson WJ, A. M. Erler, S. L. Nasarabadi, E. W. Skowronski, and P. M. Imbro, (2005). A multiplexed PCR-coupled liquid bead array for the simultaneous detection of four biothreat agents. *Mol. Cell Probes* 19:137-44.

Wong, CW, T. J. Albert, V. B. Vega, J. E. Norton, D. J. Cutler, T. A. Richmond, L. W. Stanton, E. T. Liu, and L. D. Miller, (2004). Tracking the evolution of the SARS coronavirus using high-throughput, high-density resequencing arrays. *Genome Research.* 14: 398-405.

Xiong J, W. M. Fischer, K. Inoue, M. Nakahara, and C. E. Bauer, (2000). Molecular evidence for the early evolution of photosynthesis. *Science* 289: 1703–1705.

Yang L, D. K. Tran, and X. Wang, (2001). BADGE, Beads Array for the Detection of Gene Expression, a high-throughput diagnostic bioassay. *Genome Res* 11: 1888-98.

Ye, F., M. S. Li, J. D. Taylor, Q. Nguyen, H. M. Colton, W. M. Casey, M. Wagner, M. P. Weiner, and J. Chen, (2001). Fluorescent microsphere-based readout technology for multiplexed human single nucleotide polymorphism analysis and bacterial identification. *Hum Mutat* 17: 305-16.

In: Encyclopedia of DNA Research (3 Volume Set)
Editors: S. J. Duncan and P. H. Wiley
ISBN 978-1-61324-305-3

Chapter 30

DNA SEQUENCING: METHODS, STRATEGIES AND PROTOCOLS IN MOLECULAR BIOLOGY RESEARCH HORIZONS

***Cimaglia Fabio*[1], *Assab Emanuela*[1], *D'Urso Oscar Fernando*[1] *and Poltronieri Palmiro*[*2]**

[1] Department of Environment and Biology Sciences,
via Monteroni km 7, I-73100 Lecce, Italy
[2] CNR, Institute of Sciences of Food Productions,
via Monteroni km 7, I-73100 Lecce, Italy

ABSTRACT

In recent years the exploitation of sequenced genomes has made possible the deepening of knowledge on how many genes are contained in the genomes of higher organisms. The identification of thousands of functional RNAs showed that sequenced genomes contain much more genes than previously sought. We cloned a library of RNAs sized 60-500 bases mouse and identified thirty small RNAs isolated from the developing embryo brain, the major number of them belonging to H/ACA and C/D box snoRNAs. Many of these sRNAs and snoRNAs are coded in introns of protein coding and non-protein coding transcripts. The small RNAs can form secondary structures with free energy ranging from -3.4 to -70 kcal/mol. Three-dimensional architectural motifs are increasingly recognized as determinants of RNA functionality. Such motifs can encode spatial information required for interaction with biomolecules. Localisation on the mouse genome using the UCLA Santa Cruz server showed a high conservation in these short sequences with overlapping regions of other genomes. Most of these new short RNAs have been identified today with an ENSEMBL identification number, but in our sequences there are 5' or 3' ends differences, probably relative to processing events and enzyme modifications.

A different DNA sequencing approach was used to identify genes from organisms with unsequenced genomes of wild species related to cultivated crops. In Solanaceae, gene duplication events produced a highly variable number of sequences coding for

* Correspondence author: Palmiro.poltronieri@ispa.cnr.it, Tel. +390832422609.

protein inhibitors targeting proteases, hydrolases and polygalacturonases. Exploiting degenerate primers and PCR amplification, we identified new Kunitz-type proteinase inhibitors of group A, group B and group C from wild Solanum species (*S. palustre* and *S. stoloniferum*), as well as additional isoforms from potato varieties. The new data allowed to design a phylogenetic tree grouping all known Kunitz-type inhibitors in Solanum species. The tree sub-roots, grouping highly related sequences conserved either in *S. palustre* and in *S. tuberosum*, could be useful as marks of gene duplication events at the basis of Solanum sub-families evolution and divergence.

This book chapter aims to provide clues for DNA sequencing projects directed to unsequenced organisms in which many transcripts wait to be discovered, either coding small RNAs or genes homologue to known genes coding for protein products.

INTRODUCTION

In recent years, the introduction of massive sequencing methods has opened a new era of DNA sequencing. These new Sequencing methods were exploited in the identification and analysis of short gene sequences and in the resequencing of completed genomes to study DNA polymorphic regions (454 Life Science pyrosequencing, the SOLID instrument, Applied Biosystem, and Solexa, Illumina). Standard DNA sequencing methods are expensive and time consuming, requiring the preparation of subtracted libraries before submission of samples to 384 capillary, medium to high-throughput DNA sequencers. At the beginning of the sequencing era, many EST sequences were deposited in GenBank. Single-step sample preparation methods did not bias the libraries toward short cDNAs as current methods of subtraction do. The FANTOM project of full-length mouse cDNAs, and the H-Invitational of full-length human cDNAs accomplished the objective of identifying poly-A transcript sequences and to locate them to corresponding genome regions. In a second step, consequent to gene identification, SAGE (LongSAGE, SuperSAGE) methods were set up with the aim to produce short tags corresponding to differentially expressed genes, able to unequivocally identify the transcripts and to quantitatively evaluate its expression levels (Winter et al. 2007). This protocol relies on the pyrosequencing approach to evaluate from ten thousands to hundred of thousands tags, each one unequivocally corresponding to a unique transcript.

The expressed transcripts can be visualised and aligned to sequenced genomes using BLAT software. One of the most used browsers is that at University of California Santa Cruz (http://genome.cse.ucsc.edu), allowing the visualisation of overlapping genes in syntenic chromosomal regions of vertebrate genomes.

Non-Protein Coding Small RNAs as Riboregulators

A large number of human transcripts has been found that do not encode for proteins, named non-protein coding RNAs (npcRNAs), often lacking high homology with other genomes, but containing secondary RNA structures, and short, highly conserved stretches, present in the overlapping regions and corresponding genes of other genomes. Today is known that the major component of the vast transcriptional output is represented by highly heterogeneous families of transcripts defined as short non-coding RNAs (sncRNAs) with no or limited protein-coding potentials, reviewed by Glinsky (Glinsky GV, 2008).

A ncRNA database, containing the FANTOM3 db (mouse), the H-invitation db (human) and the miRNA, piRNA and snoRNAs databases, was produced by the Institute for Molecular Bioscience, Queensland University http://research.imb.uq.edu.au/rnadb (Pang KC, Stephen S, Dinger ME, Engrstrom PG, Lehnard B, Mattick JS. RNAdb 2.0-an expanded database of mammalian non-coding RNAs. Nucleic Acid Res 2007; 35:D178-D182).

Various species of npcRNAs have been found differentially expressed in human tissues with homologs to primates. Actually, various small RNAs, often encoded in the brain, have been described as species-specific, rodent-specific, or primate-specific. RNAs also regulate mRNA stability and translation. Some npcRNA mimics the structures of other nucleic acids; the 6S RNA structure is reminiscent of an open bacterial promoter, and the tmRNA has features of both tRNAs and mRNAs. Other npcRNAs, such as the RNase P RNA, have catalytic functions. Many npcRNAs are associated with proteins, as the RNA activators of dsRNA-protein kinase. The heat-shock RNA switches on the transcriptional activity of heat shock factor HSF-1 by inducing trimerization and recruitment of RNA polymerase to promoter sites (Shamowsky et al. 2006). Other npcRNAs, such as the snRNAs and the SRP RNA, serve key structural roles in RNA-protein complexes. The importance of small non-coding RNAs (ncRNAs) with regulatory functions has emerged by several studies. ScaRNAs, snoRNAs piwiRNAs and microRNAs represent the mainstream output of the transcribed non-coding RNAs.

Sno-like RNAs and other classes of regulatory RNAs originate either from introns and exons of longer poly-A containing transcripts (produced by RNA Polymerase II), or transcribed (as H1, U6 or 7SK) by the RNA Pol III machinery. Pre-snRNAs require a processing machinery for maturation and final localisation. There are snoRNAs that direct RNA modification, RNAs that modulate translation by forming base pairs with specific target mRNAs, and probably most of the miRNAs are examples of this category. Sno-like RNAs and other classes of regulatory RNAs originate either from introns and exons of protein coding and npcRNA genes produced by the RNA Polymerase II transcription machinery, or transcribed (as H1, U6 or 7SK) by the RNA Pol III machinery.

MicroRNAs (miRNAs) have a role role as fine regulators of the eukaryotic gene expression. miRNAs have been shown to be crucial players of mRNA stability and protein expression. The microRNAs is the class of 21 base-long small RNAs, having a hairpin structure, that affect target genes by forming perfect matches or base pairing to highly similar sequences. In sheep, mutant Texel MSTN mRNA has mistakenly become the target of miRNAs because of its disguise using a target octamer motif borrowed from genuine target genes. This phenomenon led to muscular hypertrophy observed in the “callipyge” mutant sheep.

miRNAs are generated as a primary transcript (pri-miRNA) by RNA polymerase II (Lee et al. 2004) or by RNA polymerase III-like viral miRNAs (Pfeffer et al. 2005). Pri-miRNAs are capped, polyadenylated (Cai, et a. 2004) and subsequently they enter a microprocessor complex (500–650 kDa) consisting of a Drosha (an RNase III endonuclease) and an essential cofactor DGCR8/Pasha protein containing two double-stranded RNA binding domains (Denli et al. 2004). There, they are processed and cleaved giving rise to another precursor of 60–80 nucleotide stem- loop sequence (pre-miRNA) with a 5’ phosphate and two 3’ nucleotide overhang. The pre-miRNAs are then transported to the cytoplasm by Exportin-5, a member of the Ran transport receptor family (Yi et al. 2003). Finally, the stem–loop structure is sequentially processed by the cytosolic RNase III Dicer (Grishok et al. 2001) to yield the mature single-stranded miRNA. The single-stranded mature miRNA is incorporated into the cytosolic effector complex, called RNA-Induced Silencing Complex (RISC). Within the RISCs miRNA exerts its function of translational silencing by base-pairing the 3' UTR of the specifically targeted mRNAs (Gregory et al. 2005).

Massively Parallel Signature Sequencing (MPSS, Lynx Technologies) showed to be a sensitive and effective method of identifying short transcripts devoid of poly-A tail, as microRNAs and snRNAs. MPSS was used in the discovery of a very large number of small transcribed plant RNAs (Green et al. 2005). One negative feature in the sequencing of non poly-A RNA is that sequencing of the RNA libraries produces a high number of ribosomal RNA sequences, so that the cost of this method reported to the number of new sequences is relatively high.

Experimental Setting, Materials and Methods

A library of dsDNA corresponding to small RNAs (size 60-500 nt) from mouse brain (embryos from day 12 to 17) was kindly provided by Munster University (J. Brosius) in the frame of the EU project RIBOREG, Novel roles for non-coding RNAs in development and disease (2004-2007).

The library was suspended in 3 µl, with a final concentration of 15 ng/µl. Transformation of competent cells (One Shot TOP10, Invitrogen) was performed according to the Invitrogen protocol, producing a transformation efficiency of 0,42 x 10^9.

Starting from 2 µl of diluted DNA, 4000 colonies were individuated, that were stored in plates with numbered grids. PCR amplification using Invitrogen primers allowed the sequencing of 96 clones on an Applied Biosystems 4 capillary DNA sequencer 373.

New SnoRNAs and SnRNAs in Developing Mouse Brain

The sequencing of the 96 clones provided 26 new independent sequences, in addition to replicate sequences corresponding to ribosomal RNA. The result of DNA Sequencing of one 96 well plate of the sRNA library produced several repeated sequences reproducing the same ribosomal RNA (18S rDNA or 5S rDNA) representing multiple sequences of high abudance. In addition, many snoRNAs were identical (or different for one single base) to U2 snoRNA, U3 SnoRNA, U4 snoRNA, U5 snoRNA, A9 snoRNA, A13 snoRNA, A31 snoRNA, A40 snoRNA, D98 snoRNA, and RNAase P RNA. All these sequences, having homology to characterised regions on mouse chromosomes, and known as snoRNAs and snRNAs harbouring genes, were not deposited in GenBank database. These sequences with a difference in length of one or two nucleotides may reflect a bias in the DNA sequencing or in sequence analysis. One 96 bases long sequence variant of U5 snRNA was also not deposited. Fifteen new variants of snoRNAs and snRNAs were identified in mouse brain (developmental stages 12-17 days), deposited in GenBank, under the accession numbers FM991905-FM991919, corresponding to mouse D47 snoRNA, D81snoRNA, 7SK (U94 snoRNA homologue), U3 snRNA, A28 snoRNA, D22/D29 snoRNA, A16 snoRNA, U1 snRNA, D34 snoRNA, D12 snoRNA, D35a snoRNA, U2 snRNA, a scaRNA-like sRNA possessing high free energy content (-70,3 kcal/mol, Figure 1) in its secondary structure, a sRNA overlapping

BC0711254, and AK172090 originated snRNA, respectively. Considering the free energies of these small RNAs, considering the secondary structures calculated by software analysis, a very high free energies is freed in the secondary structure of rRNA-like short RNA (Figure 1), and FM991912, with free energy of – 47,4 kcal/mol (Figure 2).

All the sequenced genes, produced in 2006, were not present at that time in the UCSC genome browser at the time of this work, since the ENSEMBL annotations were added recently.

Many small RNAs in mouse are produced from unspliced mRNA. Two snoRNAs sequenced in this work localise in introns of the *gas5* gene, a growth arrest induced non-protein coding RNA that produces 9 snoRNAs (Raho et al. 2000). AK009175, a mouse gene with high level of expression in brain, was recently found to contain three intronic snoRNAs (A16, A44 and A61) belonging to the MBI-420 group, highly conserved in all mammalian genomes. The human AK092096 gene, AK009175 homologue, codifies for the neuronal cell differentiation protein AAK00754. FM991911 is a 161 basis long variant of SNORA16. In respect to the 134 base long A16 snoRNA, FM991911 possesses –34,3 kcal/mol free energy in its secondary structure, thus having high stability as the classical snoRNAs. The finding that FM991911 is expressed in brain, support the brain co-expression of the processed mRNA and the intronic snoRNA.

FM991909, 115 basis long, is produced from GI:47474960, 605 bases long sequence that originates SNORA28, 127 basis long. It seems that this snoRNA has an alternative processing, conserving a high free energy in its secondary structure (-14,2 kcal/mol).

FM991907, 115 basis long, originates from the 331 basis long 7SK gene. The processed RNA has a high similarity with SNORU94, thus possibly being a new functional small RNA.

While the shortened variants could be functional, processed variants or degradation products of snoRNAs, it seems interesting to note that few new small RNAs have additional bases that are not reported in the corresponding sequences in databases, as in the case of FM991912. Furthermore, we identified two variants with two additional bases in respect to the corresponding SNORA31 and SNOR40 sequences. It may be possible that some snoRNA and snRNA is processed with addition of new bases due to activity of modification enzymes.

A second library produced by Munster University, corresponding to small RNA with size 20-50 bases, was also used to transform competent cells and to store each independent clone. However these DNAs were not sequenced, since the sequencing costs were too high. Recently other sequencing methods have been exploited for very short RNAs, as microRNAs and piwiRNAs. One of these methods consist in the production of concatenamers, grouping several different DNAs, to an average length of 800 bp. However, when trying to sequence a library of concatenamers corresponding to small RNAs from *Medicago truncatula* (a collaborative work with Martin Crespi and Gary Stacey), only one third of the library was sequenced. The main drawback of this method consists in low reproducibility of purification of plasmids from *E. coli* cells, especially when working in automation on 384 well plates, and robotic extraction of DNA.

In this work, we showed the feasibility to monitor the expressed small RNAs in a tissue-specific and development-stage specific frame, including new sequences having variation in the length and showing processing steps and modification (base extension, splicing) at their extremities. These data were produced in january 2007, for the final RIBOREG report to the scientific officers of the EU Community, two years before the present annotation of transcripts and their localisation on the mouse genome, that includes the new ENSEMBL *Mus musculus* genes for small RNAs. This work showed the feasibility of DNA sequencing projects in unsequenced organisms, in which thousands of transcripts coding for small RNAs are awaiting to be discovered.

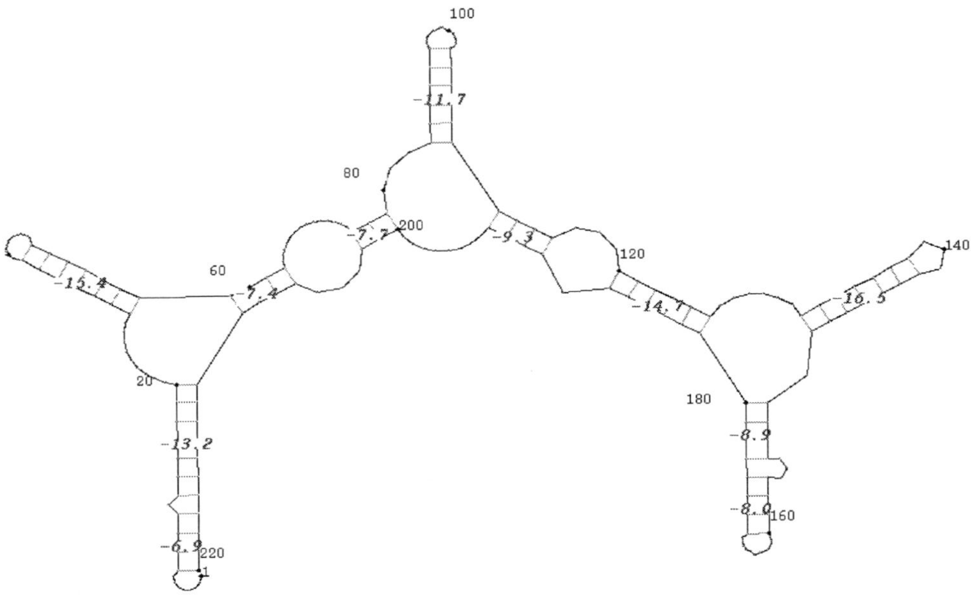

Figure 1. FM991917, a rRNA-like short RNA possessing high free energy (-70,3 kcal/mol). The high free energy entrapped in its secondary structure may produce a more stable small RNA.

DNA Sequencing of Hortologue Genes from Wild Species Related to Cultivated Crops

The Solanaceae is the third most valuable crop family, that includes highly different plants and many edible species (as coffee, pepper, eggplant, tomato and potato). The Solanaceae family is unique in that there have been no large-scale duplication events (e.g. polyploidy) early in the radiation of this family. The polyploidy events (e.g. tetraploid potatoes and tetraploid tobacco) are all recent events and diploid forms of both of these species are still in existence. As a result, microsynteny conservation amongst the genomes of tomato, potato, pepper and eggplant is very high. Previously, we sequenced several potato protein inhibitors of proteases and polygalacturonases (Speransky et al. 2007; Krinitsina et al. 2006). Solanum species contain a large array of proteinase inhibitors, of which Kunitz-type inhibitors are present in high number and may function in defense against pathogens and animal antifeedants (Santino et al. 2005). Kunitz-type inhibitors are classified in three groups, A, B and C, based on protease specificity and sequence identity (Heibges et al. 2003a, 2003b). KPIs belonging to the same homology group exhibit distinct features in their specificity towards target proteases (Heibges et al. 2003b). The KPI-B group includes inhibitors of serine proteases as trypsin, chymotrypsin and elastase (Heibges et al. 2003b). KPIs of group A, are considered aspartic protease (cathepsin D) inhibitors, with few exceptions, as tomato jasmonic-induced protein JIP21, a group-A KPI that inhibits chymotrypsin (Lison et al 2006). The KPI group C includes inhibitors of cysteine proteases (papain, ficin, bromelain, cathepsin B) or other different hydrolyzing enzymes.

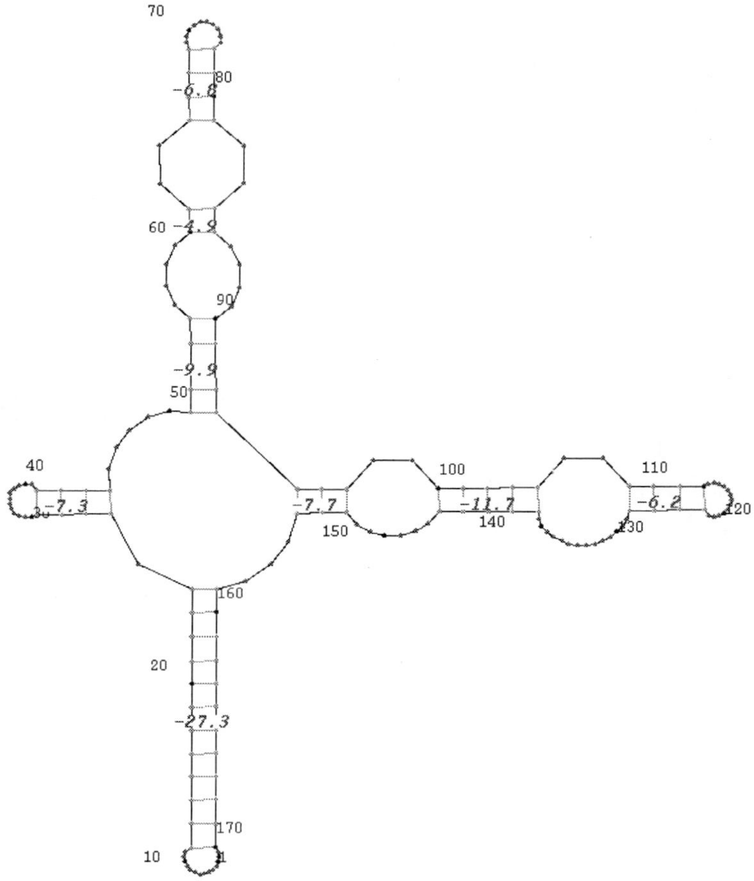

Figure 2. FM991912, a 169 base long variant of U1 snRNA, has a very high free energy that may produce a stable U1-like small RNA.

In different potato cultivars more than 80 *KPI* cDNA and genomic sequences have been identified, most of them expressed in the tuber, while other with leaf-specific expression (Kang et al. 2002). The potato tuber proteome has been extensively studied and a phylogenetic tree of potato KPIs has been made (Bauw et al. 2006). At present, 32 *PKPI* group-B gene variants have been sequenced from the genomes of 18 potato cultivars, namely Bintje (van den Broek and Jongsma), Ulster (Strukelj et al. 1990), Superior (Hannapel 1993) Danshaku (Ishikawa et al. 1994), Saturna and Provita (Heibges et al. 2003a), Keszthelyi 855-Whyte Lady (Banfalvi et al. 1996), Elkana (Pouvreau et al. 2003), Pentland Squire (Maganja et al. 1992), Kuras (Bauw et al. 2006), Desiree (Hermosa et al. 2006), Jopung (Park et al. 2005), Golden Valley (Kim et al. 2006) Agata (Ledoigt et al. 2006), Shepody (Flinn et al. 2005), Rishiri (Nakame et al. 2003) and Istrinskii (Speranskaya et al. 2005). Potato is a tetraploid species, so several deposited KPI sequences are isoforms and alleles of the same gene. Thirteen sequences corresponding to *KPI-B* genes were identified in tubers of tetraploid *S. tuberosum* cv. Provita (Heibges et al 2003a). Assuming maximum heterozygosity, this would indicate the presence of at least four *PKPI-B*-specific loci. A similar estimate has been done for KPI group-A for KPI group-C, with at least three loci for each group in the variety Istrinskii.

Genomic Gene PCR Cloning

During the project NATO-Russia (JSTC.RCLG 980102), numerous *KPI* genes were identified using a PCR-based cloning strategy in *Solanum palustre* (syn. *S. brevidens*), a wild, diploid, potato relative non-tuberous South-American species belonging to the subsection *Estolonifera* of the section *Petota*. These gene sequences were deposited in GenBank under the accession numbers AY945740, AY945741, AY945742, AY945743, AY945744 (Speransky et al 2007). A similar strategy was used in the cloning of polygalacturonase inhibitor (PGIP) genes from *S. tuberosum* and *S. palustre* (Krinitsina et al. 2006). Four sequences resembling those of other plant polygalacturonase inhibitors (PGIP) were identified in S. *palustre*. The sequences were deposited in GenBank under accession numbers AY6626809 (full length PGIP precursor), AY662680 (PGIP-Sbr-1), DQ185391 (PGIP-Sbr-2), DQ185392 (PGIP-Sbr-3), while DQ185394 (PGIP-Sbr-4) having a stop codon at +270 was considered a pseudogene. Alignment of the deduced amino acid sequences of four: *S. palustre* PGIPs with PGIP from potato (AY662681) and tomato *(*AAA53547). DQ185393, DQ185392 and DQ185391 represent true genes located in at least two independent genomic loci.

Material and Methods

Total genomic DNA template was added to the reaction mixture at a final concentration of 10 ng/μl. PCR conditions were: (1) 1 cycle: 94°C – 1 min; (2) 5 cycles: 94°C – 30 sec, 56°C – 10 sec, 72°C – 10 sec and (3) 25 cycles: 94°C – 5 sec, 60°C – 5 sec, 72°C – 30 sec. The amplified DNA fragments were analyzed by electrophoresis in a 1% agarose gel and purified using a DNA extraction kit. The eluted fragments were cloned into pGEM-T Easy (Promega) using an *E. coli* BMH 71-18 strain. DNA purifications and restriction mapping were carried out as described in the literature. Standard T7-promoter, SP6 (Promega) and KPI-F and KPI-R primers were used for DNA sequencing on both strands (AbiPrism 3130 DNA Sequencer, Applied Biosystems). The obtained nucleotide sequences were processed by the BioEdit 7.0.1 software package for alignment. Clustering of individual sequences was performed with Dendroscope using Neighbour-Joining, Maximum likelihood (with bootstrap values) and Maximum parsimony methods.

Results

A new set of PKPI genes was isolated by a PCR-based strategy from the genomes of *S. tuberosum* cv. Istrinskii (tetraploid), *S. palustre* (diploid, non-tuberous), *S. andigenum* and *S. stoloniferum*. KPI group A, group B and group C sequences represent genes or alleles in closely related species of *Solanum* genus.

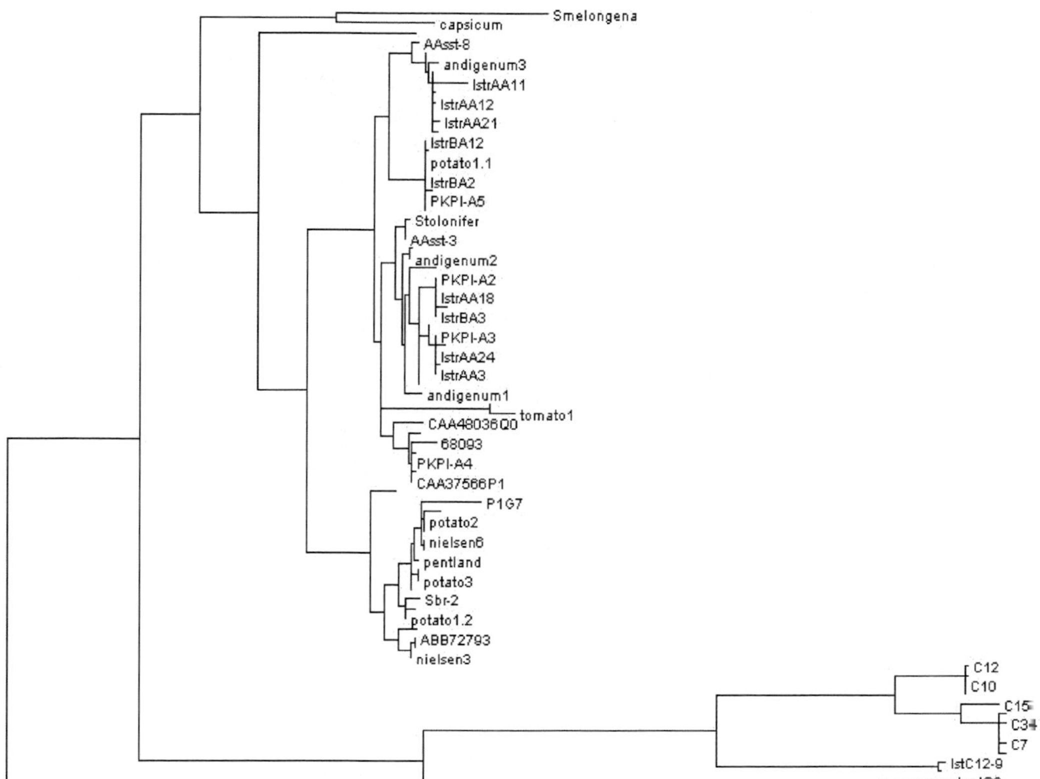

Figure 3. Phylogenetic tree using the neighbour joining method. In the bottom, two sub-clusters show the relatedness between group-C KPIs and miraculin sequences.

Twelve new KPI group A sequences (four from *S. palustre,* five from *S. tuberosum* and two from *S. stoloniferum*) and 10 new KPI group C sequences (from *S. tuberosum* and *S. palustre*) were identified, but not have been deposited yet in GenBank.

Sequences corresponding to 9 KPI-A and KPI-B genes in *S. tuberosum* and in *S. palustre* demonstrate that in diploid genomes there are at least 3 loci for KPI-B and two fro KPI-A genes.

In addition, new KPI-C sequences allowed the identification of seven genes/alleles in *S. tuberosum* cv Instrinskii summing up to three genomic loci for KPI-C. Two PKPI-C were similar to an invertase inhibitor cDNA reported from cv. Provita (99% identity) and trypsin inhibitor from cv. Bintje (98%). Five other genes were original, four showing 89-92% identity with known PKPI-C from potato, and one being 98% homologous to S9C11 from cv Provita (Heibges et al., 2003).

After alignment of all KPI related sequences (including miraculin), a phylogenetic tree was constructed, comparing the KPI family in the wild and cultivated species in the *Solanum* genus. The tree sub-roots, grouping highly related sequences conserved either in *S. palustre* and in *S. tuberosum*, could be useful as marks of gene duplication events at the basis of Solanum sub-families evolution and divergence. In particular, it is clear that miraculin sequences cluster near the group-C KPIs (Figure 3, 4, 5 and 6).

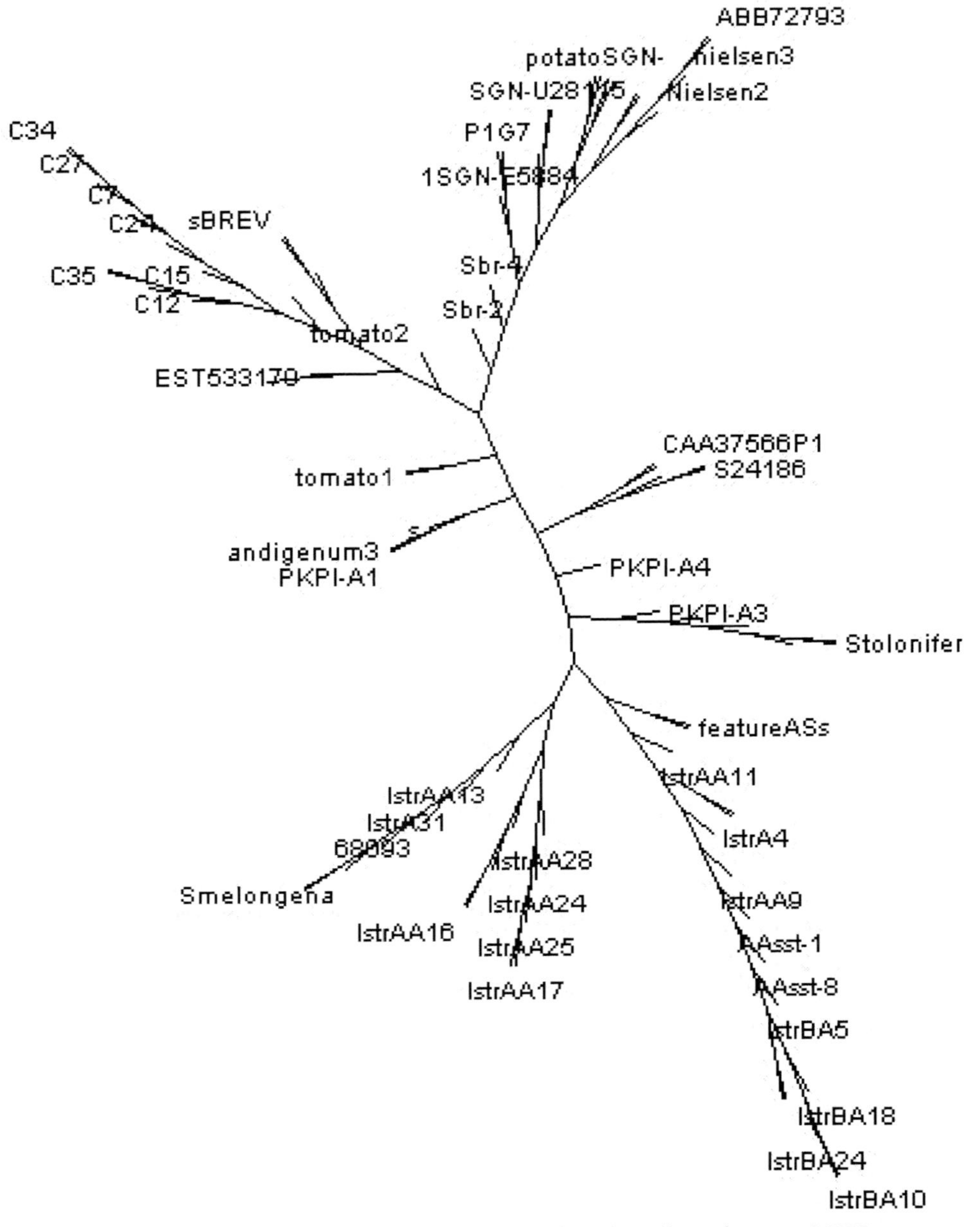

Figure 4. Maximum parsimony method allowed to cluster the miraculin and group-C KPI sequences in a divergent branch of the phylogenetic tree.

CONCLUSION

Two gene cloning strategies were exploited to cope with different levels of difficulty in DNA sequencing methods. In the case of short sequences coding for small RNAs, the library approach showed to be more suitable. The new pyrosequencing methods combined with ordinate clones on plates may allow high-throughput analysis of tissue-specific or development–specific RNA transcripts. In the case of homologue genes present in high copy number of wild species, closely related to cultivated corps, but for which no genomic data are available, a different strategy, genomic gene PCR cloning, showed to be a rapid, cheap and reliable protocol producing useful gene information.

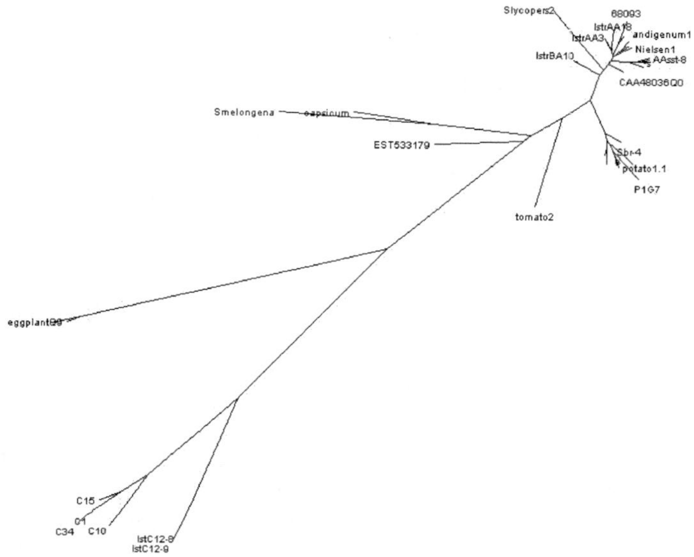

Figure 5. This Neighbour Joining-based phylogenetic tree shows the distances that exist between group-C KPI sequences and the miraculin cluster (eggplant Q9 and *S.brevidens/S. palustre* sequences).

Figure 6. This Maximum parsimony-based phylogenetic tree shows the high number of KPI sequences originated from S. tuberosum and S. palustre, either from group B or group A clusters. In several cases it is possible to make a comparison with more distant species, as tomato or *S. stoloniferum*.

REFERENCES

Lu, C; Tej, SS; Luo, S; Haudenschild, CD; Meyers, BC; Green, PJ. Elucidation of the small RNA component of the transcriptome. *Science,* 2005, 309, 1567-9.

Glinsky, GV. Phenotype-defining functions of multiple non-coding RNA pathways. *Cell Cycle*, 2008, 7(11) in press.)

Mallardo, M; Poltronieri, P; D'Urso, OF. Non-protein coding RNA biomarkers and differential expression in cancers. A review. *J Exp. Clin. Cancer Res.,* 2008, 27, 19.

Bauw, G; Nielsen, HV; Emmersen, J; Nielsen, KL; Jorgensen, M; Welinder, KG. Patatins, Kunitz-type protease inhibitors and other major proteins in tuber of potato cv Kuras. *FEBS J.*, 2006, 273, 3669-3684.

Cai, X; Hagedorn, CH; Cullen, BR. Human microRNAs are processed from capped, polyadenylated transcripts that can also function as mRNAs. *RNA*, 2004, 10, 1957-1966.

Denli, AM; Tops, BB; Plasterk, RH; Ketting, RF; Hannon, GJ: Processing of primary microRNAs by the microprocessor complex. *Nature*, 2004, 432, 231-235.

Gregory, RI; Chendrimada, TP; Cooch, N; Shiekhattar, R. Human RISC couples microRNA biogenesis and posttranscriptional gene silencing. *Cell*, 2005, 123, 631-640.

Grishok, A; Pasquinelli, AE; Conte, D; Li, N; Parrish, S; Ha, I; Baillie, DL; Fire, A; Ruvkun G; Mello CC. Genes and mechanisms related to RNA interference regulate expression of the small temporal RNAs that control *C. elegans* developmental timing. *Cell,* 2001, 106, 23-34.

Heibges, A; Glaczinski, H; Ballvora, A; Salamini, F; Gebhardt, C; 2003a. Structural diversity and organization of three gene families for Kunitz-type enzyme inhibitors from potato tubers (*Solanum tuberosum L.*). *Mol. Gen. Genom.,* 269(4), 526-534.

Heibges, A; Salamini, F; Gebhardt, C. Functional comparison of homologous members of three groups of Kunitz-type enzyme inhibitors from potato tubers (*Solanum tuberosum L*) *Mol. Gen. Genomics*, 2003, 269, 1215-1221.

Kang, SG; Choi, JH; Suh, SG. A leaf specific 27 kDa protein of potato Kunitz-type protease inhibitor is induced in response to abscisic acid, ethylene, methyl jasmonate and water deficit. Mol. Gen. *Genetics*, 2002, 13, 144-154.

Krinitsina, AA; Speransky, AS; Poltronieri, P; Santino, A; Bogacheva, AM; Buza, NL; Protsenko, MA; Shevelev, AB. Cloning of Polygalacturonase Inhibitor Protein Genes from *Solanum brevidens Fill. A*. *Genetica,* 2006, 42, 477-486.

Lee, Y; Kim, M; Han, J; Yeom, KH; Lee, S; Baek, SH; Kim, VN. MicroRNA genes are transcribed by RNA polymerase II. *EMBO J.,* 2004, 23, 4051-4060.

Lehesranta, SJ; Davies, HV; Shepardt, LV; Koistinen, KM; et al. Proteomic analysis of the potato tuber life cycle. *Proteomic,* 2006, 6, 6042-6052.

Lison, P; Rodrigo, I; Conejero, V. A novel function for the cathepsin D inhibitor in tomato. *Plant Physiol.,* 2006, 142, 1329-1339.

Machinandiarena, MF; Olivieri, FP; Daleo, GR; Oliva, CR; Isolation and Characterization of a Polygalacturonase-Inhibiting Protein from Potato Leaves: Accumulation in Response to Salicylic Acid, Wounding and Infection, *Plant Physiol. Biochem,* 2001, 39. 129-136.

Nakane, E; Kawakita, K; Doke, N; Yoshioka, H. Elicitation of primary and secondary metabolism during defence in the potato. *J. Gen. Plant Pathol.,* 2003, 69, 378-384.

Pfeffer, S; Sewer, A; Lagos-Quintana, M; Sheridan, R; Sander, C; Grässer, FA; van Dyk, LF; Ho, CK; Shuman, S; Chien, M; Russo, JJ; Ju, J; Randall, G; Lindenbach, BD; Rice, CM; Simon, V; Ho, DD; Zavolan, M; Tuschl, T. Identification of microRNAs of the herpesvirus family. *Nat Methods*, 2005, 2, 269-276.

Raho, G; Barone, V; Rossi, D; Philipson, L; Sorrentino, V. The gas 5 gene shows four alternative splicing patterns without coding for a protein. *Gene,* 2000, 256, 13-17

Rawlings, ND; Tolle, DP; Barrett, AJ. Evolutionary families of peptidase inhibitors. *Biochem J.,* 2004, 378, 705-716.

Rossi, A; D'Urso, OF; Gatto, G; Poltronieri, P; Ferracin M, Remondelli, P; Negrini M, Caporaso MG, Bonatti, S; Mallardo, M. Non-coding RNAs change their expression profile after Retinoid induced differentiation of the promyelocytic cell line NB4. BMC Research Notes, 2009, submitted.

Santino, A; Poltronieri, P; Mita, G. Advances on plant products with potential to control toxigenic fungi. A review. Food Addit. *Contam.,* 2005, 22, 389-395.

Shamovsky, I; Ivannikov, M; Kandel, ES; Gershon, D; Nudler, E. 2006. RNA-mediated response to heat shock in mammalian cells. *Nature,* 440, 556-560.

Speranskaya, AS; Krinitsina, AA; Poltronieri, P; Fasano, P; Santino, A; Shevelev, AB; Valueva, TA. Molecular cloning of Kunitz-type proteinase inhibitor group B genes from potato. *Biochemistry* (Moskow), 2005, 70, 292-299.

Speransky, AS; Cimaglia, F; Krinitsina, AA; Poltronieri, P; Fasano, P; Bogacheva, AM; Valueva, TA; Halterman, D; Shevelev, AB; Santino, A. Kunitz-type protease inhibitors group B from *Solanum palustre*. *Biotechnol. J.,* 2007, 2, 1417-24.

Stotz, HU; Contos, JJ; Powell, AL; et al; Structure and Expression of an Inhibitor of Fungal Polygalacturonases from Tomato, *Plant. Mol. Biol.,* 1994, 25(4), 607-617.12.

Yi, R; Qin, Y; Macara, IG; Cullen, BR; Exportin-5 mediates the nuclear export of premicroRNAs and short hairpin RNAs. *Genes Dev.,* 2003, 17, 3011-3016.

In: Encyclopedia of DNA Research (3 Volume Set)
Editors: S. J. Duncan and P. H. Wiley
ISBN 978-1-61324-305-3

Chapter 31

RETRO ARRAY—A COMPREHENSIVE DIAGNOSTIC DNA CHIP FOR RAPID DETECTION AND IDENTIFICATION OF RETROVIRUSES, RETROVIRAL CONTAMINANTS, AND MISTAKEN IDENTITY OF CELL LINES

Wolfgang Seifarth[1], Oliver Frank[1], Julia Schreml,[1] and Christine Leib-Mösch[1,2]

[1] III. Medizinische Klinik, UMM – Universitätsmedizin Mannheim, University of Heidelberg, Germany

[2] Helmholtz Zentrum München, German Research Center for Environmental Health, Institute of Virology, Neuherberg, Germany

ABSTRACT

Retroviruses not only represent some of the most dangerous pathogens, but also constitute, as remnants of former infections that happened millions of years ago, a large fraction of the human genome. We have established a fast and reliable DNA chip-based assay (RetroArray) for detection and identification of a wide variety of human and other vertebrate exogenous and endogenous retroviruses in biological/clinical samples. The assay combines a pan-retrovirus multiplex polymerase chain reaction (PCR) using fluorochrome-modified primers and DNA chip hybridization. Using RetroArray, distinct transcription profiles of human endogenous retroviruses (HERVs) have been established for a variety of human tissues. Using paired samples (normal vs. disease) this method can be applied to examine HERV activity in human tumors and can help to identify retrovirus-derived tumor antigens. In addition, RetroArray has been designed to detect human exogenous retroviruses such as human immunodeficiency virus (HIV) and human T-cell leukemia virus (HTLV), and putative cross-species transmissions of animal retroviruses in human samples. Species specificity and sensitivity of the assay was demonstrated by detecting traces of pig endogenous retrovirus (PERV) DNA down to ~25 copies in human cDNA samples. Furthermore, retroviral transcripts may be identified in particle preparations from cell culture supernatants. This makes the assay a valuable technique for monitoring packaging cell lines and vector preparations commonly used in human gene therapy applications to exclude cotransfer of replication competent retroviruses (RCRs) or endogenous retroviruses (ERVs) into target cells. Therefore, RetroArray could improve significantly the safety of human gene therapy, tissue engineering, xenotransplantation and production of therapeutic polypeptides in cell culture. Mistaken identity of human cell lines and frequently observed laboratory contaminations with cells of other species, as well as infection with polytropic animal retroviruses such as squirrel monkey retrovirus (SMRV) or murine leukemia virus (MLV) can influence experimental results and may lead to invalid conclusions. The RetroArray technique is an excellent tool for testing purity and homogeneity of cell lines. Characteristic HERV transcription profiles can be used to assess the cell type and to monitor cell lines for contaminating cells.

INTRODUCTION

Retroviruses represent an extremely wide range of virus-host interactions: among them there are deadly pathogens such as HIV and HTLV. For others, coevolution with the host has occurred in a unique way: these so-called ERVs have achieved becoming an integral part of their hosts' genome and thus may persist millions of years as controlled parasites. In a few cases they actually became symbionts, which have adopted essential cellular tasks during evolution [Rawn and Cross, 2008; Conley *et al.*, 2008].

HERVs comprise about 8–9 % of the genome [Lander *et al.*, 2001]. They are thought to be relicts of retroviral germ line infections that were genetically fixed during primate evolution [for review see: Gifford and Tristem, 2003; Mager and Medstrand, 2003; Sverdlov, 2005]. During millions of years they have amplified and spread throughout the primate genome by re-infection and/or retrotransposition [Belshaw *et al.*, 2005; Katzourakis *et al.*, 2005]. Most HERVs have been silenced by mutations or epigenetic control mechanisms such as DNA methylation, chromatin inactivation, RNA editing enzymes like APOBEC and RNA interference [Schulz *et al.*, 2006; Lee *et al.*, 2008; Hakim *et al.*, 2008]. Epigenetically silenced HERVs may be reactivated by various environmental conditions such as chemicals, radiation or exogenous viruses [Hohenadl *et al.*, 1999; Zeilfelder et al., 2007; Garrison 2007; Ruprecht 2006]. In animals recombination between different activated ERVs or exogenous retroviruses and ERVs has been shown to generate novel pathogenic viruses causing leukemia and other tumors [Weiss, 2006]. So far, an infectious HERV or recombinant virus has not been detected in humans. However, a subgroup of insertionally polymorphic HERVs present in only a proportion of the human population has been described recently [Moyes *et al.*, 2007]. Several attempts have been made to link such polymorphic HERV integrations with human diseases. The majority of HERVs contains functional regulatory sequences like promoters, enhancer elements and polyadenylation signals, which could affect the expression of neighbouring genes [van de Lagemaat *et al.*, 2003; Conley *et al.*, 2008]. Such genomic interaction could, for example, lead to an activation of oncogenes or inactivation of tumor suppressor genes. Furthermore, many HERV elements code for individual retroviral gene products [de Parseval *et al.*, 2003; Galli *et al.*, 2005; Denne *et al.*, 2007] that could be pathological. Transcriptionally active HERVs have been found in all human tissues investigated so far [Stauffer *et al.*, 2004; Seifarth *et al.*, 2005] and some of the expressed

HERV proteins have been associated with tumors, neurological diseases and autoimmunity [Nelson *et al.*, 2003; Bannert and Kurth, 2004; Voisset *et al.*, 2008].

Occasionally, retroviruses may jump between species [Martin *et al.*, 1999], and in most cases when they are passed to a new host, they become more pathogenic [Silvestri, 2005]. The acquired immunodeficiency syndrome (AIDS), e.g., is a zoonotic infection caused by primate lentiviruses that crossed species barriers on multiple occasions and have generated HIV-1 and -2 [Sharp *et al.*, 2005]. In particular ERVs, which are often xenotropic and do not re-infect their own host cells, may infect other species *in vivo* and *in vitro*. This ability to move between species has caused doubts over the safety of xenotransplantations or the use of animal cell lines for therapeutic applications [Weiss, 2006]. Exogenous and endogenous retroviruses of other species may further play a role as potential laboratory contaminants. Many human cell lines commonly used for research have proved to be contaminated by animal retroviruses such as SMRV [Oda *et al.*, 1988] or MLV [Takeuchi *et al.*, 2008], although these retroviruses are normally not infectious for humans. The lack of adequate defense mechanisms (e.g. immune response) in cell culture may facilitate infections by retroviruses from other species. Furthermore, silencing of ERVs may be impaired more readily in cell culture than in intact organisms [Bromham, 2002]. Therefore, the risk of zoonoses could be increased by the use of animal cell cultures in medical research, production of biologic medicines such as monoclonal antibodies, and in human gene therapy. Similar concerns apply to xenografts for transplant recipients with a compromised immune system.

To date, commercial microarrays used in medical research and diagnosis disregard ERVs, although these sequences represent a large fraction of mammalian genomes. HERVs, for example, count for about half a million elements including more than 3,000 relatively complete proviral structures in the human genome [Sperber *et al.*, 2007]. Therefore, we have developed a microarray-based comprehensive retrovirus assay (RetroArray) that allows for simultaneous detection and identification of most relevant RT-encoding retroviruses in biological samples. Employing this technique, the study of HERV elements and their biological role in gene expression regulation, cell differentiation and genome evolution, as well as modulation of immunological functions and tumorigenesis will become more feasible and may help to achieve new insights in the important field of retrovirology. Furthermore, Retroarray may be useful to improve safety in biotechnology and therapeutic applications.

RETROARRAY DESIGN

For many applications in basic science or routine lab work and particularly for applications with relevance to human or livestock health, a retrovirus assay should warrant a comprehensive detection and identification of retroviruses in biological and clinical samples. Irrespective to the disposition of the target, e.g. proviral DNA, retroviral transcripts or RNA contained in virions, it should map any retroviral genetic material with the highest possible sensitivity and specificity. Targeting of unknown retroviruses or a quick adaptation to novel retroviruses would be advantageous. Moreover, operation should be easy, starting from routine lab sample preparations using standard sampling procedures with minimal hands-on-time and the capability of robotic automation (standardization of all methodological steps).

Qualitative results should be clear without the need for expensive and complex software solutions.

Table 1. Human mixed oligonucleotide primers (MOP) used for pan-retrovirus multiplex PCR

betaretrovirus primer cocktail (MOP-ABD)

Target ID	Source	Orientation	5' Modification	Nucleotide sequence	Degenerations
HML-1	U35102	forward reverse	- Cy3	GGAGAAAAGTACTGCCACAAGGC GGAGAACAGAATATCATCCATGTA	- -
HML-2	M14123	forward reverse	- Cy3	GGAGAAAAGTGTTACCTCAGGGA GGAGAATAAAATATCATCAATATA	- -
HML-3	U35236	forward reverse	- Cy3	GGAGAAAAGTGTTGTCACAGGGC GGAGAAGTATATATCATCCATATA	- -
HML-4	AF020092	forward reverse	- Cy3	GGAGAAAAGTCCTACCACAAGGC GGAGAAGAGGAGATCATCCATGTA	- -
HML-5	U35161	forward reverse	- Cy3	GGAGAAGTGCTTCCTGAAGGGATG GGAGAATAAAATATCATCCATAAA	- -
HML-6	U60269	forward reverse	- Cy3	GGAGAAAGAGTTTTACCCCAAGGC GGAGAAAAGAATATCATCCATATA	- -
HML-7	AP003171	forward reverse	- Cy3	GGAGAAGTTTTACCTCAAGGAATG GGAGAACAGTATATCATCCATATA	- -
HML-8	AL513321	forward reverse	- Cy3	GGAGAAGTACTTCCTCAGGGAATG GGAGAATAAAATATCATCAATATA	- -
HML-9	AC025569	forward reverse	- Cy3	GGAGAAGTTCTACCCCAAGAGATG GGAGAACAAAATATCATCCACATA	- -
HML-10	U07856	forward reverse	- Cy3	GGAGAAAAGTTTTGCCCCAGGGT GGAGAATCTACTTTTTGCTGCACA	- -

Gammaretrovirus Primer Cocktail (MOP-C)

Target ID	Source	Orientation	5' Modification	Nucleotide sequence	Degenerations
gamma retrovirus	Shih et al., 1989	forward	-	GGAGAATGGAAAGTGYTRCCMCARGG	8
		reverse	Cy3	GGAGAACAGCAGSAKGTCATCCAYGTA	8
HERV-I HERV-L	M92067, \$ G895836, \$	forward reverse	- Cy3	GGAGAATKKACMSKMYTRCYYCARGGG GGAGAAAKMWKRYCATCMAYRTAMTG	1072 8192
HERVH	AF026252	forward reverse	- Cy3	GGAGAAGTACTGCTGCAAAGCTTCA GGAGAACACACGATCGGCAGGGAGA	- -
RGH2	D11078	forward reverse	- Cy3	GGAGAAGTAATGCTGCAAGGTTTC GGAGAAGAAGAACATCACCAATATA	- -
HERV-F	AC000378, \$ Z94277	forward reverse	- Cy3	GGAGAARTMCTMCMYCARGGGTT GGAGAAAAGGAGGTCATCTAGATAT	64 -
HIV-1	K02013	forward reverse	- Cy3	GGAGAAGTGCTTCCACAGGGATGG GGAGAAATACAAATCATCCATGTA	- -

Housekeeping gene primer cocktail (MOP- HKG)

Target ID	Source	Orientation	5' Modification	Nucleotide sequence	Degenerations
beta-actin	E01094	forward reverse	- Cy3	ATGATGATATCGCCGCGCTCG CATGTCGTCCCAGTTGGTGACG	- -
Ubiquitin	U49869	forward reverse	- Cy3	GTTGGCTTTGTTGGGTGAGCTTG AAGAGTACGGCCATCTTCCAGCTG	- -
RPL19	NM_000981	forward reverse	- Cy3	CCCGAATGCCAGAGAAGG CTTCCTTGGTCTTAGACCTG	- -
GAPDH	NM_002046.1	forward reverse	- Cy3	AGTCAACGGATTTGGTCGTATTGGG ACGTACTCAGCGCCAGCATCG	- -
HPRT	NM_000194	forward reverse	- Cy3	GTGATGATGAACCAGGTTATGACCTTG CTACAGTCATAGGAATGGATCTATCAC	- -

$ = degenerate primers derived from various members of the taxon = does not apply

Abbreviations: HML, human MMTV-like; HFV, human foamy virus; GAPDH, glycerole-aldehyde-3-phosphate-dehydrogenase; HPRT, hypoxanthine-phosphoribosyl-transferase; RPL19, ribosomal protein L19.

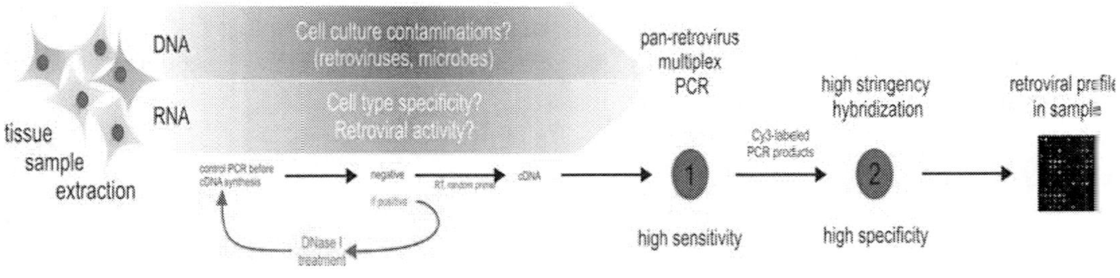

Figure 1. Experimental flow of a typical RetroArray DNA chip hybridization experiment. Retroarray is a two-step assay (1, 2) that combines the high sensitivity of a multiplex PCR targeting retroviral RT genes by means of mixed oligonucleotide primers (MOP) with the specificity of a highly stringent glass chip hybridization. It is to emphasize that expression profiling of RNA will be compromised by any traces of DNA within RNA preparations. Therefore, all RNA samples have to be treated with RNase-free DNase I until they proof DNA-negative in control PCRs using MOP primers. In contrast, for detection and identification of xenotropic or contaminating retroviruses, simple DNA preparations make appropriate starting material.

We established RetroArray, a combined PCR/oligonucleotide microarray-based assay that meets all the mentioned requirements. Within one single experiment, it allows for parallel detection and identification of a wide variety of human and other vertebrate exogenous and endogenous retroviruses in all types of samples. The assay combines both the advantages of multiplex PCR (high sensitivity) and high stringency DNA chip hybridization (high specificity) (Figure 1). RetroArray is designed as an open modular assembly system that allows target customization of the test with respect to various scientific goals.

Targeting Retroviruses With a *pol*-Specific Multiplex PCR

To achieve simultaneous coverage of as many different retroviruses as possible, either a specific target sequence has to be chosen for each individual retrovirus of interest or the selection of one (best choice) or a few common targets shared by the genomes of all retroviruses. The latter strategy has been adopted for the design of RetroArray, since the *pol* gene, in particular the RT region, is the most conserved part of the retroviral replicative machinery and therefore represents an appropriate target sequence for multiplex PCR-based amplification.

Studies on the evolutionary relationship of retroviruses and other retroelements have shown that all RTs share several highly conserved core homology regions [Xiong and Eickbush, 1990]. Two of the most conserved amino acid regions are the VLPQG and YV/MDD motifs (Figure 2). The spacing between both motifs is consistently about 90 base pairs and this region shows considerably less homology (50-85%) within different retrovirus families than the flanking sequences. According to a general principle initially outlined by Shih and coworkers [1989] these motifs were used to derive universal pan-retrovirus PCR primers (MOP, mixed oligonucleotide primers) which allow for amplification and detection of most relevant retroviral RT-related DNA templates.

After comparison of RT core homology regions of human and other vertebrate endogenous and exogenous retroviral sequences available in Genbank database (for accession numbers see Table 1), two sets of mixed *pol*(RT) oligonucleotide primers were designed. One primer set, MOP-ABD, was optimized for amplification of β-retroviruses including human HERV-K(HML) taxa, whereas the second primer set (MOP-C), based on degenerate primers described by Shih et al. [1989], was selected for favored priming γ–retrovirus-related templates. Additionally, specific primer pairs for amplification of HERV-H sequences as well as exogenous human and vertebrate retroviruses such as HIV, HTLV, and human foamy retrovirus (HFV) were added in equimolar amounts to the MOP-C cocktail (for primer compositions also see Table 1). In general, a six-base extension featuring a clamp (nucleotide sequence GGAGAA) was incorporated at the 5' end of each synthetic oligonucleotide primer. Since this sequence extension has a stabilizing effect on primer-template binding kinetics, the products generated after the first PCR cycle are amplified more efficiently in the remaining cycles. Therefore, the amplification reaction can be considered as pan-retroviral multiplex PCR under moderate primer-template annealing conditions. Since the annealing step of this PCR is carried out under relaxed conditions (50 °C for 1 min), retroviral DNA templates in the reaction mixture can be amplified sufficiently with MOP-ABD and MOP-C primers even when an exactly matching primer is not available.

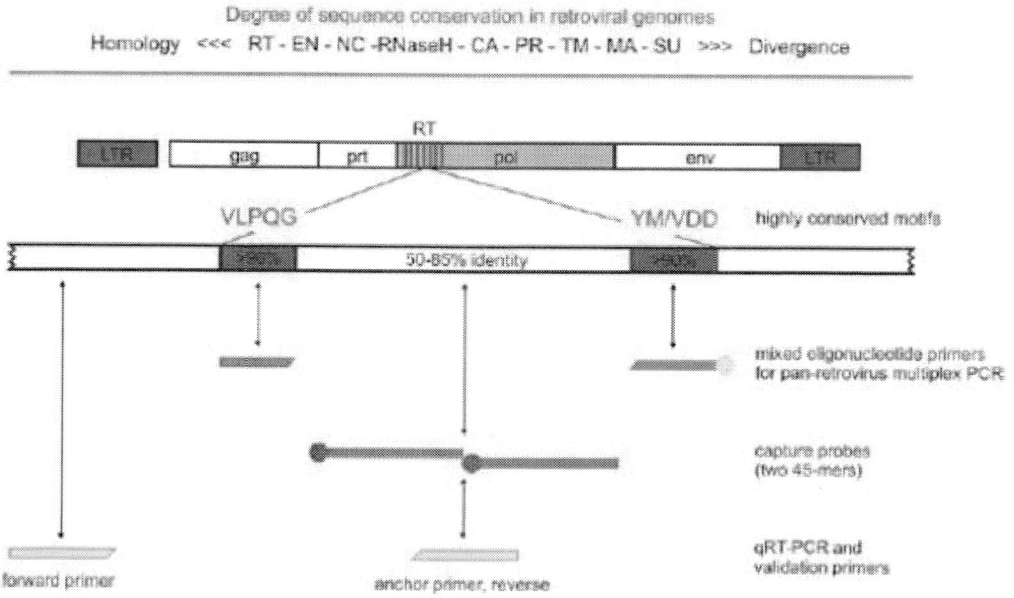

Figure 2. Localization of oligonucleotides used for target amplification (MOP, shown in green), and for microarray spotting (capture probes, in blue). Of all retroviral gene products, the RT encoded by the *pol* gene is the most conserved region among all known retroviruses. As established previously [Xiong and Eickbush 1990], seven core regions (shown as red blocks) of high amino acid identity exist in all known RTs. Two domains (blocks 4 and 5) featuring the amino acid motifs VLPQG and YM/VDD, were used for design of MOP sets that serve for amplification of all possible RT sequences. For RetroArray data corroboration by qRT-PCR, a primer pair generating overlapping amplification products is employed (yellow bars). One primer of the pair serves as anchor primer and is preferentially located between both MOP binding sites separated by 90 base pairs the other primer matches up- or downstream within a distance of about 150 bases. Spiky ends of oligonucleotides (3' end) point towards direction of PCR elongation. Yellow circles at the 5' end of reverse MOP denote fluorochrome dye (Cy3) molecules. Red circles at the 5' ends of the 45-mer capture probes represent amino linkers including C6 spacers. Two optimized primer sets, one for favored amplification of A-, B-, D-type, the other for type C retroviruses were established (for details see text).

Thus, a broad coverage of most RT sequences is achieved, albeit two separate PCR reactions proved advantageous due to the highly competitive reaction kinetics. The resulting amplification products of both reactions are then mixed and used as probe for chip hybridization. All antisense primers are 5'-modified with Cy3 fluorochrome and therefore result in terminally labelled amplification products [Seifarth *et al.*, 2000; Seifarth *et al.*, 2003; Seifarth *et al.,* 2002].

For the design of animal retrovirus-specific primers the same strategy was applied as for HERV specific primers using Genbank and the mouse genome database (http://www.ensembl.org/mus_musculus/). DNA sequences of preferably complete retroviral genomes were selected to design primers for exogenous and endogenous animal retroviruses of the avian genera *Columba* and *Gallus*; the mammalian genera *Equus*, *Canis*, *Bovis*, *Rattus*, *Sus*, *Ovis*, *Mustela*, *Felis*, and *Mus*; and the primate genera *Macaca*, *Hylobates*, and *Papio* (Table 2). In order to optimize target coverage for detection of avian, rodent, or nonrodent mammalian genera pilot experiments with distinct subsets of oligonucleotides were performed [Frank *et al.*, 2005].

As internal standard for assessing RNA quality and for data normalization of densitometric analyses, a primer set for parallel amplification of five housekeeping genes (HKG) was designed including ubiquitin, glycerinaldehyde-3-phosphate dehydrogenase (GAPDH), alpha-tubulin, beta-actin, hypoxanthine-phosphoribosyl-transferase (HPRT) (see Table 1 for Genbank accession numbers). These HKG primers are added to each MOP-PCR.

Table 2. Selective *pol* primers to be combined for detection of non-human retrovirus/retroelements

No.	Target ID	Source	Orientation	5' Modification	Nucleotide sequence	Degenerations
1	FIV	M59418	forward reverse	- Cy3	*GGAGAATAG*AGCTTACCACAGGG *GGAGAA*ATATATATATCATCCACATA	- -
2	FIV2	U56928	forward reverse	- Cy3	*GGAGAATAG*GTTTACCACAAGGA identical with reverse primer 1	- -
3	FeLV	L06140	forward reverse	- Cy3	*GGAGAATAG*CGCCTTCCHCAAGGG *GGAGAA*CAGCAGTARGTCATCTACATA	3 2
4	FeSFV	U78765	forward reverse	- Cy3	*GGAGAATAG*GTTTTACCACAAGG identical with reverse primer 1	- -
5	feline ERV-L	AJ233664	forward reverse	- Cy3	*GGAGAATAG*GTCCTACCTCAGGG not done	-
6	canine ERV-L	AJ233669	forward reverse	- Cy3	*GGAGAATAG*GTTCTTCCTCAGAAG *GGAGAA*AAGAGGATGTCATCCATGTA	- -
7	EiAV	M87581	forward reverse	- Cy3	identical with forward primer 2 GGAGAACACGAACAAATCATCCATATA	- -
8	BLV	K02120	forward reverse	- Cy3	identical with forward primer 5 *GGAGAA*GAAAAGGATATCGTCCATATA	- -
9	BoEV	X99924	forward reverse	- Cy3	*GGAGAATAG*CGCTTCCCACAAGGA *GGAGAA*ACCAAGAGGTCGTCCACATA	- -
11	BIV	M32690	forward reverse	- Cy3	*GGAGAATAG*GTTCTACCACAAGGA *GGAGAA*AATCAACAAATCATCCATATA	- -
12	BFV	AY134750	forward reverse	- Cy3	*GGAGAATAG*GTTCTCCCBCAGGG *GGAGAA*CACATAGACGTCATCCACATA	3 -
13	bovine ERV-L	AJ233662	forward reverse reverse	- Cy3 Cy3	*identical with reverse primer 5* *GGAGAACATAATGTCATCCAGGTA* *GGAGAAGTCATCCAGGTAATCCAG*	- - -
14	JSRV	A27950	forward reverse	- Cy3	*identical with forward primer 12* *identical with reverse primer 1*	3 -
15	OvEV-2	X99932	forward reverse	- Cy3	*GGAGAATAGCAACTTCCACAAGGA* *GGAGAACAGTAGTATGTCATCCACGTA*	- -
16	PERV	AF038600	forward reverse	- Cy3	GGAGAATAGCGACTGCCCCAAGGG GGAGAACAGAAGCAGGTCATCCACGTA	- -
17	MuRVY	X87639	forward reverse	- Cy3	*GGAGAATAG*AGGTCGCGACAGAGG *GGAGAA*AAGGATTAAATCATCTCCATA	- -
18	MuRRS	X02487	forward reverse	- Cy3	*GGAGAATAG*TGGTAACCCCAGGAG *GGAGAA*AAGGAGCAAGTCATCAACATA	- -
19	VL30	AF053745	forward reverse	- Cy3	*GGAGAATAG*AGACTACCACAGGGG *GGAGAA*GATTAAAAGATCATCTACGTA	- -

No.	Target ID	Source	Orientation	5' Modification	Nucleotide sequence	Degenerations
20	VL30-like	AL844168	forward reverse	- Cy3	*GGAGAATAG*AGACTCCCACAGGG identical with reverse primer 9	- -
21	IAP-1	X87638	forward reverse	- Cy3	identical with forward primer 5 *GGAGAA*AATCAAAATGTCATCCATATA	- -
22	IAP-2	U58494	forward reverse	- Cy3	*GGAGAATAG*GTCTTACCACAGGGA identical with reverse primer 6	- -
23	MMTV	M15122	forward reverse	- Cy3	*GGAGAATAG*GTTTTGCCCCAGGGT identical with reverse primer 6	- -
24	MMTV-like	AL606472	forward reverse	- Cy3	*GGAGAATAG*GTTCTTTCTTAAGAC identical with reverse primer 6	- -
25	MLV	J02255	forward reverse	- Cy3	identical with forward primer 20 *GGAGAA*CAGCAGTAAGTCATCYACGTA	- 2
26	mRE-CX	AL672245	forward reverse	- Cy3	GGAGAATAGGTCTGTAGTCAGATG GGAGAAATAAAAAATATCACCCATACT	- -
27	mRE-C17	AC087217	forward reverse	- Cy3	GGAGAATAGGTTCTGCTGGAGGTT GGAGAAAATCAAAATATCATCCATACT	- -
28	mRE-C4	NT_039260.7	forward reverse	- Cy3	GGAGAATAGTTTCTACCTCAAGGA GGAGAAGAGAAGGACATCATCTACATA	- -
29	mRE-C1	NT_039185.7	forward reverse	- Cy3	GGAGAATAGAGACTTCCACAAGGA GGAGAACAAAAGAAGGTCATCTATATA	- -
30	mRE-C9	NT_039474.7	forward reverse	- Cy3	*GGAGAATAG*GTTCTGCCTCGGGGA *GGAGAA*ACAGAAGACATCATCTACATA	- -
31	mRE-C2	NT_039207.7	forward reverse	- Cy3	*GGAGAATAG*GTTCTACCCCAACTC *GGAGAA*GAGAAGAATATCATCTATATA	- -
32	murine ERV-L	Y12713	forward reverse	- Cy3	*GGAGAATAG*GTTTTGCCTCAAGTA *GGAGAA*CAGCATAATGTCATCAATATA	- -
33	MusD1	AF246632	forward reverse	- Cy3	identical with forward primer 12 *GGAGAA*CATCAAGACATCATCTGTGAA	3 -
34	MusD2	AF246633	forward reverse	- Cy3	GGAGAATAGTATCTCCCACAGGGC identical with reverse primer 33	-
35	RLV	M77194	forward reverse	- Cy3	identical with forward primer 1 *GGAGAA*TAGAAGAAGGTCATCAACAAA	- -
36	rat ERV-L	AJ233604	forward reverse	- Cy3	*GGAGAATAG*GTTTTGCCTCGAGGA not done	-
37	MiEV-1	X99931	forward reverse	- Cy3	identical with forward primer 1 identical with reverse primer 15	- -
38	RVwp	AJ236133	forward reverse	- Cy3	identical with forward primer 3 *GGAGAA*TAGCAACAAATACCATTG	3 -
39	ALV	Z46390	forward reverse	- Cy3	*GGAGAATAG*GTCTTGCCCCAAGGG *GGAGAA*TAGCAAAAGATCATCCATATA	- -
40	EAV-HP	AJ292966	forward reverse	- Cy3	identical with forward primer 4 identical with reverse primer 39	- -
41	MPMV	M12349	forward reverse	- Cy3	identical with forward primer 4 identical with reverse primer 6	- -
42	Mac-ERV - ERVK	Han *et al.*, 2007	forward reverse	- Cy3	§ §	§ §

Table 2. (Continued)

No.	Target ID	Source	Orientation	5' Modification	Nucleotide sequence	Degenerations
43	GaLV	M26927	forward	-	*GGAGAATAG*CGGCTACCACAAGGG	-
			reverse	Cy3	identical with reverse primer 9	-
44	BaEV	D10032	forward	-	identical with forward primer 3	3
			reverse	Cy3	*GGAGAA*CAAGAGGAGGTCATCTACATA	-

§ = For amplification, the human MOP cocktails (see Table 1) are to be used.

Abbreviations: MacERV1-6 and MacERVK1-2, Macaque endogenous retrovirus types. FIV, feline immunodeficiency virus; FIV2, feline immunodeficiency virus type 2; FeLV, feline leukemia virus; FeSFV, feline syncytium-forming virus; ERV-L, endogenous retrovirus type L; EiAV, equine infectious anemia virus; BLV, bovine leukemia virus; BoEV, bovine endogenous retrovirus; BIV, bovine immunodeficiency virus; BFV, bovine foamy virus; JSRV, Jaagsiekte retrovirus; OvEV-2, ovine endogenous retrovirus; MuRVY, murine endogenous retrovirus from chr Y; MuRRS, murine retroviral-related sequence; VL30, murine virus-like element encoding 30S RNA; VL30-like, murine virus-like element for 30S RNA-like; IAP-2, intracisternal particle type 2 ; IAP-1, intracisternal particle type 1; MMTV, mouse mammary tumor virus; MMTV-like murine retroelement; MLV, murine leukemia virus; mRE-CX, unclassified murine retroelement on chromosome X; mRE, unclassified murine retroelement; MusD (1/2), type D-like murine endogenous retrovirus; RLV, rat leukemia virus; MiEV-1, mink endogenous retrovirus; RVwp, wood pigeon retrovirus; ALV, avian leukosis virus; EAV-HP, endogenous avian retrovirus type HP; MPMV, Mason Pfizer monkey virus; GaLV, gibbon ape leukemia virus; BaEV, baboon endogenous retrovirus; PERV, porcine endogenous retrovirus, types A, B, C.

Capture Probe Design and Modular Assembly of RetroArray

As capture probes retrovirus-specific synthetic oligonucleotides are deposited on aldehyde-activated glass slides. For design of these capture probes, a similar strategy as for the design of the pan-retrovirus primer cocktails was applied. Available databases were initially screened for RT-related sequences. RT-sequences of exogenous and endogenous retroviruses were then classified according to the current nomenclature and further subgrouped with respect to their degree of nucleotide homology (Figure 3). Finally, representative members for prominent retroviruses/HERV taxa were selected with special emphasis on full-length proviral genomes and retroviral sequences that have been associated previously in literature with any biological activities and/or human diseases.

The 90 bp stretch between the primer binding motifs VLPQG and YV/MDD is considerably less conserved than the primer motifs (Figure 2) and allows for discrimination between HERVs, which are identical within the primer regions. Therefore, this sequence was used for synthesis of a pair of oligonucleotides each 45 nucleotides in length. All oligonucleotides were NH_2(C6)-modified (aminolinker) at their 5' ends for covalent glass slide coupling via Schiff base reaction. Thus, each spotted dot represents an equimolar mixture of two 45-mer oligonucleotides covering 90 bp of the RT sequence of a distinct retrovirus or a representative subgroup of a retrovirus family.

Currently, the RetroArray chip comprises up to 118 spots covering human and other vertebrate retroviral sequences.

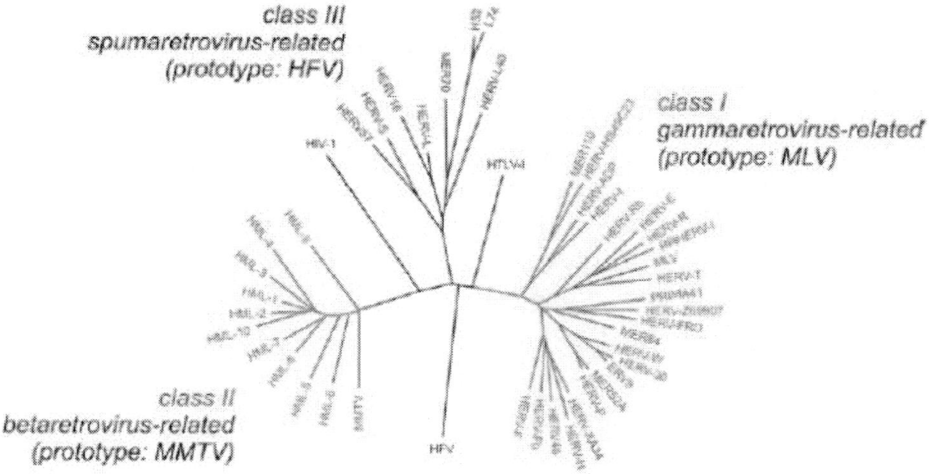

Figure 3. Classification of retroviral taxa covered by RetroArray (adapted from Mager and Medstrand, 2003). The unrouted tree is based on a 550 bp *pol* region encoding RT. For abbreviations see text and tables 1 and 2.

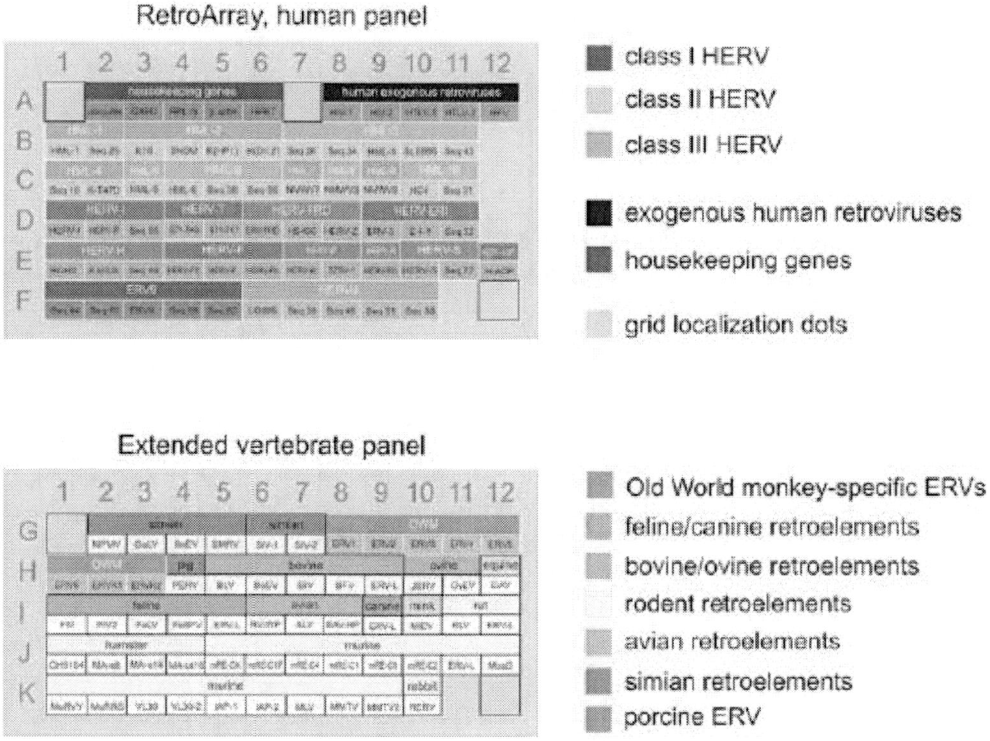

Figure 4. Potential layout of RetroArray capture probes arranged according to their origin (human, upper panel; other vertebrates, lower panel). Since RetroArray was designed as an open modular assembly system, the researcher can choose from the list of currently 118 available retroviral capture probes and five housekeeping genes according to the scientific question of interest.

In addition to exogenous retroviruses and ERVs common for humans and Old World monkeys, RetroArray may cover representatives of the eight macaque-specific endogenous retroviruses identified by the macaque genome sequencing project [Han *et al.*, 2007]. Furthermore, an extended vertebrate panel with 49 capture probes of rodent and of bovine, ovine, porcine, avian, simian, feline and canine origin is available (Figure 4). For source and identity of human and non-human capture probes see Tables 3 and 4, respectively. As internal controls, five housekeeping genes (for details see Figure 4 and Table 3) and several grid localization dots are spotted on the chip. The latter are short random oligonucleotides that are modified at 5' and 3' ends with aminolinkers and Cy3, respectively. They serve for the optical localization of the spotted array during the scanning procedure. For some experimental settings it may be useful to include further non-retroviral capture probes in addition to housekeeping genes, such as sequences of common cell culture contaminants like mycoplasma or reporter genes of retroviral vectors [Zeilfelder et al., 2007].

Table 3. Origin and classification of human retrovirus-specific capture probes

Retrovirus class	Family / group	RepBase* Name	Sequence source (Genbank accession number)
Class I	HERV-I	HERVI	HERV-I (M92067)
retroviruses		HERVIP10F	HERV-IP-T47D (U27241)
(gammaretrovirus-like)			Seq65 (AP000842)
	HERV-T	HERVS71	S71pCRTK6 (U12969)
			S71pCRTK1 (U12970)
	HERV-FRD	MER50I	HERV-FRD (U27240)
			HS49C23 (Z93019)
			HERV-Z (Z69907)
	HERV-E	HERVE	E4-1 (M10976)
			Seq32 (AC010636)
	HERV-H	HERVH	RGH2 (D11078)
			HERV-H (AF026252)
			Seq66 (AL359740)
	HERV-F	HERVH48I	HERV-F2 (AC002416)
		HERVFH19I	HERV-F (Z94277)
		HERVFH21	HERV-Fb (AC000378)
	HERV-W	HERV17	HERV-W (AF009668)
			SZRV-1 (AF135487)
	HERV-R	HERV-R	ERV3 (AC004609)
			HERV-Rb (AC004045)
	HERV-S	HERV18	HERV-S (Z84470)
			Seq77 (NT_022171)
	HERV-ADP	HERVP71A_I	HERV-ADP (AC005741)
	ERV9	HERV9	Seq64 (AC005253)
		HERV17	Seq63 (AC018926)
		HERV9	ERV9 (X57147)
		HERV9	Seq59 (AC006397)

Retrovirus class	Family / group	RepBase* Name	Sequence source (Genbank accession number)
		HERVFH19I	Seq60 (AL135749)
Class II	HML-1	HERVK14I	HML-1 (U35102)
retroviruses			Seq29 (S77579)
(betaretrovirus-like)	HML-2	HERVK	HERV-K10 (M14123)
			HERV-K2.HOM (U87592)
			HERV-KHP1 (U87588)
			HERV-KD1.2 (U87595)
	HML-3	HERVK9I	Seq26 (AC073115)
			Seq34 (AL592449)
			HML-3 (U35236)
			HERV1 (S66676)
			Seq43 (AF047595)
	HML-4	HERVK13I	Seq10 (AF047591)
			HERV-K-T47D (AF020092)
	HML-5	HERVK22I	HML-5 (U35161)
	HML-6	HERVK3I	HML-6 (U60269)
			Seq38 (AC010328)
			Seq56 (AC018558)
	HML-7	HERVK11DI	NMWV7 (AP003171)
	HML-8	HERVK11I	NMWV3 (AL513321)
	HML-9	-	NMWV9 (AC025569)
	HML-10	HERVKC4	HERV-KC4 (U07856)
			Seq31 (AL162734)
Class III retroviruses	HERV-L	HERVL	HERV-L (G895836)
(spumavirus-like)			Seq39 (AC091914)
			Seq45 (AC006971)
			Seq51 (AL353741)
			Seq58 (AL590730)
Human exogenous			HIV-1 (K02013)
retroviruses			HIV-2 (J04542)
			HTLV-1 (M81248)
			HTLV-2 (M10060)
			HFV (Y07725)
Human housekeeping			ubiquitin (X04803)
genes			GAPDH (NM_002046.2)
			RPL19 (NM_000981)
			beta-actin (E01094)
			HPRT (NM_000194)

* = according to RepBase: Genetic Information Research Institute, Sunnyvale, CA, USA (www.girinst.org), [Jurka et al., 2005].

Since RetroArray was designed as an open modular assembly system, the researcher can choose from the list of all retroviral and housekeeping gene capture probes. Capture probes for further retroviruses can be easily integrated into the microarray (add-on system) as soon as a novel retrovirus of potential interest is described and the *pol* sequence is available in databases. Of note, RetroArray by itself has the potential to detect novel retroviruses. This is possible by the use of pan-retrovirus primers and by controlled cross-hybridization after lowering the stringency of hybridization conditions. DNA hybridizing to the covalently bound oligonucleotides can be eluted from the glass slides by alkaline denaturation and reamplified after neutralization to provide sufficient double-stranded DNA for cloning and subsequent sequence analysis [Frank *et al.*, 2005].

Oligonucleotides for qRT-PCR and Data Validation

For corroboration of microarray data it has become generally accepted to employ quantitative RT-PCR (qRT-PCR), for example LightCycler technology, in combination with target-specific amplification primers for testing the same nucleic acid preparation [Frank *et al.*, 2005; Frank *et al.*, 2008]. However, for RetroArray data validation, specific qRT-PCR amplification using the same pan-retrovirus amplification primers (MOP) is hindered because the primer cocktails MOP-ABD and MOP-C have been designed to amplifiy many targets simultaneously. Primers derived from the highly conserved VLPQG and YV/MDD region will amplify more than one single target. To overcome this impairment, sets of HERV-specific primers for qRT-PCR were designed in such a way that for each HERV subgroup one primer anneals within the capture probe sequences spotted on the chip, whereas a second primer is located 100 to 150 bp upstream or downstream of the first primer (Figure 2). To obtain complementary information e.g. on the expression of spliced *env* transcripts, which lack the complete *pol* gene, or for a more subtle differentiation between HERV taxa, specific primers from more divergent regions of the retroviral genome, such as *gag* or *env* coding sequences can be used [de Parseval *et al.*, 2003; Muradrasoli *et al.*, 2006; Frank *et al.*, 2005; Frank *et al.*, 2008]. For normalization of qRT-PCR, the use of standard plasmids of cloned housekeeping genes and *pol* fragments of the respective retroviral target can be advantageous in standard SYBR Green-based qRT-PCR experiments [Frank *et al.*, 2008].

For further characterization of retroviral amplicons, eluting captured MOP products from hybridized RetroArrays and subsequent reamplification and cloning for DNA sequencing according to standard methods are possible (see section "Limitations, potential pitfalls and caveats"). For example, the obtained sequence data can be used to pinpoint the genomic locus of integrated proviruses/HERVs using BLAT at the Human Genome Browser database (http://genome.ucsc.edu/) as demonstrated recently [Flockerzi *et al.*, 2007; Frank *et al.*, 2008].

Experimental Procedure and Technical Aspects

Preparation of Microarrays

The standardized experimental flow [Seifarth *et al.*, 2003] is shown in Figure 1. For preparation of DNA chips, probably any standard robotic device will provide satisfactory results. We employ an Affymetrix GMS 417 Arrayer (Affymetrix, Santa Clara, CA, USA).

Corresponding pairs of 45-mer synthetic oligonucleotides are diluted to 100 mM in 3x SSC (1x SSC = 0.15 M NaCl plus 0.015 M sodium citrate) and mixed in equimolar amounts. Capture probes are spotted (three hits per dot) onto CSS-100 aldehyde-activated glass slides (TeleChem International Inc., Sunnyvale, CA, USA). Generally, to rule out local defects in the slide surface (e.g. scratches) and for hybridization reproducibility, two replicas are spotted onto the same slide. After spotting, slides are incubated in a humid chamber over night, then air dried, washed and treated with sodium borohydride ($NaBH_4$) as recommended by the glass slide manufacturer. Slides are stored in the dark at room temperature until hybridization.

Sample Preparation

Any kind of biological samples containing nucleic acids are suited for RetroArray testing. Since PCR is the first step of the assay (Figure 1), DNA or cDNA (reverse transcribed RNA) is required as starting template. While DNA can be prepared according to regular standard protocols, RNA preparations require special attention in regard to qualitiy and purity. Since integrated retroviral sequences do not contain introns, and ERVs may occur in hundreds of copies in the host genome, expression profiling of RNA will be compromised by any traces of genomic DNA contaminating RNA preparations. Therefore, all RNA samples have to be treated with RNase-free DNase I until they proof DNA-negative in control PCR experiments using MOPs and omitting the cDNA synthesis step (Figure 1). This is contrary to most other commercial retroarrays (e.g. Affymetrix Gene Chips) that monitor the transcriptional activity of regular cellular genes using intron-spanning capture probes. For detection and identification of exogenous retroviruses, such as xenotropic or cell culture contaminating retroviruses, DNA preparations or simple cell extracts represent an appropriate starting material, since retroviruses normally integrate in their host's genome.

For RNA preparation, commercial kits such as RNeasy (Qiagen, Hilden, Germany) or Trizol (Biozol, Eching, Germany) might be employed. Alternatively, total RNA can be extracted from ground tissue by guanidinium isothiocyanate-cesium chloride ultracentrifugation [Sambrook and Russell., 2001]. To remove genomic DNA contaminations, it is mandatory to treat all RNA samples with RNase-free DNase I. Subsequently, 25 ng of each RNA preparation should be tested by conventional PCR using the MOP cocktail and omitting the reverse transcription step. Only RNA preparations negative for amplification products (expected size 150 bp) are appropriate for subsequent reverse transcription and pan-retrovirus multiplex PCR [Seifarth *et al.*, 2003].

For cDNA synthesis, any commercial cDNA synthesis first strand kit can be employed. However, it is mandatory to use random priming instead of oligo(dT) priming because the latter will overrepresent the 3' ends of RNA molecules. Since the *pol* sequence targeted by MOP PCR is located in the middle of transcripts (Figure 2) spanning up to 9 kb in length random priming is more efficient in copying this region into cDNA. It should be noted that commercial RT preparations can contain DNA traces of RT expression plasmids and can lead to false positive results in RetroArray experiments as will be addressed in more detail in the section "Limitations, potential pitfalls and caveats" of this chapter.

RetroArray Hybridization

RetroArray hybridization is carried out according to a standardized protocol as initially described by Seifarth and coworkers [2003]. It includes the following steps:

1. Prehybridization
2. Hybridization
3. Washing
4. Drying of hybridized glass slides

In general, all handling steps involving fluorochromes are performed under dimmed light conditions, incubations are carried out in the dark. Prehybridization of glass microarrays is performed within Falcon blue cap tubes (Greiner GmbH, Nürtingen, Germany) in 6x SSC, 0.5% sodium dodecyl sulfate (SDS), 1% bovine serum albumin (DNase-free) at 50 °C for at least 4 h. Slides are then washed five times (1 min each) in bidest. water under vigorous agitation. Subsequently, slides are dried by a short centrifugation step (500x g for 1 min at 22 °C). For hybridization, a 22x22 mm glass lifter slip (Erie Scientific Co., Portsmouth, NH, USA) is taped to the microarray. Subsequently, 25 µl of the denatured (5 min at 95 °C) Cy3-labeled probe (containing 3x SSC, 0.1x SDS, 50% formamide) is applied to the small space between cover slip and microarray and sealed with Fixogum rubber cement (Marabu GmbH, Tamm, Germany). Sealed slides are incubated in a humid chamber for at least 16 h at 42 °C. Hybridized slides are then washed three times under vigorous agitation (10 min each) at 22 °C in 1x SSC, 0.5x SSC and 0.1x SSC, respectively. After a brief centrifugation step (500x g for 1 min at 22 °C), the hybridized microarrays were stored in the dark at 22 °C until evaluation by scanning.

To adjust the hybridization conditions for discrimination of related HERV taxa pilot experiments using 25 pmol per reaction of synthetic Cy3-labeled oligonucleotides complementary to the corresponding capture probe were performed [Frank *et al.*, 2008]. Varying hybridization temperature and/or amount of formamide in the hybridization solution to achieve higher or lower stringency conditions could be exploited to discriminate between closely related HERV taxa or to detect yet unknown retroviruses by cross-hybridization. For sensitivity, as few as 25 pig endogenous retrovirus (PERV) DNA copies were detectable in spiking experiments when defined amounts of cloned proviral PERV DNA were diluted with human mRNA-derived cDNA from healthy blood donors [Seifarth *et al.*, 2003].

Table 4. Origin and classification of non-human retrovirus-specific capture probes

Animal Genus	Sequence source
	(Genbank accession number)
Felis	FIV (M59418)
	FIV2 (U56928)
	FeLV (L06140)
	FeSFV (U78765)
	ERV-L (AJ233664)
Canis	ERV-L (AJ233669)

Animal Genus	Sequence source
Equus	EiAV (M87581)
Bovis	BLV (K02120)
	BoEV (X99924)
	BIV (M32690)
	BFV (AY134750)
	ERV-L (AJ233662)
Ovis	JSRV (A27950)
	OvEV-2 (X99932)
Sus	PERV (AF038600)
	MuRRS (X02487)
	VL30 (AF053745)
	VL30-like (AL844168)
	IAP-1 (X87638)
	IAP-2 (U58494)
	MMTV (M15122)
	MMTV-like (AL606472)
	MLV (J02255)
	mRE-CX (AL672245)
	mRE-C17 (AC087217)
	mRE-C4 (NT_039260.7)
	mRE-C1 (NT_039185.7)
	mRE-C9 (NT_039474.7)
	mRE-C2 (NT_039207.7)
	ERV-L (Y12713)
	MusD (1/2) (AF246632/3)
Rattus	RLV (M77194)
	ERV-L (AJ233604)
Mustela	MiEV-1 (X99931)
Columba	RVwp (AJ236133)
Gallus	ALV (Z46390)
	EAV-HP (AJ292966)
Macaca	MPMV (M12349)
	MacERV1 (Han *et al.*, 2007)
	MacERV2 (Han *et al.*, 2007)
	MacERV3 (Han *et al.*, 2007)
	MacERV4 (Han *et al.*, 2007)
	MacERV5 (Han *et al.*, 2007)
	MacERV6 (Han *et al.*, 2007)
	MacERVK1 (Han *et al.*, 2007)
	MacERVK2 (Han *et al.*, 2007)
Hylobates	GaLV (M26927)
Papio	BaEV (D10032)

Signal Evaluation, Densitometry and Data Analysis

For acquisition of RetroArray raw data, the hybridized glass slides are scanned using an Affymetrix Scanner GMS 418, and the resulting images (16-bit TIFF) are visually evaluated (yes/no decision) or are subjected to densitometric analysis (HERV profiling) using ImaGene 4.0 software (Biodiscovery Inc., Los Angeles, CA, USA). For image presentation on computer monitors or in printed matter, false-color output gives best results as a virtual color scale will be used to efficiently represent the possible 65.536 grey scale signal increments.

Due to the systematic limitation of the assay, densitometric data should be judged with caution. It is important to emphasize that RetroArray is primarily a qualitative screening and detection technique. The use of multiplex PCR with closely related primer sequences combined with relaxed primer–template binding conditions may lead to preferential amplification of distinct "high copy" or a few "best fit" templates, whereas others stay underrepresented. In contrary, with increasing cycle numbers availability of certain primers may be reduced leading to an early decrease of reaction products [Seifarth *et al.*, 2000]. Thus, no more than 30 rounds of PCR should be performed. The pan-retrovirus multiplex PCR employed for unpaired single samples does not allow an internal standardization except for RNA quality, overall hybridization efficacy and signal gain. Therefore, further experiments such as qRT-PCR or Northern blotting are highly recommended for exact quantification.

A semi-quantitative evaluation based on densitometric analysis may be undertaken when paired samples are available. These could for example be RNA preparations from cells before and after treatment with any agents, or paired samples of nonmalignant and tumor tissue derived from the same organ of the same patient (Figure 5). Such samples should be analyzed preferentially in one single experiment using the same chip lot and one common PCR mastermix to minimize the main interfering factors [Frank *et al.*, 2008]. For densitometric analysis, relative signal intensities (signal mean) are corrected by subtracting the corresponding signal background values (background mean). To discriminate positive signals from background, an arbitrary cutoff value corresponding to twofold background intensity values of the respective chip is used. This proved to be in best agreement with the optical appearance of raw images when observed on a color-calibrated monitor in a darkened room [Frank *et al.*, 2008]. To account for the influence of RNA quality, signals are normalized to RNA levels of an appropriate housekeeping gene that can be selected from the HKG capture probe panel. So far, the HPRT gene showed the most consistent transcript levels in brain and solid tumor samples. Thus, microarray-based relative HERV transcript levels can be expressed as a ratio of HERV to HPRT [Frank *et al.*, 2005; Frank *et al.*, 2008].

Limitations, Potential Pitfalls and Caveats

RetroArray was designed to simultaneously detect and identify a wide variety of human and other vertebrate exogenous and endogenous retroviruses in biological samples and clinical specimen. To achieve this, MOPs derived from a *pol*(RT) region highly conserved in all known retroviruses have been employed. The 90 bp internal sequences of the amplicon, which were used for design of the capture probes spotted on the chip, are less conserved and allow for discrimination of single HERV families or subgroups (Figure 2). Cross-hybridization of HERV elements with less than 20% divergence within the hybridizing region cannot be excluded under standard hybridization conditions. HERV subgroups were therefore defined by at least 20% nucleotide difference and a representative member of each subgroup was selected to design the capture probes. While this requirement is met by all exogenous retroviruses and also by most HERV taxa, it cannot be used to discriminate between HML-2 subgroups, which are more closely related to each other than it applies to older HERV families. However, members of the HML-2 family constitute the most active HERVs with high biological significance and pathogenic potential. Therefore, we decided to include at least four different HML-2 elements, as they may be discriminated under appropriate hybridization conditions. Hybridization conditions may be optimized in particular cases by pilot experiments using specific oligonucleotides as hybridization probes [Frank et. al., 2008].

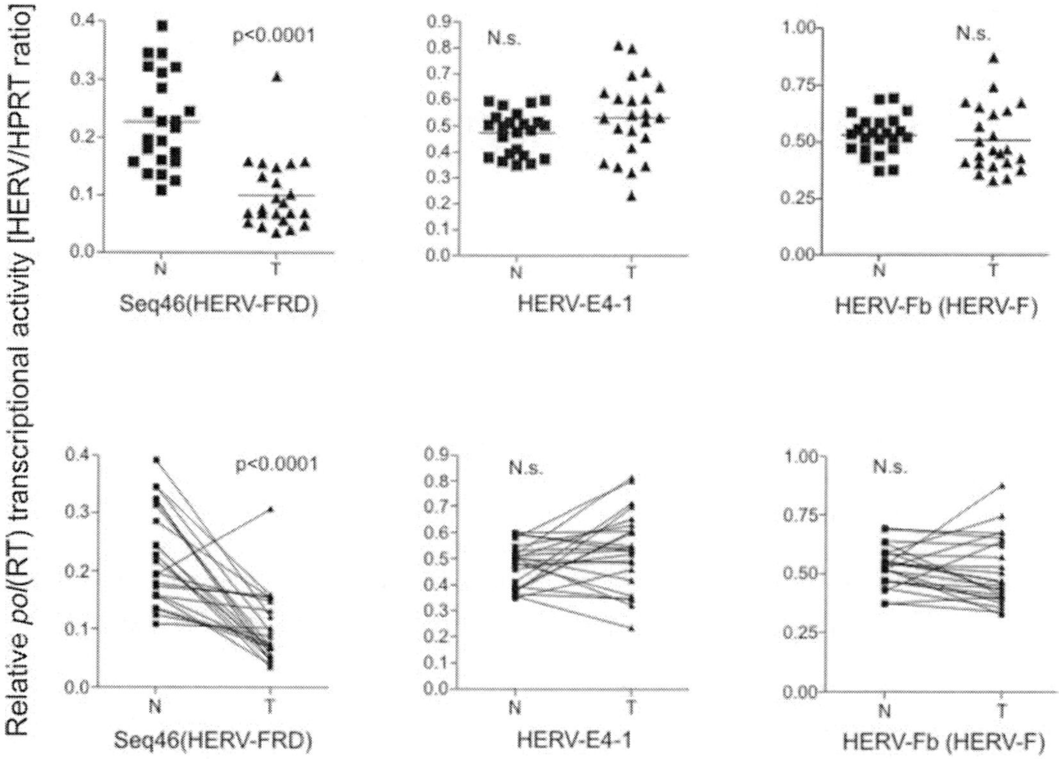

Figure 5. Quantification of RetroArray data by densitometry in experimental settings using matched sample pairs, e.g. normal tissue (N) and tumor tissue (T) derived from the same patient. After normalization of densitometric data to HPRT, scatter plotting may be the best choice to visualize increases or decreases of HERV transcript levels between tissue sample pairs. For example, while HERV-E4-1 and HERV-Fb show quite similar transcriptional levels in all paired mammary carcinoma samples (n=23), HERV-FRD signals show a significant ($p<0.0001$) decrease in tumor tissues when compared with corresponding normal cells. For complete data sets see Frank and coworkers [Frank *et al.*, 2008]. Abbreviations: N.s., not significant.

The copy numbers of HERV elements belonging to a given HERV family/taxon may vary between a few and several hundreds in the human genome. Therefore, RetroArray cannot indicate whether an observed signal is produced by one or a few closely related HERV elements that are highly expressed, or by many HERV elements with low transcriptional activities. To obtain such information, hybridizing cDNAs may be eluted from the chip to analyze the composition of transcripts.

For this purpose, a microarray may be spotted only with capture probe oligonucleotides binding to the HERV sequences of interest and the hybridized sequences may be eluted, reamplified with MOP or specific primers and sequenced. This method can also be used to identify novel, yet unknown retroviruses and to differentiate between cross-hybridizing contaminations and retroviral infections.

To identify single active loci of HERV families which have been found to be conspicuous in RetroArray experiments, it is recommended to perform in addition PCR experiments with HERV family-specific primers located in retroviral genes such as *gag* or *env* that are less conserved than the *pol* gene, and to amplify sequences that are long enough for a definite discrimination of closely related proviruses [Flockerzi *et al.*, 2008; Frank *et al.*, 2008]. Therefore, whenever interesting RetroArray signals are observed, we recommend either the use of specific qRT-PCR to amplify sequences from different retroviral genes or cloning and DNA sequencing of RetroArray amplicons. For specific goals, e.g. detection of species-specific retroviruses, MOP cocktails may be simplified to achieve less competition of primers and better primer-target binding kinetics. This can improve performance of the assay and shortened annealing times may speed up the RetroArray workflow.

Due to the high sensitivity of the RetroArray, not only template preparations (RNA) have to be free of undesired nucleic acid contaminants (environmental DNA), but also all the other ingredients that are used within the workflow (Figure 1). Since the genomes of all vertebrates including primates and humans possess high copy numbers of retroviral sequences, any environmental DNA traces contaminating the experiment may account for false positive signals. Therefore, it is recommended to perform all wet lab hand-on steps of the pan-retrovirus PCR under a clean bench in an air filtered environment.

Considerable problems may be caused by commercial kits or ingredients thereof that have been produced in recombinant pro- or eukaryotic expression systems. While investigating retroviral activities in samples from brains of patients with schizophrenia and bipolar disorders, we repeatedly observed signals related to MLV even in the water controls [Frank *et al.*, 2005]. Cloning by ligation-mediated PCR and sequencing of the amplicons revealed that MLV RT expression vector molecules occurred in the assay. These were obviously introduced by a commercial MLV RT preparation used for cDNA synthesis. Thus, commercial enzyme preparations may contain traces of bacterial expression plasmids causing false-positive signals with RetroArray. However, this problem may be circumvented by use of RT from other sources, e.g. from avian myeloblastosis retroviruses.

APPLICATIONS

HERV Expression Profiling and Identification of Exogenous Retroviruses in Human Diseases

RetroArray is suitable for generating comprehensive HERV transcription profiles for any specimen of interest. Using RetroArray for qualitative RNA profiling of normal human tissues (Figure 6) we are able to establish tissue-specific HERV expression signatures [Seifarth *et al.*, 2003; Seifarth *et al.*, 2005; Frank *et al.*, 2005; Frank *et al.*, 2008].

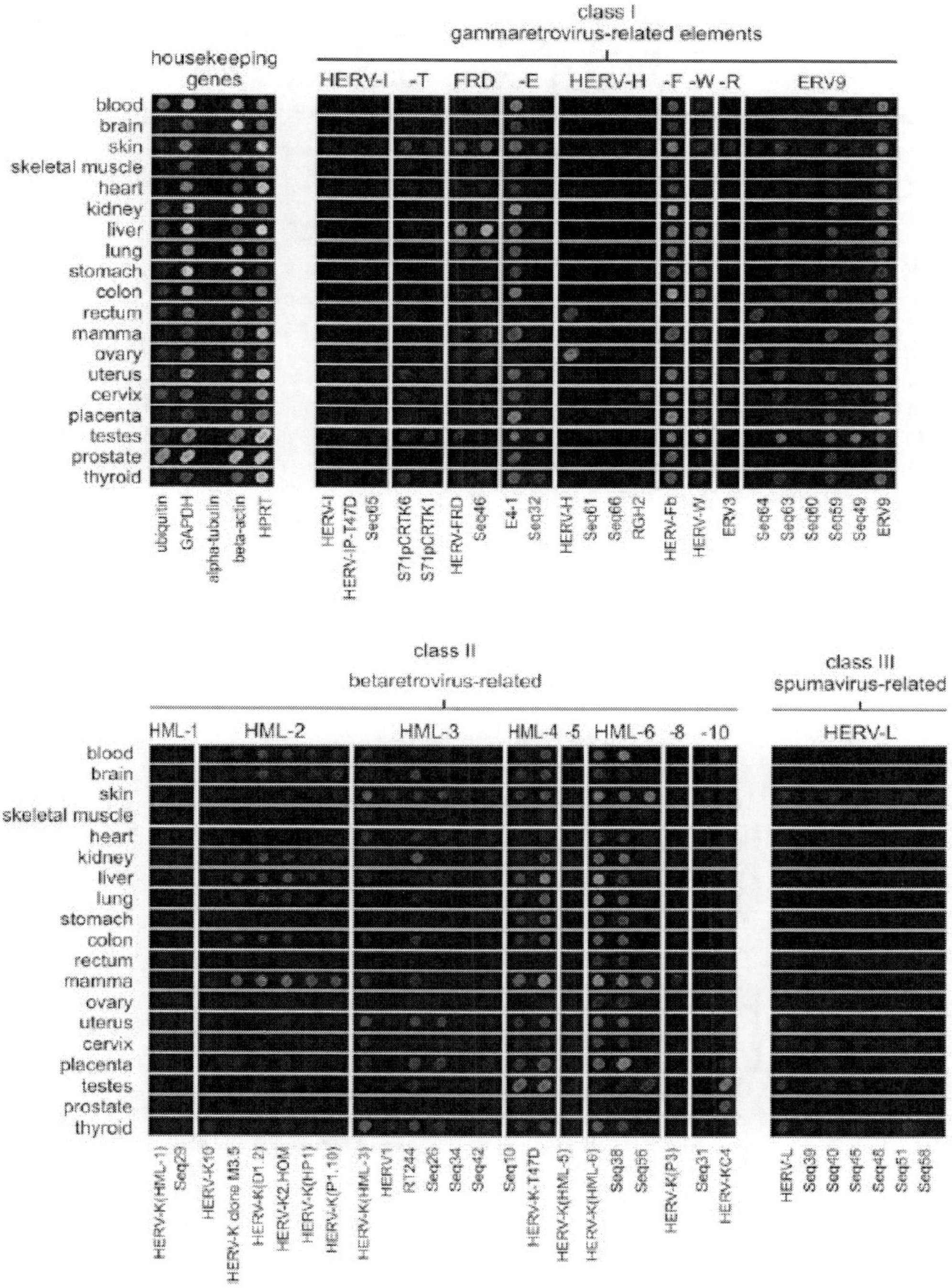

Figure 6. Transcriptional activity of HERV class I, II, and III elements observed in 19 normal human tissues. False-color chip data sets obtained by RetroArray hybridization were aligned as described [Seifarth *et al.* 2005]. A housekeeping gene panel served as an internal control for mRNA integrity. Underneath the image alignment, HERV capture probes used in this chip hybridization experiment are indicated. For IDs see Table 3.

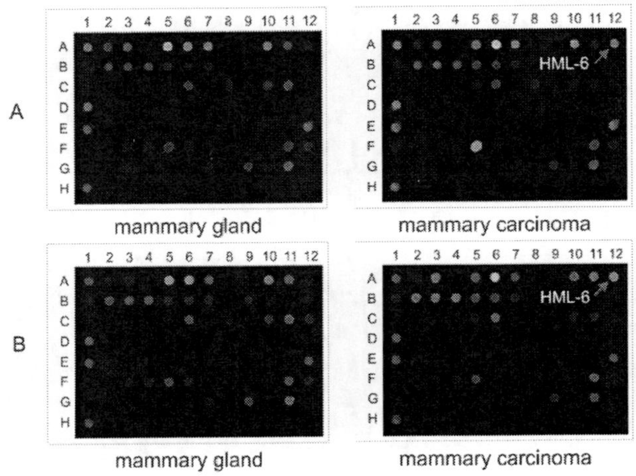

Figure 7. Comparison of HERV transcriptional pattern in paired samples (unaffected tissue vs. tumor tissue), each pair derived from the same patient (A, B). Conspicuous pattern alterations are emphasized by red arrows (= HERV-K(HML6)). For detailled analysis see Frank and coworkers [Frank *et al.*, 2008].

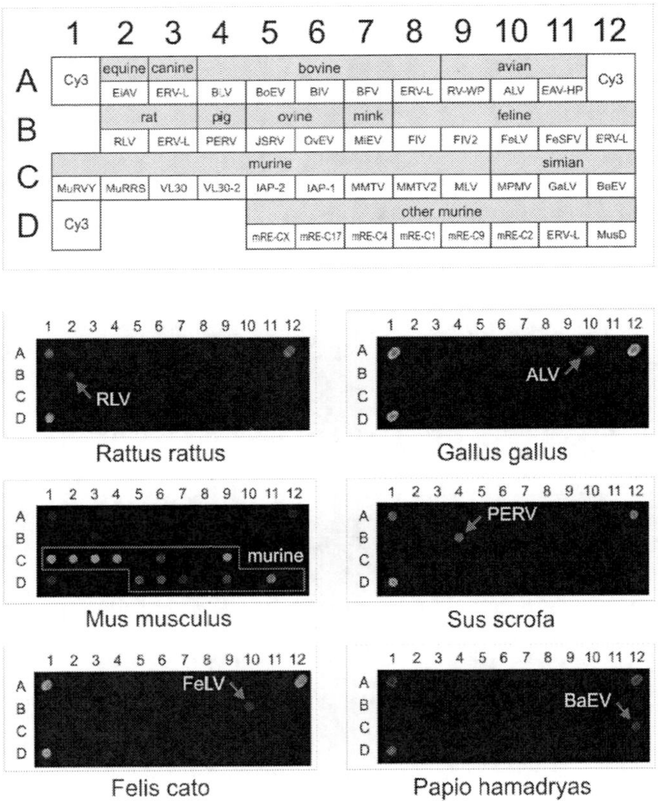

Figure 8. Species specificity of the animal retrovirus-specific microarray (pet chip). Genomic DNAs (100 ng) derived from avian (*Gallus gallus domesticus*) and mammalian (*Rattus rattus*, *Mus musculus*, *Felis cato*, *Sus scrofa domestica*, and *Papio hamadryas*) genera were amplified with non-human MOP using a standardized amplification protocol. No significant inter-species cross-reactions are detectable. A schematic of the array is shown in the top panel. Grid locators represented by a spotted Cy3-labeled arbitrary oligonucleotide are designated Cy3. For abbreviations see text and tables 1 and 2.

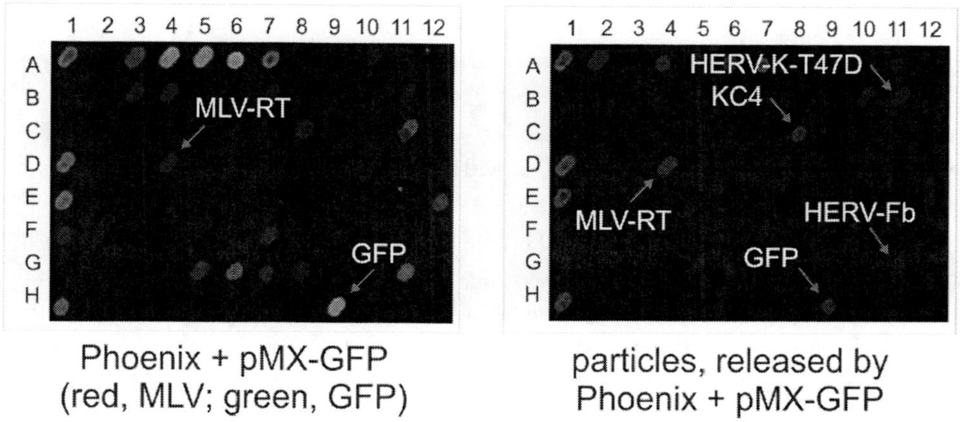

Figure 9. Co-packaging of HERV transcripts and/or RCR in retroviral vector preparations. Human Phoenix-GP cells harbouring MoMLV*gag-pol* were profiled using RetroArray (left panel) after transfection with a pMX-GFP construct. Comparatively, the corresponding particle preparations were analyzed (right panel) for packaging of retroviral transcripts. Signals with increased intensity in particle preparations point to accumulation of distinct retroviral transcripts in vector particles (red arrows). Abbreviation: GFP, green fluorescent protein.

Constitutively transcribed HERVs can be identified for each human tissue comparing HERV expression profiles obtained from the same organ such as brain [Frank *et al.*, 2005] or mammary gland [Frank *et al.*, 2008] of different individuals. In addition, this setting enables the detection of individual variations in HERV expression (variably transcribed HERVs) that may reflect the particular genetic background of a person. Having determined the basic or "core" HERV transcriptional signature of a given human tissue, more extensive biomedical investigations can be performed e.g. by monitoring disease-related differences in HERV activity (Figure 7).

Depending on the study design and/or the availability of patient samples two different approaches are conceivable:

1. Qualitative analysis can be performed comparing large series of matched healthy and diseased tissues from different patients as has been shown for example for brain samples from patients with neuropsychiatric diseases [Frank *et al.*, 2005].
2. A second approach is to use paired samples (healthy vs. diseased) derived from the same patient. In this case, RetroArray allows for semiquantitative evaluation based on densitometric analysis of HERV transcription. In this experimental setting, detection of disease-associated changes in retroviral activity may be feasible even with a relatively small series of patients as has been demonstrated for human breast cancer patients in a recent study (Figure 5 and 7) [Frank *et al.*, 2008]. The obtained findings might serve as a starting point for further investigating the role of differentially transcribed HERVs as candidates for potential molecular markers.

Beyond it, also transmission of animal retroviruses might be of relevance for oncogenesis due to potential insertional mutagenesis. Examples for such suspected zoonotic infections are a xenotropic MLV-like virus (XMRV) that was recently discovered in human prostate cancer [Urisman *et al.*, 2006], and a retrovirus with 96% sequence identity to the mouse mammary tumor virus (MMTV) that has been associated with human breast cancer [Liu *et al.*, 2001]. Underlying infections could arise from polytropic retroviruses derived from domesticated animals (pets) or pest animals that are in close contact with humans. Based on the same principle and technology as the HERV microarray, we established an animal retrovirus chip for detection of 41 exogenous and endogenous retroviruses from various domestic animals [Frank *et al*., 2005]. The general applicability of the pet chip as a screening tool has been demonstrated by using animal DNAs that served as targets in PCRs (see Figure 8).

We used this chip to elucidate the existence of the controversially debated human MMTV-like virus (HMLV) [Mant and Cason, 2004] in the aforementioned breast cancer survey. No evidence was found for occurrence of exogenous retroviruses, neither in tumor samples nor in corresponding nonmalignant mammary gland tissues of patients with breast cancer, although the microarray has been proven to be reliable in pilot experiments using synthetic oligonucleotides [Frank *et al*., 2008]. These data indicate that an exogenous retrovirus such as MMTV or HMLV is not present in breast cancer patients from South West Germany.

Monitoring of Undesired Retroviral Gene Transfer in Therapeutic Applications

Polytropic retroviruses may also cause problems when cell lines are used for the production of recombinant therapeutics such as peptide hormones, or when therapeutic monoclonal antibodies are produced by murine hybridoma cells. Retroarray may be used to screen such cell lines for virus release and to control biomedical products for contaminating retroviral genetic material. For example, HERV transcripts of different origin were demonstrated in particles released by T47D human mammary carcinoma cells [Seifarth *et al.*, 2003]. Further conceivable applications are survey of pig-derived xenotransplants before transplantation as well as long-term follow-up of transplant recipients. RetroArray is also suitable for monitoring packaging cell lines and vector preparations with the goal to exclude transfer of copackaged RCRs or ERVs into target cells. Evidence that the latter phenomenon occurs regularly was found when HERV expression profiles of packaging cell lines were compared with HERV transcripts associated with the corresponding vector particle preparations (Figure 9). For example, a selective enrichment of certain HERV sequences was observed in particle fractions produced by human Phoenix-GP cells compared to the overall HERV transcription pattern of the packaging cell line itself [Zeilfelder *et al.*, 2007]. Moreover, copackaging appears to depend on the vector system, since an HIV-1-derived vector system showed significantly less copackaged HERV sequences than MLV-based vectors. Thus, RetroArray could contribute to safety in a variety of medical applications and protect patients against undesired iatrogeneous transmission of retroviral genetic material.

Detection of Cell Culture Contaminations

Viral contaminations of cell lines are an persisting problem not only in biomedical applications but also in basic research [Lacroix, 2008]. Unknown presence of retroviruses in cell culture can introduce an additional biohazard to the experimenter and adulterate experimental outcomes. Polytropic animal viruses such as MLV or gibbon ape leukemia virus, for example, have been detected in human cell lines commonly used in HIV research and may have caused phenotypic mixing of HIV and gammaretroviruses [Takeuchi *et al.*, 2008]. Interferon samples prepared from the human lymphoid Namalwa cell line have been found to be positive for SMRV sequences [Pienkowska and Seth, 1998]. Namalwa cells were shown to harbor several copies of SMRV proviral DNA [Oda *et al.*, 1988]. Such contaminations can be easily detected and identified with the animal retrovirus-specific version of RetroArray as is shown for a SMRV infected human melanoma cell line (A375) in Figure 10.

Further important problems leading to confounded or falsified experimental results are cellular cross-contaminations and misidentified cell lines [Lacroix, 2008]. It is estimated that as much as 20 – 30% of all cell lines used today are cross-contaminated either with cells from other species (interspecies contamination) or with unrelated cells from the same species (intraspecies contamination) [Hughes *et al.*, 2007]. Due to the highly specific HERV activity profiles of particular cells of human origin (see above) RetroArray can be used to assess purity and homogeneity of human cell lines. Laboratory contamination with cells of other species will be easily detected by RetroArray analysis due to the species-specificity of the assay. For example mouse/human hybridoma cells display a retroviral pattern that is characteristically composed of both underlying cell fractions (Figure 10).

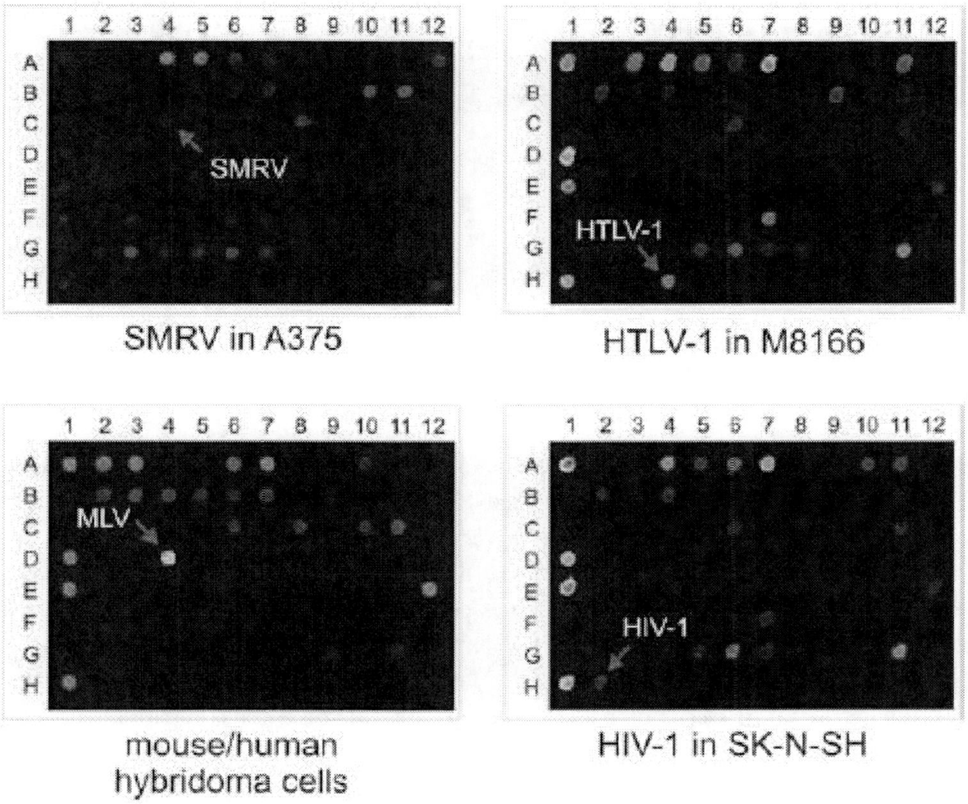

Figure 10. Retroviral profiling allows species identification (mouse/human hybridoma cells) and detection of exogenous retroviruses in human cell cultures (A375, M8166, SK-N-SH). A375, M8166, and SK-N-SH are human melanoma, lymphocytic, and neuroblastoma cell lines, respectively. For abbreviations see text and tables 1 and 2.

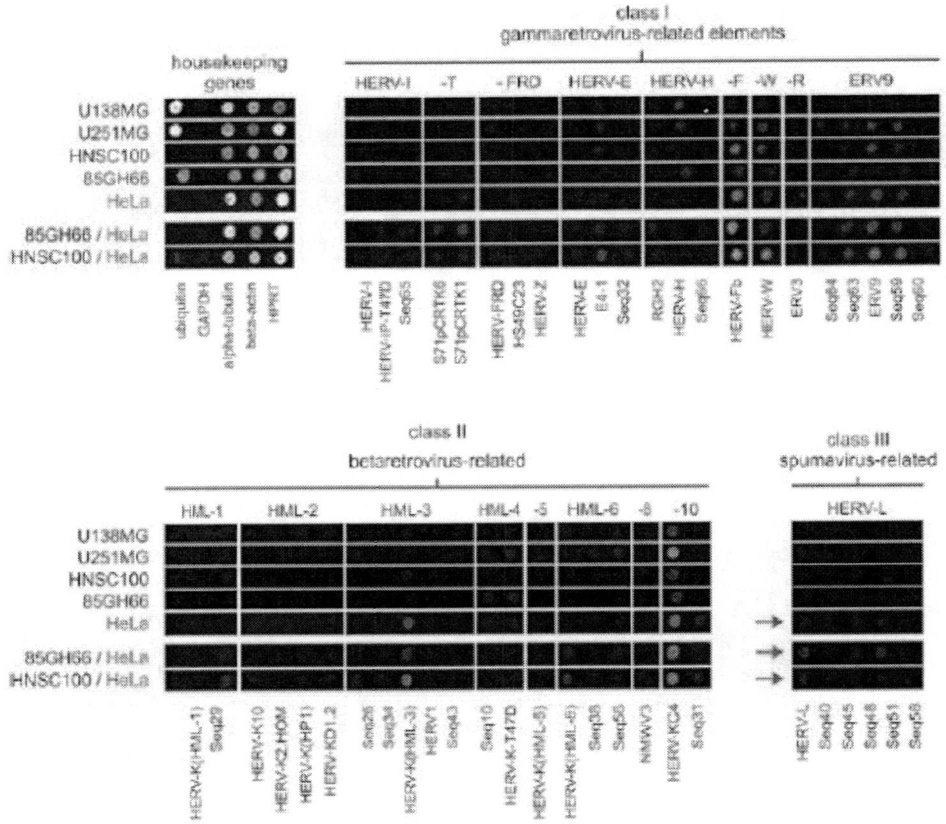

Figure 11. Mistaken identity of human cell cultures detected by HERV expression profiling. Contaminations of the human brain-derived cell lines 85GH66 and HNSC100 with fast growing HeLa cells were detected due to the tissue-specific activity of HERV-L. Red arrows denote HeLa-specific HERV-L transcription patterns. Abbreviations: 85GH66, U138MG, U251MG are astrocytoma-derived cells; HNSC100 is a human neuronal stem cell line.

HeLa cells are a very common contamination; the American Type Culture Collection ATCC presently lists at least 10 misidentified HeLa contaminants that were undetected in laboratory use for years. HeLa cells display a very characteristic HERV transcription pattern, since they express class III elements (HERV-L) that are normally active only in human skin, thyroid gland and in reproductive organs such as cervix, uterus, placenta and testes [Seifarth *et al.*, 2005]. We therefore compared the HERV expression profiles from several human brain-derived cell lines with those from HeLa and HeLa-contaminated cells. As demonstrated in Figure 11 the retroviral transcription pattern characteristic for HeLa cells, in particular HERV-L expression, can be easily recognized in contaminated human neural stem cells (HNSC-100) and astrocytoma cells (85HG66). Therefore, RetroArray analysis represents a rapid screening tool that may indicate occurrence of cross-contaminated or misidentified cell lines.

CONCLUSION

Over the last couple of years, microarrays (DNA chips) have emerged as powerful and indispensable research tools for detection of microorganisms, for gene expression profiling, mutation and polymorphism analysis in cells and tissues. Employing this technology in combination with a one-step PCR labelling procedure, we set up a high throughput assay for rapid and precise qualitative analysis of retrovirus activity in biological samples. In general, nucleic acids prepared from any biological source, i.e. all types of body fluids, tissues, cells, and cell culture supernatants, can be tested for the presence of retroviral genetic material. One of the advantageous features of RetroArray is that the test is almost unlimited with respect to the number of retroviral *pol*(RT) sequences that can be addressed. Thus, capture probes corresponding to novel *pol*(RT)-encoding sequences can be easily added without having to modify the experimental design. Furthermore, the assay allows the identification of novel so far unknown exo- and endogenous retroviruses. Retroviral elements with only weak homologies to well characterized families should be detectable after lowering hybridization stringency conditions. Recent publications demonstrate that RetroArray can serve as a powerful tool for various applications in the field of basic science and modern biomedicine.

REFERENCES

[1] Bannert, N; Kurth, R. Retroelements and the human genome: new perspectives on an old relation. *Proc. Natl. Acad. Sci.*, U S A 2004, 101 Suppl 2, 14572-14579.

[2] Belshaw, R; Katzourakis, A; Paces, J; Burt, A; Tristem, M. High copy number in human endogenous retrovirus families is associated with copying mechanisms in addition to reinfection. *Mol. Biol. Evol.*, 2005, 22(4), 814-817.

[3] Bromham, L. The human zoo: endogenous retroviruses in the human genome. *Trends Ecol. Evol.*, 2002, 17, 91-97.

[4] Conley, AB; Piriyapongsa, J; Jordan, IK. Retroviral promoters in the human genome *Bioinformatics*, 2008, 24(14), 1563-1567.

[5] de Parseval, N; Lazar, V; Casella. JF; Benit, L; Heidmann, T. Survey of human genes of retroviral origin: identification and transcriptome of the genes with coding capacity for complete envelope proteins. *J. Virol.*, 2003, 77(19), 10414-10422.

[6] Denne, M; Sauter, M; Armbruester, V; Licht, JD; Roemer, K; Mueller-Lantzsch, N. Physical and functional interactions of human endogenous retrovirus proteins Np9 and rec with the promyelocytic leukemia zinc finger protein. *J. Virol.*, 2007, 81(11), 5607-5616.

[7] Flockerzi, A; Maydt, J; Frank, O; Ruggieri, A; Maldener, E; Seifarth, W; et al. Expression pattern analysis of transcribed HERV sequences is complicated by ex vivo recombination. *Retrovirology*, 2007, 4, 39.

[8] Flockerzi, A; Ruggieri, A; Frank, O; Sauter, M; Maldener, E; Kopper, B; et al. Expression patterns of transcribed human endogenous retrovirus HERV-K(HML-2) loci in human tissues and the need for a HERV Transcriptome Project. *BMC Genomics*, 2008, 9, 354.

[9] Frank, O; Giehl, M; Zheng, C; Hehlmann, R; Leib-Mösch, C; Seifarth, W. Human endogenous retrovirus expression profiles in samples from brains of patients with schizophrenia and bipolar disorders. *J. Virol.*, 2005, 79(17), 10890-10901.

[10] Frank, O; Verbeke, C; Schwarz, N; Mayer, J; Fabarius, A; Hehlmann, R; et al. Variable transcriptional activity of endogenous retroviruses in human breast cancer. *J. Virol.*, 2008, 82(4), 1808-1818.

[11] Galli, UM; Sauter, M; Lecher, B; Maurer, S; Herbst, H; Roemer, K; et al. Human endogenous retrovirus rec interferes with germ cell development in mice and may cause carcinoma in situ, the predecessor lesion of germ cell tumors. *Oncogene*, 2005, 24(19), 3223-3228.

[12] Garrison, KE; Jones, RB; Meiklejohn, DA; Anwar, N; Ndhlovu, LC; Chapman, JM; et al. T cell responses to human endogenous retroviruses in HIV-1 infection. *PLoS Pathog.*, 2007, 3(11), e165.

[13] Gifford, R; Tristem, M. The evolution, distribution and diversity of endogenous retroviruses. *Virus Genes.*, 2003, 26(3), 291-315.

[14] Hakim, ST; Alsayari, M; McLean, DC; Saleem, S; Addanki, KC; Aggarwal, M; et al. A large number of the human microRNAs target lentiviruses, retroviruses, and endogenous retroviruses. *Biochem. Biophys. Res. Commun.*, 2008, 369(2), 357-362.

[15] Han, K; Konkel, MK; Xing, J; Wang, H; Lee, J; Meyer, TJ; et al. Mobile DNA in Old World monkeys: a glimpse through the rhesus macaque genome. *Science*, 2007, 316(5822), 238-240.

[16] Hohenadl, C; Germaier, H; Walchner, M; Hagenhofer, M; Herrmann, M; Stürzl, M; et al. Transcriptional activation of endogenous retroviral sequences in human epidermal keratinocytes by UVB irradiation. *J. Invest. Dermatol.*, 1999, 113(4), 587-594.

[17] Hughes, P; Marshall, D; Reid, Y; Parkes, H; Gelber, C. The costs of using unauthenticated, over-passaged cell lines: how much more data do we need? *Biotechniques*, 2007, 43(5), 575, 577-8-581-2.

[18] Jurka, J; Kapitonov, VV; Pavlicek, A; Klonowski, P; Kohany, O; Walichiewicz, J. Repbase Update, a database of eukaryotic repetitive elements. *Cytogenet. Genome Res.*, 2005, 110(1-4), 462-467.

[19] Katzourakis, A; Rambaut, A; Pybus, OG. The evolutionary dynamics of endogenous retroviruses. *Trends Microbiol.*, 2005, 13(10), 463-468.

[20] Lacroix, M. Persistent use of "false" cell lines. *Int J Cancer*, 2008, 122(1), 1-4.

[21] Lander, ES; Linton, LM; Birren, B; Nusbaum, C; Zody, MC; Baldwin, J. et al. Initial sequencing and analysis of the human genome. *Nature*, 2001, 409(6822), 860-921.

[22] Lee, YN; Malim, MH; Bieniasz, PD. Hypermutation of an ancient human retrovirus by APOBEC3G. *J. Virol.*, 2008, 82(17), 8762-8770.

[23] Liu, B; Wang, Y; Melana, SM; Pelisson, I; Najfeld, V; Holland, JF. et al. Identification of a proviral structure in human breast cancer. *Cancer Res.*, 2001, 61(4), 1754-1759.

[24] Mager, D; Medstrand, P. Retroviral repeat sequences. In: Cooper D, editor. *Nature encyclopedia of the human genome*. London, United Kingdom: *Nature Publishing Group*, 2003, 57-63.

[25] Mant, C; Cason, J. A human murine mammary tumour virus-like agent is an unconvincing aetiological agent for human breast cancer. *Rev. Med. Virol.*, 2004, 14(3), 169-177.

[26] Martin, J; Herniou, E; Cook, J; O'Neill, RW; Tristem, M. Interclass transmission and phyletic host tracking in murine leukemia virus-related retroviruses. *J. Virol.*, 1999, 73(3), 2442-2449.
[27] Moyes, D; Griffiths, DJ; Venables, PJ. Insertional polymorphisms: a new lease of life for endogenous retroviruses in human disease. *Trends Genet.*, 2007, 23(7), 326-333.
[28] Muradrasoli, S; Forsman, A; Hu, L; Blikstad, V; Blomberg, J. Development of real-time PCRs for detection and quantitation of human MMTV-like (HML) sequences HML expression in human tissues. *J. Virol. Methods*, 2006, 136(1-2), 83-92.
[29] Nelson, PN; Carnegie, PR; Martin, J; Davari, EH; Hooley, P; Roden, D. et al. Demystified. Human endogenous retroviruses. *Mol. Pathol.*, 2003, 56(1), 11-18.
[30] Oda, T; Ikeda, S; Watanabe, S; Hatsushika, M; Akiyama, K; Mitsunobu, F. Molecular cloning, complete nucleotide sequence, and gene structure of the provirus genome of a retrovirus produced in a human lymphoblastoid cell line. *Virology*, 1988, 167(2), 468-476.
[31] Pienkowska, M; Seth, A. Detection of squirrel monkey retroviral sequences in interferon samples. *J. Hepatol.*, 1998, 28(3), 396-403.
[32] Rawn, SM; Cross, JC. The evolution, regulation, and function of placenta-specific genes. *Annu. Rev. Cell Dev. Biol.*, 2008, 24, 159-181.
[33] Ruprecht, K; Obojes, K; Wengel, V; Gronen, F; Kim, KS; Perron, H. et al. Regulation of human endogenous retrovirus W protein expression by herpes simplex virus type 1: implications for multiple sclerosis. *J. Neurovirol.*, 2006, 12(1), 65-71.
[34] Sambrook, J; Russell, D. Molecular Cloning: A Laboratory Manual. 3rd ed. Cold Spring Harbor, NY, USA: Cold Spring Harbor Laboratory Press, 2001.
[35] Schulz, WA; Steinhoff, C; Florl, AR. Methylation of endogenous human retroelements in health and disease. *Curr. Top Microbiol. Immunol.*, 2006, 310, 211-250.
[36] Seifarth, W; Krause, U; Hohenadl, C; Baust, C; Hehlmann, R; Leib-Mösch, C. Rapid identification of all known retroviral reverse transcriptase sequences with a novel versatile detection assay. *AIDS Res. Hum. Retroviruses*, 2000, 16(8), 721-729.
[37] Seifarth, W; Leim-Mösch, C; Baust, C. Method for specifically detecting and identifying retroviral nucleic acids/retroviruses in an item to be examined. PCT/DE2000/001071, WO 2000/068435.
[38] Seifarth, W; Spiess, B; Zeilfelder, U; Speth, C; Hehlmann, R; Leib-Mösch, C. Assessment of retroviral activity using a universal retrovirus chip. *J. Virol. Methods*, 2003, 112(1-2), 79-91.
[39] Seifarth, W; Frank, O; Zeilfelder, U; Spiess, B; Greenwood, AD; Hehlmannm R. et al. Comprehensive analysis of human endogenous retrovirus transcriptional activity in human tissues with a retrovirus-specific microarray. *J. Virol.*, 2005, 79(1), 341-352.
[40] Sharp, PM; Shaw, GM; Hahn, BH. Simian immunodeficiency virus infection of chimpanzees. *J. Virol.*, 2005, 79(7), 3891-3902.
[41] Shih, A; Misra, R; Rush, MG. Detection of multiple, novel reverse transcriptase coding sequences in human nucleic acids: relation to primate retroviruses. *J. Virol.*, 1989 63(1), 64-75.
[42] Silvestri, G. Naturally SIV-infected sooty mangabeys: are we closer to understanding why they do not develop AIDS? *J. Med. Primatol.*, 2005, 34(5-6), 243-252.
[43] Sperber, GO; Airola, T; Jern, P; Blomberg, J. Automated recognition of retroviral sequences in genomic data--RetroTector. *Nucleic Acids Res.*, 2007, 35(15), 4964-4976.

[44] Stauffer, Y; Theiler, G; Sperisen, P; Lebedev, Y; Jongeneel, CV. Digital expression profiles of human endogenous retroviral families in normal and cancerous tissues. *Cancer Immun*, 2004, 4, 2.

[45] Sverdlov, ED. Retroviruses and primate genome evolution. Georgetown, Texas: Landes Bioscience, 2005.

[46] Takeuchi, Y; McClure, MO; Pizzato, M. Identification of gammaretroviruses constitutively released from cell lines used for human immunodeficiency virus research. *J. Virol.*, 2008, 82(24), 12585-12588.

[47] Urisman, A; Molinaro, RJ; Fischer, N; Plummer, SJ; Casey, G; Klein, EA. et al. Identification of a novel Gammaretrovirus in prostate tumors of patients homozygous for R462Q RNASEL variant. *PLoS Pathog*, 2006, 2(3), e25.

[48] van de Lagemaat, LN; Landry, JR; Mager, DL; Medstrand, P. Transposable elements in mammals promote regulatory variation and diversification of genes with specialized functions. *Trends Genet.*, 2003, 19(10), 530-536.

[49] Voisset, C; Weiss, RA; Griffiths, DJ. Human RNA "rumor" viruses: the search for novel human retroviruses in chronic disease. *Microbiol. Mol. Biol. Rev.*, 2008, 72(1), 157-196.

[50] Weiss, RA. The discovery of endogenous retroviruses. *Retrovirology*, 2006, 3, 67.

[51] Xiong, Y; Eickbush, TH. Origin and evolution of retroelements based upon their reverse transcriptase sequences. *Embo. J.*, 1990, 9(10), 3353-3362.

[52] Zeilfelder, U; Frank, O; Sparacio, S; Schön, U; Bosch, V; Seifarth, W. et al. The potential of retroviral vectors to cotransfer human endogenous retroviruses (HERVs) from human packaging cell lines. *Gene.*, 2007, 390(1-2), 175-179.

In: Encyclopedia of DNA Research (3 Volume Set)
Editors: S. J. Duncan and P. H. Wiley
ISBN 978-1-61324-305-3

Chapter 32

BIOGENOMIC PATHWAYS IN BREAST AND OVARIAN MALIGNANCIES: CLINICAL AND SURGICAL WORKFLOW FOCUSED ON TAILORING PROCEDURES

A. Tinelli*[*1], A. Malvasi[1], V. Lorusso[2], E. Chiuri Vincenzo[2], S, Cinieri[3], D. Vergara[4], R. Martignago[5], U. Vergari[6], M. Pisano[7] and G. Leo[7]

[1]Department of Gynecology and Obstetric, Vito Fazzi Hospital, Lecce, Italy
[2] Department of Oncology, Oncological Hospital, Lecce, Italy
[3] Department of Oncology, Perrino Hospital, Brindisi, Italy
[4] National Nanotechnology Laboratory (NNL), CNR-INFM, University of Salento, Leece, Italy
[5] Department of Biological and Environmental Sciences and Technologies, Laboratory of Human Anatomy, University of Salento, Leece, Italy
[6] International Center of Bioethics and Human Rights, University of Salento, Lecce, Italy
[7] Molecular Biology and Experimental Oncology Lab, Oncological Hospital, Lecce, Italy

ABSTRACT

Breast and ovarian cancers are the most common worldwide gynaecological cancers and connected to germline mutations in the tumour suppressor genes BRCA1 and BRCA2, both involved in different cellular mechanism, such as DNA damage recognition, double-strand break repair, checkpoint control, transcription regulation and chromatin remodelling. The inactivation of BRCA-related pathways is probably associated with significantly higher percentages of these gynaecological cancers.

In this elaborate are discussed the biogenomic perspectives in gynecological surgery and the current opportunity to treat breast and ovarian malignancies by tailoring

* Corresponding author: Dr. Tinelli Andrea, MD,Department of Gynecology and Obstetric, Vito Fazzi Hospital, Piazza F. Muratore, 73100, Lecce, Italy, Cellular Number 0039/339/2074078, Fax Number 0039/0832/661511, E-mail: andreatinelli@gmail.com.

minimally invasive procedures, basing on molecular genetic analysis and by using of prophylactic surgery.

The authors reviewed the clinical and genetic evaluation for germline mutations to monitor the women predisposed to cancer developments and to choose the risk reduction strategies, based on a more precise estimate of individual risk; in the high risk women, the authors focused on the tailoring minimally invasive gynaecological procedures, a pre-emptive prophylactic surgery based on molecular genetic analysis, targeting to an organ or a part of its prior to malignant transformation or while the cancer is in situ.

Clinical and surgical workflow in high risk patients for breast and ovarian cancers include highly sensitive screening, starting at a young age, and the use of tailoring minimally invasive operations on the ovaries or on the breast; the tailoring surgery, based on molecular genetic analysis and targeting a low tumor burden, have the potential to achieve greater cure rates than extensive resections required for clinically evident disease and, moreover, it is as an acceptable surgical approach for smaller incisions, less postoperative pain, and shorter hospital stays.

Keywords: gynecological tumors, breast cancers, ovarian cancers, BRCA 1 mutation, BRCA 2 mutation, minimally invasive surgery, prophylactic surgery.

Introduction

The appearance of genomic studies represented an unprecedented opportunity in medical and biological research, with the aim to investigate genes and their functions into the cells.[1]

The information generated by genomics have the potential to revolutionize the medicine, but despite significant scientific advances, only few genomics-based tests or treatments have reached consumers [2] and the comparison of genomes has resulted in some surprising biological discoveries[3].

In the current genomic era, the appropriate use of minimally invasive therapy for treatment of diseases, led to the possibility of prophylactic surgical interventions, based on molecular genetic analysis; in fact, predictive genetic tests are enabled to identify patients with high risk to develop a malignancy of a specific organ. Unlikely, the actually predictive tests and prophylactic surgeries are applicable only to patients with select hereditary cancer syndromes.

The goal of prophylactic surgery is to reduce the risk of cancer, by removing the organ(s) carrying mutated genes before the occurrence of clinical disorders.

In this review, we expose about the biogenomic perspectives in gynecological surgery and the current opportunity to treat some gynecological diseases by tailoring minimally invasive procedures and how these new approaches will lead to novel treatment strategies and targets relevant for patients with a broad range of sporadic malignancies.

Breast and Ovarian Cancer: Gynaecological Malignancies Related to Genomic Mutations

Breast cancer is the most common malignancy among U.S. women, with more than 200,000 new cases diagnosed annually; the breast cancer mortality trend in the USA is declined with the introduction of more widespread screening, which have led to an earlier detection, as well as advances in the adjuvant treatment of early-stage disease. [4]

Ovarian cancer is the most lethal gynecologic malignancy; on a worldwide basis, it is estimated that over 140,000 cases are diagnosed annually, representing more than 4% of all cancer cases in women and in the USA, incidence and mortality estimates for 1990 indicated that approximately 20,500 new cases were diagnosed and 12,000 women died for it; overall, the probability that a woman will develop an ovarian cancer in her lifetime is 1 in 70. The incidence rate is currently 46% higher in whites than in blacks and the incidence of ovarian cancer seems to have decreased slightly since 1973, but the survival rate has not changed much. [5]

Survival is related to age: during the period 1981-86, women diagnosed prior to age 65 had a 5-year relative survival rate of 48% compared to 24% for those diagnosed at age 65 and older; the age-adjusted mortality rates are 1.69, 3.02, 7.04, and 11.02 per 100,000 for Japan, Italy, the USA and Denmark, respectively. [6]

It is widely accepted that the appropriate use of adjuvant chemotherapy and endocrine therapy improves the disease-free and overall survival of patients with early-stage breast cancer; therefore, it is a standard clinical practice to administer adjuvant systemic therapy to patients with node-positive and high-risk, node-negative breast cancer. [7]

Epithelial carcinomas account for 80-90% of ovarian neoplasms; the remaining ovarian tumors originate from germ or stromal cells.

Epithelial tumors are classified in benign, malignant (invasive), and tumors of low malignant potential (or borderline malignancy): the peak incidence of benign epithelial tumors occurs in the 20-40-year-old group, whereas the peak incidence for malignant neoplasms occurs in the 40-70-year-old group [8].

Cancers of the breast and ovary appear to share some common etiologic factors [9] : women with breast cancer have twice expected risk of ovarian cancer and women with ovarian cancer have a three-to-fourfold increased risk of subsequent breast cancer. [10,11]

Germline mutations account for 5% to 10% of breast cancer diagnosed in the United States each year [12]; because women who develop cancer related to such mutations have a relatively young age, these mutations account for a disproportionate share of life-years lost due to cancer. [13]

Germline mutations in the *BRCA1* and *BRCA2* genes confer ovarian and breast cancer susceptibility; mutations of genes encoding ATM and CHEK2 or mutations leading to EMSY amplification also affect BRCA pathways. [14] These alteration are present more frequently than *BRCA1* or *BRCA2* mutations in the general population, suggesting that inactivation of BRCA-related pathways is associated with significantly higher risk of cancers [15]; the *BRCA1* gene (chromosome 17) and *BRCA2* gene (chromosome 13) genes were characterized in 1994 and 1995. [14,15]

Data obtained in many thousands of high-risk individuals who were not identified as mutation carriers are available, which were also eligible for mutation testing [16]; these data would include all mutations affecting BRCA pathways, with some studies providing a built-in control inasmuch as they also include those who were not eligible for mutation testing. [17-19]

The data published in more than 30 epidemiologic studies (meta-analyses) show that the silencing of *BRCA* genes provides growth or survival advantages to a broader spectrum of tumors, namely stomach, pancreas, prostate, and colon [20, 21], leading to suppose that

BRCA attends to different cellular pathways. [21-23] Thus, a better understanding of BRCA pathways may provide new targets for chemotherapy or chemoprevention. [22-25]

In the general population, ~0.1% to 0.2% are carriers of *BRCA1* or *BRCA2* mutations having a risks for breast cancer exceeding 80%, respect to mutation-free individuals. [26] In particular, risks of ovarian cancer for BRCA1- and BRCA2-mutation carriers are greater than 40% and 20%, respectively. [27]

Kauff and collaborators estimated that for a female mutation carrier, the risk of developing breast cancer by age of 70 is between 60% and 85% and the risk of developing ovarian cancer by age of 70 is between 15% and 65%.[28]

This variation of the estimate risks depends upon the kind of *BRCA* mutation (or different mutations in the same gene), family history (strong or weak), as well as genetic and other factors that interfere with predisposition.

BRCA1 and/or BRCA2 proteins are involved in different cellular mechanism, like DNA damage recognition, double-strand break repair, checkpoint control, transcription regulation, and chromatin remodelling; these functions are essential and important for all cell types. [21-23]

Despite the general nature of BRCA functions, tumors predominantly target breast and ovary; thus, a major problem is that the broad theoretical importance of BRCA-related pathways conflicts with such specific cancers targeting. [26]

Most of breast and ovarian cancers are considered sporadic, with no known genetic link; nevertheless, about 10% of women with breast cancer have a genetic predisposition, so that family history has been recognized as a major risk factor for breast and ovarian cancers.

Hereditary breast and ovarian cancer is more likely to strike at a younger age than sporadic cancer, and in some carriers' bilateral breast cancer or both breast and ovarian cancer develop.[29] Most, but not all, women with inherited mutations that affect one allele of either *BRCA1* or *BRCA2*, will develop breast and/or ovarian cancer; this risk varies considerably between 40% to 85% for breast cancer and 11% to 65% for ovarian cancer. [30]

Data available suggest that among women with *BRCA* mutations, those with a strong family history have a higher risk than those without. [31]

Genetic Testing for Mutations in Ovarian and Breast Cancer

Genetic tests for *BRCA* mutations identification offers the opportunity to choose risk reduction strategies based on a more precise estimate of individual risk with respect to the past.[32]

However, these tests are complex because there is a wide spectrum of known *BRCA* mutations; most pathogenic mutations are deleterious and result in non functional products. [21-23]

The no-deleterious missense mutations cannot be clinically interpreted since they may represent neutral polymorphic variants rather than pathogenic mutations; at present, commercial genetic tests protocols are pointed to five specific large genomic rearrangements in *BRCA1* gene. [27,28]

The mutation detection rate using this technique is 88% at best (Myraid Genetics Inc. Salt Lake City, UT, USA); among those who have a negative test, approximately 12% are found to actually carry a large genomic deletion or duplication in a *BRCA* gene when further tested by multiplex ligation dependent probe amplification (MLPA).

These intricacies of genetic testing underscore the need for proper genetic counseling. [39,40]

A means to detect disease must exist in order to determine the efficacy of target surgical interventions; in terms of breast cancer, the most widely recommended strategy entails frequent, intense surveillance that combines annual mammography, annual or semiannual clinician breast examination, and monthly breast self-examination, beginning at age 25 to 35 years. [33]

The addition of magnetic resonance imaging (MRI), to the clinical breast examination and annual mammography, could increase the rate of early detection, but the rate of tumors of more than 1 cm is 43% and only 63% of patients have node-negative disease.

Theoretically, more frequent mammographic and MRI screening, for example every six months, could decrease the risk of late diagnosis, and increase the unneeded additional investigations and biopsies 31; unfortunately, due to their low specificity, there are no much evidences. Whereas heightened surveillance cannot prevent the development of breast cancer, it increases the likelihood of early detection and effective treatment; however, the sensitivity of mammograms has been shown to be lower among younger women and for *BRCA* mutation carriers the risk for early-onset breast cancer is increased up to 20-fold. [34]

It has been estimated that ~25% of young women, high-risk women that undergo close surveillance will develop breast cancer and eventually die from the disease [35] ; moreover, there are concerns that women may not comply with the schedule of regular examinations or that the radiation incurred by mammography may slightly increase the risk of breast cancer in healthy BRCA mutation carriers, who by the nature of their mutations have defective DNA repair mechanisms. [34]

There is considerable variability within and between countries in risk reduction strategies utilized by healthy BRCA1 or BRCA2 mutation carriers. Genomics-based diagnostic techniques, such as gene-array analysis and research focused on evaluating molecular pathways, will help to identify those *BRCA* carriers who will develop breast or ovarian cancer. [36]

Exciting findings using gene-expression profiles based on tissue-array analysis in clinically evident breast cancer are already available [37]: in current clinical practice, tailoring minimally invasive surgery is, sometimes, clearly superior to surveillance [38], because the high protective index, over 90% by surgery, are resulted in an increasing acceptance of surgical prevention among both oncologists and women with *BRCA* mutations. [39]

Clinical Options for Women at High Risk for Breast and Ovarian Germline Mutations

Clinical options for women with high genetic risk of breast cancer include screening starting at a young age, the use of highly sensitive detection methods and prophylactic surgeries of the ovaries or breast. [40]

Tamoxifen, an antiestrogenic drug, can reduce breast cancer risk, but it is effective only in estrogen-receptors (ER) positive tumors and probably also in BRCA-associated tumors tamoxifen chemoprevention reduced breast cancer incidence only in *BRCA2* carriers but not in those with *BRCA1* mutations; research is ongoing to confirm these findings and to determine whether other ER modulators and aromatase inhibitors are more effective than tamoxifen. [41]

Other factors may modify the genetic risk: pregnancy, physical exercise, adolescence, healthy weight and breast-feeding may delay the onset or even reduce the risk of cancer [42] and, since breast tumorigenesis is causatively linked to estrogen, it is not surprising that bilateral salpingo-oforectomy (BSO) reduces the risk of both ovarian and breast cancer. [43]

This protective effect of BSO has been suggested by previous studies and decision analysis. [44]

Because prophylactic surgeries are highly effective in reducing risk, but also highly invasive, it is particularly important to distinguish mutation carriers from noncarriers with similarly severe family histories. [45]

Women with *BRCA1* or *BRCA2* mutations are possible candidates for such surgeries. [46]

Molecular and genetic tests can identify deleterious *BRCA1* and *BRCA2* mutations in as yet unaffected women with severe family histories of breast or ovarian cancer and they have become an integral part of clinical practice in many communities [47]: however, mutation screening of *BRCA1* and *BRCA2* is technically challenging, because each gene harbors more than 1,000 different disease-associated mutations, the vast majority of which are individually rare. [47]

With the exception of founder alleles such as those in the Ashkenazi Jewish population, mutations detection in *BRCA1* and *BRCA2* requires the screening of the entire gene, through short-range polymerase chain reaction (PCR) followed by genomic sequencing. [48] The evident of the failure response to tamoxifen chemoprevention suggest the urgent need for surgical prophylaxis [49-51]: women with BRCA2 mutation face a lower probability of developing cancer in the breast (45–82%) or ovary (11–23%), compared with BRCA1 carriers. [52]

But this risk remains high, much higher than in general population, and despite the tamoxifen effectiveness – most BRCA2 are ER positive tumors – the uncertainty about cancer mortality persists [44]; in a recent analysis of preventive strategies, it was reported that prophylactic surgery or chemoprevention could lead to better survival and quality-adjusted survival than surveillance alone for women with a positive test for *BRCA1* or *BRCA2* mutations. [53]

Surgical and Non Surgical Preventive Options in Breast and Ovarian Germline Mutations: The Prophylactic Procedures

Surgical procedures in patients with *BRCA1* or *BRCA2* mutations are, substantially, prophylactic procedures and are codified in medical-legal issues; prophylactic surgery includes bilateral mastectomy (BM), BSO and resection of both breasts and ovaries.

By removing the *BRCA*-targeted organ(s), it is possible to dramatically reduce an overall cancer risk; indeed, although BRCA mutated genes are found in every cell in the body of carriers, they target nearly exclusively breast tissue, and to a lesser extent, the ovary and fallopian tube. [46]

Prophylactic BM in women with family history of breast cancer had been suggested, but rarely approved in scientific community; with the availability of genetic tests, this surgery is performed only to women with evident *BRCA* mutations. [44] Ideally, surveillance may result in lifelong preservation of target organs or its surgical resection only when cancer becomes clinically evident; unfortunately, individual women who will benefit from close screening and/or chemoprevention cannot be identified. [44]

Oophorectomy is associated with reduced risk of ovarian and fallopian tube cancer in high-risk women, although there is a substantial residual risk for peritoneal cancer in *BRCA1* and *BRCA2* mutation carriers following prophylactic BSO. [54] The ovarian and breast cancer risk reductions by prophylactic surgery are substantial, but the approaches are drastic: prophylactic removal of the breasts or ovaries bring to think for a better way to fend off the diseases. [55]

Several studies in mutation carriers indicate that it could provide a 90% reduction in breast cancer risk, but researchers agree that, despite the dramatic risk reduction, other options are needed: ovaries removal can reduce risk for both breast and ovarian cancers. [40]

At 2002 annual meeting of the American Society of Clinical Oncology, Kauff presented 4-year follow-up data for BRCA mutation carriers who underwent removal of the fallopian tubes and ovaries, called: risk-reducing bilateral salpingooophorectomy (RR-BSO). [56]

The surgery was associated with a 70% lower risk of breast cancer and 83% lower risk of ovarian cancer and prophylactic BSO is elected far more than prophylactic mastectomy; ovarian cancer screening is less effective than breast cancer screening so, for women with a family history of ovarian cancer, BSO may be considered the best option for ovarian cancer risk management, because the level of cancer risk reduction associated with prophylactic BSO has been estimated to be as high as 95%.[56]

In some Oncological Centers where *BRCA* screening is routinely effective, 65% of women with *BRCA* mutations elect to undergo RR-BSO, only one-third elect prophylactic mastectomy and the majority of women who elect mastectomy also choose RR-BSO.

For *BRCA1* mutations, ovarian cancer risk appears during a woman's late 30s and for a woman with a *BRCA2* mutation, risk doesn't really appear until her late 40s, early 50s, so the screening into the late 40s, early 50s, may not be unreasonable.

The surgical survival benefit for women who had prophylactic mastectomy with BSO at 30 years of age was 4.9 years, but it increased to 7.8 years, if the penetrance rates of breast and ovarian cancer were 85% and 63%, respectively. [57]

A recent important report confirmed that oophorectomy is associated with reduced risk of ovarian and fallopian tube cancer in high-risk women, although there is a substantial residual risk for peritoneal cancer in *BRCA1* and *BRCA2* mutation carriers following prophylactic salpingo-oophorectomy. [54]

To estimate the risk of ovarian cancer after a first diagnosis of breast cancer among women with a *BRCA1* or *BRCA2* mutation and to identify host and treatment-related factors that might modify the risk, a group of authors performed an important study; this research concluded that oophorectomy should be recommended in female *BRCA1* and *BRCA2* mutation carriers with a diagnosis of breast cancer, especially those in stage I disease. Breast cancer systemic therapy did not significantly alter the risk of ovarian cancer. [58]

Risk-reducing salpingo-oophorectomy is often considered by carriers of *BRCA* mutations who have completed childbearing; however, there are limited data supporting the efficacy of this approach.

In women with *BRCA* mutations, some authors compared the effect of risk-reducing salpingo-oophorectomy with that of surveillance for ovarian cancer, on the incidence of subsequent breast cancer and *BRCA*-related gynecologic cancers and they concluded that BSO can decrease the risk of breast cancer and BRCA-related gynecologic cancers[59].

A study conducted to estimate the risk of contra-lateral breast cancer in *BRCA1* and *BRCA2* carriers after diagnosis and to determine which factors are predictive of the risk of a second primary breast cancer, reveled that the risk of contra-lateral breast cancer in women with a *BRCA* mutation is approximately 40% at 10 years and that it is reduced in women who take tamoxifen or who undergo an oophorectomy. [60]

Inherited mutations in the *BRCA1* or *BRCA2* genes are associated with a greatly increased risk of breast and ovarian cancers and a modestly increased risk of several other cancer types. Several case reports of endometrial carcinoma in women with a *BRCA* mutation and this have led to speculate about the contributing of these genes on the risk of endometrial cancer. [61]

A study tested that the hypothesis that germline mutation of a *BRCA* gene is associated with an increased risk of endometrial carcinoma (based on 199 women in 12 years of observation), concluded that the risk of endometrial carcinoma is not increased. [62]

An interesting study defined the rate of neoplasia in prophylactic surgical specimens with the use of a careful surgical and pathologic protocol in a prospective research of high-risk women who were undergoing risk-reducing BSO; indeed, neoplasia was identified in surgical specimens and clinical outcomes of women who were undergoing risk-reducing BSO. The conclusions were that age ≥ 45 years and mutations in *BRCA1* or *BRCA2*, could identify occult neoplasia in women who undergo risk-reducing BSO; it can be hypothesized that older age and *BRCA* mutation would be predictors for tubal or ovarian neoplasia and that a careful surgical and pathologic protocol would lead to a low rate of subsequent primary peritoneal cancer. Furthermore, a thorough pathologic and surgical protocol at the time of risk-reducing BSO may improve the risk of subsequent primary peritoneal cancer. [63]

Minimally Invasive Gynaecologic Treatment in Breast and Ovarian Germline Mutations

It is difficult to evaluate the optimal surgical procedure: BSO for the prevention of ovarian and fallopian-tube cancers in carriers of *BRCA1* and *BRCA2* mutations is widely recommended, but support for this approach comes from retrospective studies in which participants were not genotyped and in some cases were included in the analysis after self-selection for genetic testing years after the preventive surgery. [63]

Of three surgical approaches available, radical removal of both breast and ovaries provides the highest cancer protection but as a consequence of its major invasiveness and complications, it is preferred by few women.

Laparoscopic BSO is associated with low morbidity (4%) and have all the benefits of a minimal-access surgery; the efficacy of BSO to reduce risk of both breast cancer of 50% and ovarian cancer of 90%, the more extensive surgery and increased morbidity with BM and reconstruction, the advantages of laparoscopic BSO and the absence of effective ovarian-cancer screening, strongly convince for the superiority of laparoscopic BSO. [63]

The high efficacy, the minimal invasiveness and non effective surveillance explain an increasing trend towards BSO in recent years, even because intensive screening with ultrasonography and CA-125 in *BRCA* carriers' at early stage has been failed. [51]

But for BSO, the decisions don't end with the surgery, because of most ovarian cancers in women with inherited mutations are high-grade serous cancers and these are infrequently screen in an early stage. [64]

The removal of the ovaries can result in menopause and symptoms that include hot flashes, vaginal dryness and night sweats; hormone replacement therapy (HRT) can help to treat these symptoms but, in some cases, HRT can increase a woman's risk of breast cancer. [61]

The use of HRT after BSO arises some concerns, although HRT-associated adverse effects, observed in general population– increased risk of breast cancer, coronary disease and thromboembolism – were not found among *BRCA* carriers in a recent analysis. [65]

Climacteric symptoms can be handled with local hormonal therapies, such as low-dose vaginal estrogens, which have less systemic absorption, with the goal of short term therapy by low dose of HRT. [61]

Moreover, if someone has had prophylactic mastectomy but doesn't want to remove the ovaries, oral contraceptive use is probably "a good option" for ovarian cancer protection, because oral contraceptive users have substantially reduced risk, 50% lower, for ovarian cancer. [62]

Conclusions

With the advent of direct DNA analysis to detect germline mutations characteristic of a specific cancer linked to genomic mutations, the use of tailoring therapy (prophylactic surgery and minimally invasive target treatment) have assumed major importance, since it can be defined as the pre-emptive operative removal of an organ prior to malignant transformation.

Tailoring minimally invasive gynecological surgery should only be considered if the benefits of removing a gynecological non-cancerous organ from an asymptomatic individual outweigh the risks of the operation, but several conditions need to be considered.

First, there should be either near-complete penetration of the mutation causing a specific malignancy, or a near-certain risk of developing a specific hereditary cancer, as ovarian and breast tumors; second, there must be a reliable means of identifying patients destined to develop the malignancy and this typically involves genetic screening, to determine genetic susceptibility to a particular disease.

Then, there must be a sensitive method for determining whether the patient becomes disease-free and remains predisposed after mini-surgery, with an acceptable low operative morbidity.

The choice of removing organs should be restored as completely as possible, but it is not always possible with prophylactic surgery, since the bilateral ovarian and breast removing lead to a semi organ amputation, with lack of a part of female body.

So the conviction that female organs surgical removal is the best between the bad options is currently passed in clinical practice, thanks to a surgical minimally invasive tailoring approach and we hope that the risk reduction strategies will be less drastic and will not require removal of the high risk organs.

REFERENCES

[1] Miller, LD; Liu, ET. Expression genomics in breast cancer research: microarrays at the crossroads of biology and medicine. *Breast Cancer Res.,* 2007 Mar 26, 9(2), 206.

[2] Min Jou, W; Haegeman, G; Ysebaert, M; Fiers, W. Nucleotide sequence of the gene coding for the bacteriophage MS2 coat protein, *Nature*, 1972 May 12, 237(5350), 82-8.

[3] Fiers, W. et al., Complete nucleotide-sequence of bacteriophage MS2-RNA - primary and secondary structure of replicase gene, *Nature*, 260, 500-507, 1976.

[4] Sanger, F; Air, GM; Barrell, BG; Brown, NL, Coulson, AR; Fiddes, CA; Hutchison, CA, Slocombe, PM; Smith, M. Nucleotide sequence of bacteriophage phi X174 DNA, *Nature*, 1977 Feb 24, 265(5596), 6.

[5] Budd, GE. On the origin and evolution of major morphological characters. *Biol. Rev. Camb. Philos .Soc.,* 2006 Nov, 81(4), 609-28. Epub 2006 Sep 25.

[6] Van Neues, RW; Brown, JD. Distant Segments of Saccharomyces cerevisiae scR1 RNA Promote Assembly and Function of the Signal Recognition Particle. *J. .Mol. .Biol.*, 2007 May 4, 368(3), 677-90. Epub 2007 Feb 22.

[7] Garcia-Sancho, M. Mapping and sequencing information: the social context for the genomics revolution. *Endeavour.*, 2007 Mar, 31(1), 18-23. Epub 2007 Mar, 2.

[8] Sandman, T; Jakobsen, JS; Furlong, EE. ChIP-on-chip protocol for genome-wide analysis of transcription factor binding in Drosophila melanogaster embryos. *Nat. Protoc.*, 2006, 1(6), 2839-55.

[9] Zhu, T; Salmeron, J. High-definition genome profiling for genetic marker discovery. *Trends Plant Sci.,.* 2007 Apr 6.

[10] Jankovic, R; Brankovic-Magic, M; Radulovic, S. The past and the future of cancer genetics. *J. BUON.*, 2004 Oct-Dec, 9(4), 347-54.

[11] Haber, DA; Settleman, J. Cancer: drivers and passengers. *Nature*, 2007 Mar 8, 446(7132), 145-6.

[12] Mittal, V, Nolan, DJ. Genomics and proteomics approaches in understanding tumor angiogenesis. *Expert Rev. Mol. Diagn.,.* 2007 Mar, 7(2), 133-47.

[13] Huang, Y. Pharmacogenetics/genomics of membrane transporters in cancer chemotherapy. *Cancer Metastasis Rev.,.* 2007 Mar, 26(1), 183-201.

[14] Edgren, H; Kallioniemi, O. Integrated breast cancer genomics. *Cancer Cell.*, 2006 Dec, 10(6), 453-4.

[15] Cianfrocca, M; Gradishar, WJ. Controversies in the Therapy of Early Stage Breast Cancer. *Oncologist*, 2005, 10, 766-779.

[16] Holschneider, CH; Berek, JS. Ovarian cancer: epidemiology, biology, and prognostic factors. *Semin. Surg. Oncol..* 2000 Jul-Aug, 19(1), 3-10.

[17] Davidson, B. Anatomic site-related expression of cancer-associated molecules in ovarian carcinoma. *Curr. Cancer Drug Targets*, 2007 Feb, 7(1), 109-20.

[18] Bast RC; Brewer, M; Zou, C; Hernandez, MA; Daley, M; Ozols, R; Lu, K; Lu, Z; Badgwell, D; Mills, GB; Skates, S; Zhang, Z; Chan, D; Lokshin, A; Yu, Y. Prevention and early detection of ovarian cancer: mission impossible? *Recent Results Cancer Res.,.* 2007, 174, 91-100.

[19] Edlich, RF; Winters, KL; Lin, KY. Breast cancer and ovarian cancer genetics. *J. Long Term Eff. Med. Implants,.* 2005, 15(5), 533-45.

[20] Cvelbar, M; Ursic-Vrscaj, M; Rakar, S. Risk factors and prognostic factors in patients with double primary cancer: epithelial ovarian cancer and breast cancer. *Eur. J. Gynaecol. Oncol,.* 2005, 26(1), 59-63
[21] Sogaard, M; Kjaer, SK; Gayther, S. Ovarian cancer and genetic susceptibility in relation to the BRCA1 and BRCA2 genes. Occurrence, clinical importance and intervention. *Acta Obstet Gynecol Scand,.* 2006, 85(1), 93-105.
[22] Risch, H; McLaughlin, J; Cole, D; Rosen, B; et al. Prevalence of germline BRCA1 and BRCA2 mutations in a population series of 649 women with ovarian cancer. *Am. J. Hum. Genet,.* 2001, 68, 700-710.
[23] Anderson, K; Jacobson, JS; Heitjan, DF; Zivin , JG; Hershman, D; Neugut, AI; Grann, VR. Cost-Effectiveness of Preventive Strategies for Women with a BRCA1 or a BRCA2 Mutation. *Ann. Int. Med. March*, 2006, 144 (6), 397-407.
[24] Mok, SC; Elias, KM; Wong, KK; Ho, K, Bonome, T; Birrer, MJ. Biomarker discovery in epithelial ovarian cancer by genomic approaches. *Adv. Cancer Res.,.* 2007, 96, 1-22
[25] Dite, GS; Jenkins, MA; Southey, MC. et al. Familial risks, early-onset breast cancer, and BRCA1 and BRCA2 germline mutations. *J. Natl. Cancer Inst.*, 2003, 95, 448 -57.
[26] Drucker, L; Stackievitz, R; Shpitz, B; Yarkoni, S. Incidence of BRCA1 and BRCA2 mutations in Ashkenazi colorectal cancer patients: preliminary study. *Anticancer Res.*, 2000, 20, 559-561.
[27] Tinelli, A; Leo, G; Vergara, D; Martignago, R; Malvasi, A; Tinelli, R; Marsigliante, S; Maffia, M; Lorusso, V. Endometrial and ovarian cancer: beyond the clinical significance of hormonal carcinogenesis and the promising biomarkers. *J. Chinese Clinical Medicine*, 2007, 2(12), 711-720.
[28] Bermejo, JL; Hemminki, K. Risk of cancer at sites other than the breast in Swedish families eligible for BRCA1 or BRCA2 mutation testing. *Ann. Oncol.*, 2004, 15, 1834-1841.
[29] Shih, H; Nathanson, K; Seal, S; Collins, N; Stratton, M; Rebbeck, T; Weber, B. BRCA1 and BRCA2 mutations in breast cancer families with multiple primary cancers. *Clin. Cancer Res.,* 2000, 6, 4259-64.
[30] Tinelli, A; Vergara, D; Martignago, R; Leo, G; Malvasi, A; Tinelli, R; Marsigliante, S; Maffia, M; Lorusso, V. Ovarian cancer biomarkers: a focus on genomic and proteomic findings. *Current Genomics*, 2007, 8, 335-342.
[31] Semba, S; Yokozaki, H; Yasui, W; Tahara, E. Frequent microsatellite instability and loss of heterozygosity in the region including BRCA1 (17q21) in young patients with gastric cancer. *Int. J. Oncol.* 1998, 12, 1245-1251.
[32] The Breast Cancer Linkage Consortium. Cancer risks in BRCA1 mutation carriers. *J. Natl. Cancer Inst.*, 1999, 91, 1310-1316.
[33] Thompson, D; Easton, DF. Breast Cancer Linkage Consortium. Cancer incidence in BRCA1 mutation carriers. *J. Natl. Cancer Inst.*, 2002, 94, 1358-1365.
[34] Brose, M; Rebbeck, T; Calzone, K. et al. Cancer risk estimates for BRCA1 mutation carriers identified in a cancer risk evaluation program. *J. Natl. Cancer Inst.*, 2002, 94, 1359-1365.
[35] Johannsson, O; Loman, N; Moller, T; Kristoffersson, U; Borg, A; Olsson, H. Incidence of malignant tumors in relatives of BRCA1 and BRCA2 germ-line mutation carriers. *Eur. J. Cancer*, 1999, 35, 1248-1257.

[36] Wainberg, S; Husted, J. Utilization of Screening and Preventive Surgery Among Unaffected Carriers of a BRCA1 or BRCA2 Gene Mutation. *Cancer Epidemiol. Biomarkers Prev.*, Dec 2004, 13(12), 1989-1995.

[37] Draper, L. Breast cancer: trends, risks, treatments, and effects. *AAOHN J.* 2006 Oct, 54(10), 445-51.

[38] Kauff, ND; Satogopan, JM; Robson, ME. et al. Risk-reducing salpingooophorectomy in women with a BRCA1 or BRCA2 mutation. *N. Engl. J. Med.*, 2002, 346, 1609 -15.

[39] Solomon, JS; Brunicardi, CF; Friedman, JD. Evaluation and treatment of BRCA-positive patients. *Plast Reconstr. Surg.*, 1999, 105, 714- 9.

[40] Finch, A; Beiner, M; Lubinsky, J. et al. Salpingo-oophorectomy and the risk of ovarian, fallopian tube, and peritoneal cancers in women with a BRCA1 or BRCA2 Mutation. *JAMA.*, 2006 Jul, 12, 296(2), 185-92.

[41] Loman, N; Bladstrom, A; Johannsson, O; Borg, A; Olsson, H. Cancer incidence in relatives of a population-based set of cases of early-onset breast cancer with a known BRCA1 and BRCA2 mutation status. *Breast Cancer Res.*, 2003, 5, R175 - 86.

[42] Adjuvant Breast Cancer Trials Collaborative Group. Ovarian ablation or suppression in premenopausal early breast cancer: results from the international adjuvant breast cancer ovarian ablation or suppression randomized trial. *J. Natl. Cancer Inst.*, 2007 Apr, 4, 99(7), 516-25.

[43] You, YN; Lakhani, VT; Wells, SA. jr. The role of prophylactic surgery in cancer prevention. *World J. Surg.*, 2007 Mar, 31(3), 450-64.

[44] Frank, TS. The pros and cons of genetic testing for breast and ovarian cancer risk. *Int. J. Fertil.*, 1999, 44, 139 -45.

[45] Tengs, TO; Winer, EP; Paddock, S; Aguilar-Chavez, O; Berry, DA. Testing for the BRCA1 and BRCA2 breast-ovarian cancer susceptibility genes: a decision analysis. *Med. Decis. Making*, 1998, 18, 365-75.

[46] Georgiou, I; Fatouros, M; Arampatzis, I. et al. Evaluating cancer risks in BRCA mutation carriers. *Gastric Breast Cancer*, 2003, 2, 45-51.

[47] Grann, VR; Panageas, KS; Whang, W; Antman, KH; Neugut, AI. Decision analysis of prophylactic mastectomy and oophorectomy in BRCA1-positive or BRCA2-positive patients. *J. Clin. Oncol.*, 1998, 16, 979-85.

[48] Armstrong, K; Schwartz, JS; Randall, T; Rubin, SC; Weber, B. Hormone replacement therapy and life expectancy after prophylactic oophorectomy in women with BRCA1/2 mutations: a decision analysis. *J. Clin. Oncol.*, 2004, 22, 1045-54.

[49] Schrag, D; Kuntz, KM; Garber, JE; Weeks, JC. Decision analysis—effects of prophylactic mastectomy and oophorectomy on life expectancy among women with BRCA1 or BRCA2 mutations. *N. Engl. J. Med.*, 1997, 336, 1465-71.

[50] van Roosmalen, MS; Verhoef, LC; Stalmeier, PF; Hoogerbrugge, N; van Daal, WA. Decision analysis of prophylactic surgery or screening for BRCA1 mutation carriers: a more prominent role for oophorectomy. *J. Clin. Oncol.*, 2002, 20, 2092-100.

[51] Rebbeck, TR; Lynch, HT; Neuhausen, SL; Narod, SA; Van't Veer, L; Garber, JE. et al. Prophylactic oophorectomy in carriers of BRCA1 or BRCA2 mutations. *N. Engl. J. Med.*, 2002, 346, 1616-22.

[52] Hartmann, LC; Sellers, TA; Schaid, DJ; Frank, TS; Soderberg, CL; Sitta, DL. et al. Efficacy of bilateral prophylactic mastectomy in BRCA1 and BRCA2 gene mutation carriers. *J. Natl. Cancer Inst.*, 2001, 93, 1633-7.

[53] King, MC; Marks, JH; Mandell, JB. Breast and ovarian cancer risks due to inherited mutations in BRCA1 and BRCA2. *Science*, 2003, 302, 643-6.

[54] Metcalfe, KA; Lynch, HT; Ghadirian, P; Thing, N. et al. The risk of ovarian cancer after breast cancer in BRCA1 and BRCA2 carriers. *Gynecol. Oncol.*, 2005 Jan, 96(1), 222-6.

[55] Gswind, A; Fischer, OM; Ullrich, A. The discovery of receptors tyrosine kinases: targets for cancer therapy. *Nat. Rev. Cancer*, 2004, 4, 361-70, van de Vijver MJ, He YD, van 't Veer LJ et al. A gene-expression signature as a predictor of survival in breast cancer. *N. Engl. J. Med.*, 2002, 347, 1999-2009.

[56] Kauff, ND; Satagopan, JM; Robson, ME; Schener, L. et al. Risk-reducing salpingo-oophorectomy in women with a BRCA1 or BRCA2 mutation. *N. Engl. J. Med.*, 2002 May, 23, 346(21), 1609-15.

[57] Antoniou, A; Pharoah, PD. et al. Average risks of breast and ovarian cancer associated with BRCA1 or BRCA2 mutations detected in case Series unselected for family history: a combined analysis of 22 studies. *Am. J. Hum. Genet.*, 2003, 72, 1117-30.

[58] Hogg, R; Friedlander, M. Biology of epithelial ovarian cancer: implications for screening women at high genetic risk. *J. Clin. Oncol.*, 2004, 22, 1315-27.

[59] Garber, JE; Hartman, AR. Prophylactic oophorectomy and hormone replacement therapy: protection at what price? *J. Clin. Oncol.*, 2004, 22, 978-80.

[60] Armstrong, K; Sanford Schwartz, J; Randall, T. et al. Hormone replacement therapy and life expectancy after prophylactic oophorectomy in women with BRCA1/2 mutations: A decision analysis. *J. Clin. Oncol.*, 2004, 22, 1045-54.

[61] Writing Group for the Women's Health Initiative Investigators. Risks and benefits of estrogen plus progestin in healthy postmenopausal women: principal results From the Women's Health Initiative randomized controlled trial. *JAMA.*, 2002, 288, 321-33.

[62] Eisinger, F. Prophylactic mastectomy: ethical issues. *Br Med Bull*, 2007 Apr, 4.

[63] Berek, JS; Hacker, NF. eds. Practical gynecologic oncology. 2d ed. Baltimore: *Williams & Wilkins*, 1994, 285-326.

[64] Aarnio, M; Sankila, R; Pukkala, E. et al. Cancer risk in mutation carriers of DNA-mismatch-repair genes. *Int. J. Cancer*, 1999, 81, 214-8.

[65] Dunlop, MG; Farrington, SM; Carothers, AD. et al. Cancer risk associated with germline DNA mismatch repair gene mutations. *Hum. Mol. Genet.*, 1997, 6, 105-10.

In: Encyclopedia of DNA Research (3 Volume Set)
Editors: S. J. Duncan and P. H. Wiley
ISBN 978-1-61324-305-3

Chapter 33

MICROARRAY ANALYSIS OF ALTERED GENE EXPRESSION IN THE KIDNEY OF MICE EXPOSED TO RADIATION

Mitsuru Nenoi
Radiation Effect Mechanisms Research Group,
National Institute of Radiological Sciences, 9-1,
Anagawa-4-chome, Inage-ku,
Chiba 263-8555 Japan

ABSTRACT

The *kidney* is a *radiosensitive organ*, and the radiation-induced nephropathy is a serious problem for cancer patients after clinical radiotherapy. To assess the molecular mechanisms for radiation nephropathy, microarray experiments have been performed in multiple laboratories using RNA from the kidney of mice exposed to radiation. Modulation of a number of genes has been revealed after acute irradiation, including those associated with extracellular matrix synthesis/degradation and regulation of intracellular redox status. However, the risk of radiation nephropathy after chronic exposure is unclear. We previously performed continuous irradiation of C57BL/6J mice with γ-rays for 485 days at dose-rates of 0.032 μGy/min, 0.65 μGy/min, and 13 μGy/min. Gene expression profiles in the kidney from irradiated and unirradiated mice were analyzed, and differentially expressed genes were identified. An evident modulation in expression of the cell cycle-associated genes was not observed in contrast to the case of acute irradiation. However, expression of the genes involved in mitochondrial oxidative phosphorylation was elevated after the long-term irradiation at the dose-rates of 0.65 μGy/min and 13 μGy/min, suggesting that the redox condition in the kidney is likely to be modulated after chronic exposure. Furthermore, it was also found that the genetic response is highly tissue-specific by comparing the gene expression profiles in the kidney and the testis. In this article, analysis of gene expression profile after a long-term irradiation with low-dose-rate γ-rays is reviewed, and the risk of radiation nephropathy after chronic exposure is discussed.

INTRODUCTION

The *kidney* is a *radiosensitive organ*, the tolerance of which is often limiting in clinical radiotherapy. Tubular epithelial cell damage and cell loss after exposure to high doses of radiation, and glomerulosclerosis after low-dose irradiation are predominant histological features of radiation-induced nephropathy [1]. Accumulating data suggest that the pathogenesis of radiation nephropathy involves dynamic interaction between multiple cell types in the kidney [2]. Reactive oxygen species and angiotensin II were suggested to mediate this interaction [3-5]. It has been demonstrated that *in vitro* irradiation of rat mesangial cells and tubular epithelial cells causes altered gene expression, such as those of *Tgfb1* and genes involved in extracellular matrix synthesis and degradation, leading to altered cell function [6-8]. Based on the complexity of mechanisms for radiation nephropathy, the recent development of cDNA and oligonucleotide microarray technology is thought to have advantage over the conventional methods such as Nothern blot in order to better understand the molecular events initiated and sustained in the irradiated kidney. Kruse *et al.* [9] performed a microarray analysis using RNA from the C3H mouse kidney irradiated with 16 Gy of X-rays. Several genes including *TSA-1* and *Jagged 1* were identified as up-regulated genes during the period of developing nephropathy (10 through 30 weeks postirradiation). *TSA-1* was previously reported to be up-regulated in response to proteinuria, and *Jagged 1* is known to play a role in angiogenesis. In addition, Zhao *et al.* [10] have performed microarray experiments using RNA from the kidney of C57BL6/J mice after whole-body irradiation with 10 Gy of γ-rays. They found expression of a large number of genes regulating transcription and translation were changed at 8 h and 24 h after irradiation. Genes regulating intracellular redox status were also observed to be modulated. Interestingly, they observed that the gene expression patterns in kidney and brain was strikingly different, suggesting that the cells respond to radiation in a tissue-specific manner.

The risk of radiation-induced nephropathy after a long-term exposure to chronic radiation remains unresolved. This is an important issue, because it is expected, in near future, that the general public will be able to travel and stay in space for extended periods, and have a chance to be exposed continuously to low-dose-rate cosmic radiation. We recently examined gene expression in the kidney of C57BL/6J mice irradiated with γ-rays at a dose-rate of 0.032-13 μGy/min for 485 days [11]. In this article, a microarray analysis of the gene expression profile after whole-body irradiation with low-dose-rate γ-rays will be reviewed, and the risk of radiation nephropathy after chronic exposure will be discussed.

EXPERIMENTAL DESIGN FOR MICROARRAY ANALYSIS

Male SPF C57BL/6J mice, 7 to 8 weeks of age, were whole-body irradiated with ^{137}Cs γ-rays in the facility of the Institute for Environmental Sciences Japan, for 485 days (22 h/day) at dose-rates of 0.032 μGy/min, 0.65 μGy/min, and 13 μGy/min. The total doses after these dose-rate irradiation were 21mGy, 420mGy, and 8000mGy, respectively. Three mice per each dose-rate conditions including the control (total 12 mice) without organ hypertrophy, neoplasia, and/or hair loss were subjected to the microarray analysis. The number of mice per dose-rate condition (n=3) was chosen, because this is the minimum sample size needed for statistical analysis. Total RNA was extracted from the kidney and testis, and the RNA quality was assessed using an Agilent 2100 Bioanalyzer (Agilent Technologies, Santa Clara, CA). A total of 1.5 μg of biotin-labeled cRNA was hybridized for 18 h with the microarray (Sentrix Mouse-6 v1.0 Expression BeadChips, Illumina, San Diego, CA). The hybridized biotinylated cRNA was detected with Streptavidin-Cy3 (GE, Fairfield, CT) and quantitated using a BeadStation 500GX-WG Systems scanner (Illumina). Sentrix Mouse-6 v1.0 Expression BeadChips is constructed by the small beads coated with 46,000 species of oligonucleotide probes. The beads are so densely arrayed on the surface of silicon substrate that the signal from one gene is taken 30 times on average across the array.

GeneSpring 7.3.1 (Silicon Genetics, Redwood City, CA) was utilized to analyze the microarray data. Gene expression data were normalized in two ways: "per chip normalization" and "per gene normalization". For "per chip normalization", all expression data on an array were normalized to the 50th percentile of all values on that array. For "per gene normalization", the data for a given gene were normalized to the median expression level of that gene across all samples. Quality control of the data was performed as illustrated in Figure 1A. First, a whole sample showing an anomalous distribution of signal intensity was excluded from the analysis. Figure 1B shows that one of the samples in the 0.032 μGy/min group, designated by an asterisk (*), exhibited a distinct distribution of normalized signal intensity compared to those of other samples for an unknown reason. This sample was simply excluded from this study. Next, data containing large statistical errors and/or interindividual differences were excluded from the analysis. For this purpose, we used the "detection score" which is a statistical measure computed based on the Z-value of a gene relative to the Z-value of the negative controls [12]. In our analysis, the genes whose detection score was less than 0.99 either for 2 samples out of 3 in the non-irradiated control group, or for 6 samples out of 8 in the irradiated group, were excluded (one sample in the irradiated group had been removed from the analysis because of an anomalous distribution of signal intensity, as mentioned above). And, the genes for which the standard deviation within each of the groups (non-irradiated, and irradiated at 0.032 μGy/min, 0.65 μGy/min, and 13 μGy/min) exceeded approximately 10% of the mean value were excluded. As a result, 4,061 genes out of 46,000 transcripts contained on the array remained. Figure 1C shows scatter plots of signal intensity in two of the samples from the unirradiated group before the quality control (a) and after removal of outlier data (b). It can be seen that the quality control was appropriately accomplished, and that fluctuations due to statistical errors and/or interindividual differences of the remaining genes were reduced, with the relative ratio being mostly less than 1.3.

ANALYSIS OF PROFOUNDLY MODULATED GENES

Fluctuation in the signal intensity within the same irradiation condition were mostly less than 1.3 as shown in Figure 1C(b). Then, we defined those genes in which the expression level was changed by more than 1.6-fold after irradiation as "profoundly modulated genes". Out of a total of 50 profoundly modulated genes, validation of the microarray data was performed for 6 genes by the real-time PCR or Northern hybridization as shown in Figure 2A. Up-regulation of *Ndufb9*, *Ndufv2*, *Lamp2* and *Commd1* after 0.65 μGy/min and 13 μGy/min could be reproduced, although the fold-change after irradiation was generally small compared to the microarray results. Down-regulation of *Hspa8* after 0.032 μGy/min and *3010027A04Rik* after 13 μGy/min could be also confirmed. It should be noted that *Hspa8* responded to radiation at dose-rate down to only 20-times higher than the typical *level of natural background radiation.* As shown in the Venn diagram in Figure 2C, 5 genes were identified in two different dose-rate conditions: *A630033E08Rik* (0.032 μGy/min and 0.65 μGy/min); and *Lamp2*, *Ndufb9*, *Esd*, and *2410003B16Rik* (0.65 μGy/min and 13 μGy/min). 1, 18, 14 genes were up-regulated, and 4, 3, 15 genes were down-regulated after 0.032μGy/min, 0.65μGy/min, 13 μGy/min, respectively.

By a pathway analysis of the profoundly modulated genes, 4 genes (*Uqcrb*, *Ndufb9*, *Ndufv2*, and *Atp5k*) out of 50 were found to be involved in a common pathway of mitochondrial oxidative phosphorylation. Their expression was up-regulated in a dose-rate-dependent manner. Since these genes function in mitochondrial oxidative phosphorylation [13-18], the activity of mitochondrial respiration was suggested to be increased as the dose-rate becomes higher.

Analysis of Significantly Modulated Genes

Simultaneous modulation of multiple genes involved in a specific physiological process may result in an obvious biological response, even if modulation of each gene is small. Then, we examined the genes whose expression level was modulated by irradiation with the statistical significance regardless of the fold change. 621 of significantly modulated genes were selected out of 4,061 of quality-controled genes by using Welch's ANOVA with a p-value cut off of 0.05. These genes were classified into 16 clusters on the basis of expression patterns after a hierarchical cluster analysis. As shown in Figures 3A and 3B, the genes belonging to the cluster C1 were the most abundant and exhibited increased expression in a dose-rate-dependent manner. In order to examine the biological interpretation, the Gene Ontology categories significantly overlapped with each of the gene clusters were searched. We found that the cluster C1 is significantly similar to the Gene Ontology categories of "mitochondrion", "mitochondrion organization and biogenesis", and "energy pathways" (Figure 3C). Thus, it was suggested again that mitochondrial oxidative phosphorylation was modulated by low-dose-rate irradiation, particularly at 0.65 μGy/min and 13 μGy/min.

Modulation of mitochondrial oxidative phosphorylation after low-dose-rate irradiation may cause altered intracellular redox condition. In this regard, it is noteworthy that the expression of Rhodanese, which is thought to be involved in regulation of antioxidative function, was observed to increase in the liver of mice irradiated with 0.4 Gy at the dose rate of 0.65 μGy/min or 8 Gy at the dose rate of 13 μGy/min [19]. By analyzing the gene expression profile in the kidney of mice after acute irradiation, Zhao *et al.* [10] reported that a number of differentially-expressed genes were associated with regulation of intracellular redox status. It is likely that the redox condition in the kidney is modulated after irradiation with γ-rays in a broad range of the dose-rate. On the basis of a hypothesis that an oxidative stress is involved in pathogenesis of nephropathy, the risk of radiation nephropathy could be suggested also after chronic exposure to radiation.

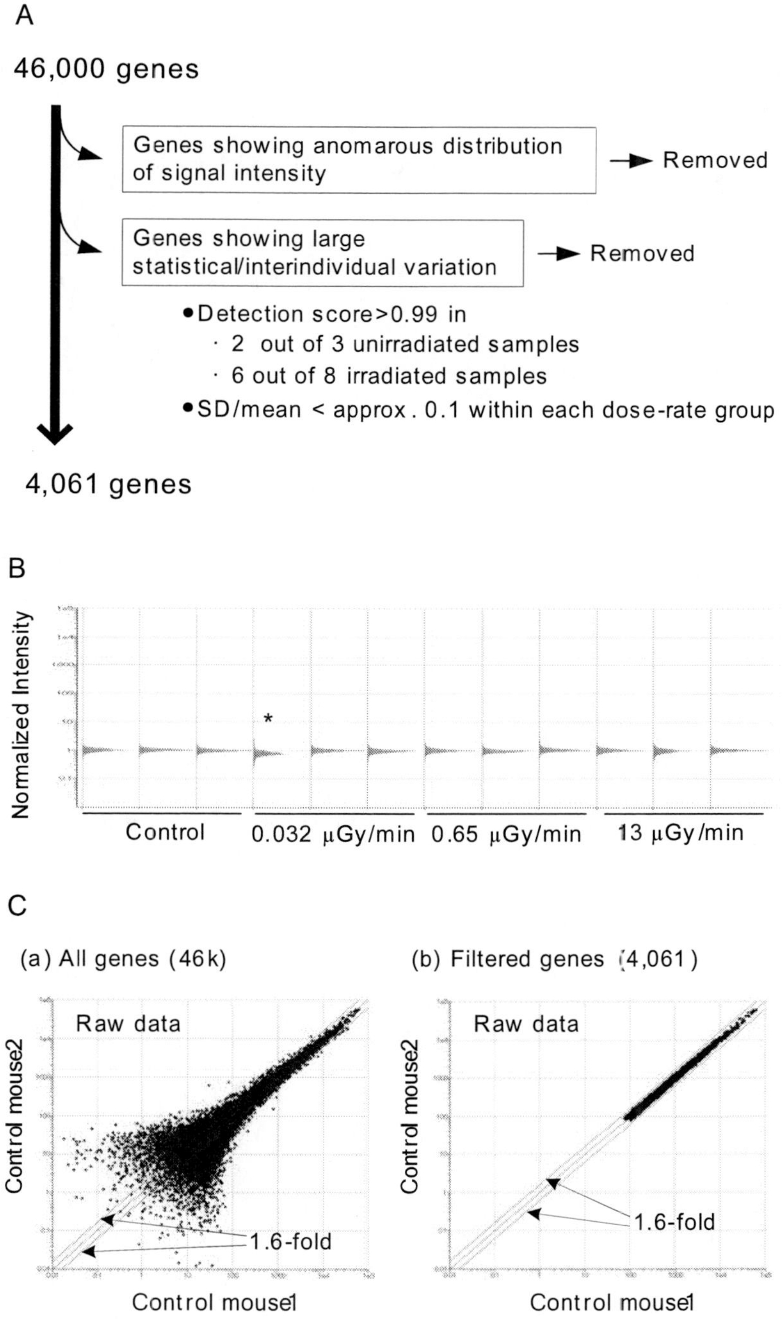

Figure 1. Quality control of microarray data. A: Schematic representation of overall procedure for quality control of the microarray data. B: Distributions of normalized signal intensity from 12 arrays, including 3 replicates from each of 4 irradiation conditions. An asterisk (*) indicates an apparently distinct distribution of signal intensities compared to those of other samples. This sample was simply excluded from subsequent analyses. C: Scatter plots of signal intensities (raw data) between 2 of the 3 control mice are shown for 46000 genes (a), and for the 4061 genes that remained after removal of outlier data (b). Lines representing a 1.6-fold difference are drawn to show that fluctuations were reduced with a relative ratio of mostly less than 1.3 after removal of outlier data. This Figure was reproduced from Ref. [11] with permission by the editorial office, Journal Radiation Research Society, Japan.

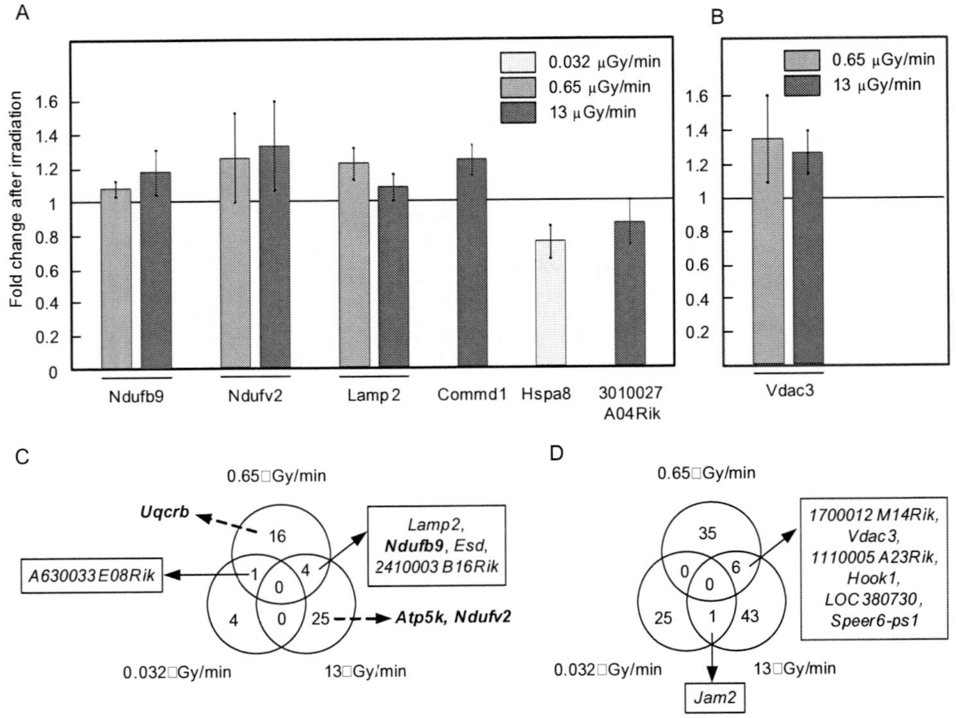

Figure 2. Genes profoundly modulated after low-dose-rate irradiation (fold-change > 1.6). A: Validation of microarray data for the kidney. Up-regulation of Ndufb9, Lamp2 and Commd1, and down-regulation of Hspa8 and 3010027A04Rik was confirmed by the real-time PCR. Up-regulation of Ndufv2 was confirmed by a northern blot analysis. B: Up-regulation of Vdac3 in the testis was confirmed by a real-time PCR. C: The number of profoundly modulated kidney genes is represented by the Venn diagram. The genes identified in two different dose-rate conditions are boxed. Genes that are commonly involved in the mitochondrial oxidative phosphorylation pathway are indicated in boldface. D: The number of profoundly modulated testis genes is represented in the Venn diagram. The genes identified in two different dose-rate conditions are boxed. This Figure was reproduced from Ref. [11] with permission by the editorial office, Journal Radiation Research Society, Japan.

Tissue Specificity

We performed another microarray analysis using RNA from testes to examine the tissue-specificity of alteration in gene expression profile. 110 of profoundly modulated genes were identified, and validation of the microarray data was performed for *Vdac3* by real-time PCR (Figure 2B). As shown in the Venn diagram in Figure 2D, 7 genes were identified in two different conditions: *Jam2* (0.032 μGy/min and 13 μGy/min); *1700012M14Rik, Vdac3, 1110005A23Rik, Hook1, LOC380730, Speer6-ps1* (0.65 μGy/min and 13 μGy/min). 1, 31, 41 genes were up-regulated, and 25, 10, 9 genes were down-regulated after 0.032μGy/min, 0.65μGy/min, 13 μGy/min, respectively.

When profoundly modulated genes were compared between the kidney and testis, we found only 2 genes (*Hspa8* and *1200016E24Rik*) commonly modulated in these organs. In addition, the direction of modulation of *Hspa8* was opposite, that is, down-regulation in the kidney but up-regulation in the testis. It should be concluded that the gene modulation after low-dose-rate irradiation is not similar between kidneys and testes. However, it should be noted that RNA was extracted from the whole organ consisting of heterogeneous cell population. Therefore, modulation of gene expression only in the major cell population was examined in this study. Evaluation of gene expression in a small number of parenchymal cells, that should be most important, remains to be solved.

Alteration of the gene expression after whole-body irradiation of the mouse with 0.2-2 Gy has been examined by Amundson *et al.* [20]. They reported that the gene regulation was highly tissue-specific on the basis of the observation that only few genes were commonly regulated in the thymus, spleen and liver. Alteration in gene expression *in vivo* after irradiation was analyzed in the brain [21], bone marrow [22], and liver [23]. Although the mouse strains and radiation doses utilized in these studies were not identical, it could be suggested that the cellular response to radiation was much different among tissues, and that genetic response to radiation is regulated largely on the epigenetic level.

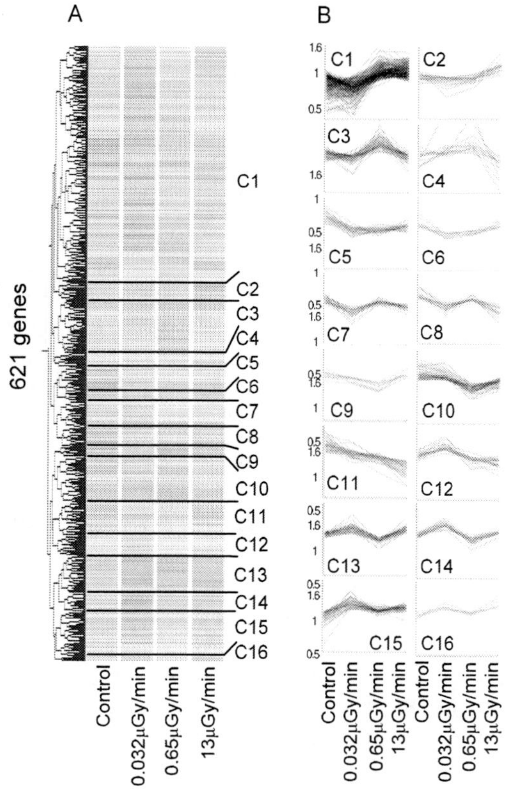

C

	Category	p-Value
C1	cytoplasm	2.8E-07
	mitochondrion	5.7E-04
	mitochondrion organization and biogenesis	0.0013
	energy pathways	0.0055
	organelle organization and biogenesis	0.0071
	cell organization and biogenesis	0.0071
	cytoplasm organization and biogenesis	0.0071
	transferase	0.031
C10	cell organization and biogenesis	1.4E-05
	cytoplasm organization and biogenesis	1.4E-05
	organelle organization and biogenesis	1.4E-05
	cytoplasm	0.0016
	mitochondrion	0.0061
	nucleosome	0.012
	mitochondrion organization and biogenesis	0.032
	GO SLIMS Cellular Component	0.040
C13	cytochrome-c oxidase activity	1.2E-05
	organelle inner membrane	6.5E-05
	mitochondrion organization and biogenesis	3.3E-04
	mitochondrion	3.3E-04
	energy pathways	7.1E-04
	cytoplasm	0.0086

Figure 3. Hierarchical cluster analysis of genes showing a statistically significant change in expression in the kidney after irradiation. A: The expression levels of the 621 significantly modulated genes are indicated by color. Green, yellow, and red represent low, medium, and high expression, respectively. B: Based on the expression pattern, genes were classified into 16 clusters. Expression profiles of genes in each of the 16 clusters are shown. C: Gene Ontology categories that overlapped with gene clusters are shown, with p values from Fisher's exact test. Only clusters C1, C10 and C13 had a statistically

significant overlap with any Gene Ontology categories. This Figure was reproduced from Ref. [11] with permission by the editorial office, Journal Radiation Research Society, Japan.

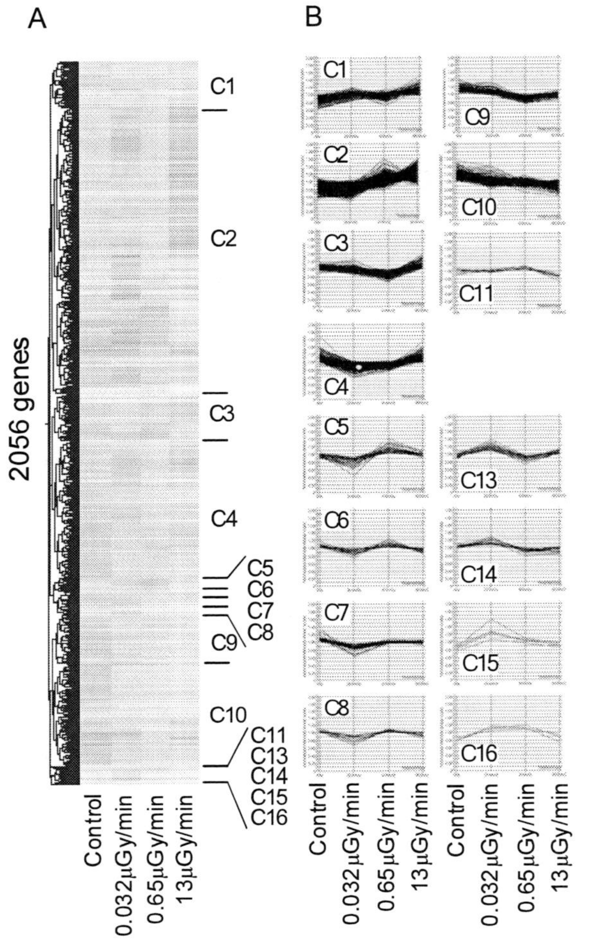

C

	Category	p-Value
C1	response to temperature	0.0013
C2	physiological process	0.0012
	protein biosynthesis	0.0088
	Tricarboxylic acid cycle intermediate metabolism	0.0077
	Chaperone cofactor dependent protein folding	0.0060
	aldehyde metabolism	0.0014
	glutathione metabolism	0.0077
	coenzyme metabolism	0.0058
	regulation of endocytosis	0.0095
	cellular physiological process	0.0003
	cytoskeleton-dependent intracellular transport	0.0049
	microtubule-based movement	0.0046
	organelle organization biogenesis	0.0078
	microtubule-based process	0.0015
	response to heat	0.0095
	response to hormone stimulus	0.0095
C9	DNA metabolism	0.0086
	DNA repair	0.0019
	response to DNA damage	0.0040
	response to endogenous	0.0049
C10	DNA replication	0.0024
	mitotic cell cycle	0.0056

Figure 4. Hierarchical cluster analysis of genes whose expression was significantly modulated in the testis after irradiation. A: Expression levels of the 2056 significantly modulated genes are indicated by color. Based on the expression pattern, genes were classified into 16 clusters. The cluster 12 contained no genes. B: Expression profiles of genes in each of the 16 clusters are shown together. C: The Gene Ontology categories that significantly overlapped with the gene clusters C1, C2, C9 and C10 are shown. This Figure was reproduced from Ref. [11] with permission by the editorial office, Journal Radiation

The hierarchical cluster analysis was performed for 2,056 of significantly modulated testis genes. Figure 4A and 4B show that the expression of genes belonging to the cluster C1 and C2 was increased with dose-rates. Gene Ontology categories significantly similar to these clusters were listed in Figure 4C. Mitochondrion-related categories were not found to be similar to the cluster C1 and C2. When the Gene Ontology categories significantly similar to the cluster C1 and C2 were compared with those similar to the cluster C1 of the kidney (Figure 3C), only one category "organelle organization and biogenesis" was commonly found. To sum up, it was confirmed that alteration of gene expression profile after low-dose-rate irradiation was different between kidneys and testes. The clusters C9 and C10 which contained genes whose expression was decreased in the testis with increase in dose-rate were significantly similar to the Gene Ontology categories "DNA metabolism", "response to DNA damage" and "DNA replication" (Figure 4). It appears that cells in testes responded to low-dose-rate radiation as if they were preparing for emergency by shutting down the general metabolisms as well as DNA repair activity.

DOSE-RATE EFFECTS

The technology to determine the gene expression changes using microarray analysis is an effective approach to understanding the complex nature of biological responses to environmental stresses. Zhao *et al.* [10] observed several cell cycle-related genes up-regulated in the kidney after acute whole-body irradiation of mice with 10 Gy of γ-rays, while we did not observed those genes profoundly modulated in the kidney after chronic irradiation even with the total dose of 8 Gy. We neither observed Gene Ontology categories of "cell cycle" which was similar to any gene clusters of significantly modulated genes in the kidney (Figure 3C). This result may suggest that modulation of the cell cycle-related genes following exposure to radiation is highly dose-rate dependent *in vivo*. The dose-rate effect has been assessed by Sokolov *et al.* [24] who observed that approximately one-third of genes differentially expressed following exposure to 1 Gy of γ-rays at the dose rate of 1 Gy/min or 0.045 Gy/min were different. Amundson *et al.* [25] investigated the dose-rate effect using human myeloid leukemia cells (ML-1) irradiated with γ-rays in the dose-rate range of 0.008-2.9 Gy/min. It was indicated that there exist two classes of low-dose-rate responders; one was the group of genes induced in a dose-rate dependent fashion, and the other was those with dose-rate independent induction. Apoptosis-related genes were preponderantly included in the dose-rate dependent gene group, and the majority of genes in the dose-rate independent gene group were involved in cell cycle regulation. These reports consistently indicate that the modulation of gene expression is dose-rate dependent

CONCLUSION

Previous reports from numerous laboratories including ours consistently suggested that the gene expression is differentially modulated after exposure to radiation in tissue-dependent and dose-rate-dependent manner. However, we found that genes regulating intracellular redox condition were modulated not only after acute irradiation but also after low-dose-rate irradiation, suggesting that the redox status in the kidney is modulated after irradiation irrespective of the dose-rate. Assuming that an oxidative stress plays an important role in pathogenesis of nephropathy, the risk of radiation nephropathy could be suggested also after chronic exposure to radiation.

ACKNOWLEDGMENTS

This study was supported in part by the Budget for Nuclear Research of the Ministry of Education, Culture, Sports, Science and Technology of Japan, based on screening by, and in consultation with, the Atomic Energy Commission of Japan. This study was performed in part under contract with Aomori Prefectural Government, Japan.

REFERENCES

[1] Robbins, ME; Wooldridge, MJ; Jaenke, RS; Whitehouse, E; Golding, SJ; Rezvani, M & Hopewell, JW. A morphological study of radiation nephropathy in the pig. *Radiat. Res.*, 1991, 126, 317-327.

[2] Jaenke, RS; Robbins, ME; Bywaters, T; Whitehouse, E; Rezvani, M & Hopewell, JW. Capillary endothelium. Target site of renal radiation injury. *Lab. Invest.*, 1993, 68, 396-405.

[3] Robbins, ME; Zhao, W; Davis, CS; Toyokuni, S; Bonsib, SM. Radiation-induced kidney injury: a role for chronic oxidative stress? *Micron*, 2002, 33, 133-141.

[4] Robbins, ME; Zhao; W. Chronic oxidative stress and radiation-induced late normal tissue injury: a review. *Int. J. Radiat. Biol.*, 2004, 80, 251-259.

[5] Moulder, JE; Fish, BL; Cohen, EP; Bonsib, SM. Angiotensin II receptor antagonists in the prevention of radiation nephropathy. *Radiat. Res.*, 1996, 146, 106-110.

[6] Zhao, W; O'Malley, Y; Robbins, ME; Irradiation of rat mesangial cells alters the expression of gene products associated with the development of renal fibrosis. *Radiat. Res.*, 1999, 152, 160-169.

[7] Wang, J; Robbins, ME. Radiation-induced alteration of rat mesangial cell transforming growth factor-beta and expression of the genes associated with the extracellular matrix. *Radiat. Res.*, 1996, 146, 561-568.

[8] Zhao, W; O'Malley, Y; Wei, S; Robbins, ME. Irradiation of rat tubule epithelial cells alters the expression of gene products associated with the synthesis and degradation of extracellular matrix. *Int. J. Radiat. Biol.*, 2000, 76, 391-402.

[9] Kruse, JJ; te Poele, JA; Velds, A; Kerkhoven, RM; Boersma, LJ; Russell, NS; Stewart, FA. Identification of differentially expressed genes in mouse kidney after irradiation using microarray analysis. *Radiat. Res*., 2004, 161, 28-38.

[10] Zhao, W; Chuang, EY; Mishra, M; Awwad, R; Bisht, K; Sun, L; Nguyen, P; Pennington, JD; Wang, TJ; Bradbury, CM; Huang, L; Chen, Z; Bar-Sela, G; Robbins, ME; Gius, D. Distinct effects of ionizing radiation on in vivo murine kidney and brain normal tissue gene expression. *Clin. Cancer Res.*, 2006, 12, 3823-3830.

[11] Taki, K; Wang, B; Nakajima, T; Wu, J; Ono, T; Uehara, T; Matsumoto, T; Oghiso, Y; Tanaka, K; Ichinohe, K; Nakamura, S; Tanaka, S; Magae, J; Kakimoto, A; Nenoi, M. Microarray analysis of differentially expressed genes in the kidneys and testes of mice after long-term irradiation with low-dose-rate γ-rays. *J. Radiat. Res.*, 2009, in press.

[12] Lenk, GM; Tromp, G; Weinsheimer, S; Gatalica, Z; Berguer, R; Kuivaniemi, H. Whole genome expression profiling reveals a significant role for immune function in human abdominal aortic aneurysms. *BMC Genomics,* 2007, 8, 237-248.

[13] Haut, S; Brivet, M; Touati, G; Rustin, P; Lebon, S; Garcia-Cazorla, A; Saudubray, JM; Boutron, A; Legrand, A; Slama, A. A deletion in the human QP-C gene causes a complex III deficiency resulting in hypoglycaemia and lactic acidosis. *Hum. Genet.*, 2003, 113, 118-122.

[14] Malaney, S; Heng, HH; Tsui, LC; Shi, XM; Robinson, BH. Localization of the human gene encoding the 13.3-kDa subunit of mitochondrial complex III (UQCRB) to 8q22 by in situ hybridization. *Cytogenet. Cell Genet.,* 1996, 73, 297-299.

[15] Lin, X; Wells, DE; Kimberling, WJ; Kumar, S. Human NDUFB9 gene: genomic organization and a possible candidate gene associated with deafness disorder mapped to chromosome 8q13. *Hum. Hered.,* 1999, 49, 75-80.

[16] Bénit, P; Beugnot, R; Chretien, D; Giurgea, I; De Lonlay-Debeney, P; Issartel, JP; Corral-Debrinski, M; Kerscher, S; Rustin, P; Rötig, A; Munnich, A. Mutant NDUFV2 subunit of mitochondrial complex I causes early onset hypertrophic cardiomyopathy and encephalopathy. *Hum. Mutat.,* 2003, 21, 582-586.

[17] Hattori, N; Yoshino, H; Tanaka, M; Suzuki, H; Mizuno, Y. Genotype in the 24-kDa subunit gene (NDUFV2) of mitochondrial complex I and susceptibility to Parkinson disease. *Genomics,* 1998, 49, 52-58.

[18] Dunnick, J; Blackshear, P; Kissling, G; Cunningham, M; Parker, J; Nyska, A. Critical pathways in heart function: bis(2-chloroethoxy)methane-induced heart gene transcript change in F344 rats. *Toxicol. Pathol.,* 34, 348-356 (2006).

[19] Nakajima, T, Taki, K, Wang, B, Ono, T, Matsumoto, T, Oghiso, Y, Tanaka, K, Ichinohe, K; Nakamura, S; Tanaka, S; Nenoi, M. Induction of rhodanese, a detoxification enzyme, in livers from mice after long-term irradiation with low-dose-rate gamma-rays. *J. Radiat. Res.,* 2009, 49, 661-666.

[20] Amundson, SA; Bittner, M; Meltzer, P; Trent, J; Fornace, AJ. Jr. Induction of gene expression as a monitor of exposure to ionizing radiation. *Radiat. Res.,* 2001, 156(5 Pt 2), 657-661.

[21] Mahmoud-Ahmed, AS; Atkinson, S; Wong, CS. Early gene expression profile in mouse brain after exposure to ionizing radiation. *Radiat. Res.,* 2006, 165, 142-154.

[22] Dai, JM; Sun, DC; Lin, RX; Yang, J; Lou, S; Wang, SQ; Microarray analysis of differentially expressed genes in mouse bone marrow tissues after ionizing radiation. *Int. J. Radiat. Biol.,* 2006, 82, 511-521.

[23] Roudkenar, MH; Li, L; Baba, T; Kuwahara, Y; Nakagawa, H; Wang, L; Kasaoka, S;Ohkubo, Y;Ono, K; Fukumoto, M. Gene expression profiles in mouse liver cells after exposure to different types of radiation. *J. Radiat. Res.,* 2008, 49, 29-40.

[24] Sokolov, M; Panyutin, IG; Neumann, R. Genome-wide gene expression changes in normal human fibroblasts in response to low-LET gamma-radiation and high-LET-like 125IUdR exposures. *Radiat. Prot. Dosimetry,* 2006, 122, 195-201.

[25] Amundson, SA; Lee, RA; Koch-Paiz, CA; Bittner, ML; Meltzer, P; Trent, JM; Fornace AJ. Jr. Differential responses of stress genes to low dose-rate gamma irradiation. *Mol. Cancer Res.,* 2003, 1, 445-452.

In: Encyclopedia of DNA Research (3 Volume Set)
Editors: S. J. Duncan and P. H. Wiley
ISBN 978-1-61324-305-3

Chapter 34

METHODS OF DATA ANALYSIS APPLIED TO SEQUENTIALLY ACCUMULATED GENE SIGNATURES

Shun Ikeda[1], Rui Ose[1, 2] and Hisashi Koga[1,*]
[1] Laboratory of Medical Genomics, Department of Human Genome Research, Kazusa DNA Research Institute, 2-6-7 Kazusa-Kamatari, Kisarazu, Chiba 292-0818, Japan
[2] Graduate School of Pharmaceutical Sciences & Faculty of Pharmaceutical Sciences, Chiba University, 1-8-1, Inohana, Chuo-ku, Chiba 260-8675, Japan

ABSTRACT

The DNA microarray is one of the standard assays in modern science. One can obtain tens of thousands of DNA microarray data sets freely through the Internet without performing exact hybridization experiments; however, several different methods of data analysis were used to gather this data, as no standard method has been established. We should therefore focus our effort on data analysis rather than data acquisition. It is promising that important evidence may be extracted from certain sequentially accumulated microarray data. Based on this possibility, we applied three different cluster analyzing methods (UPGMA, SOM, and GSEA) against the sequentially accumulated data from cytokine-treated culture cells and compared the results. In this chapter we briefly review the advantages of each method and look at the usefulness of further projections of the results in pathway databases such as KEGG.

Keywords: *EMT / UPGMA / SOM / GEDI / GSEA / KEGG*

ABBREVIATIONS

ANOVA	analysis of variance;
UPGMA	Unweighted Pair-group Method using Arithmetic Average;

* Corresponding Author: Tel: +81 438 52 3505; Fax: +81 438 52 3914; E-mail: hkoga@kazusa.or.jp.

SOM	Self-Organizing Map;
GSEA	Gene Set Enrichment Analysis;
KEGG	Kyoto Encyclopedia of Genes and Genomes;
EMT	epithelial-mesenchymal transitions;
RPE,	human retinal pigment epithelial;
GEDI	Gene Expression Dynamics Inspector;
MSigDB	Molecular Signature Database

INTRODUCTION

Sequentially accumulated gene signatures of cells after a particular type of stimulation provide us with a tremendous amount of information about the cell responses attributed to the initial transcriptional changes. However, this exploration is part of a new wave of systematic studies in cell biology, and unresolved problems regarding the data analysis remain. We compared the two most recent data analysis methods (SOM and GSEA) with conventional cluster analysis (UPGMA). For this comparison, we accumulated gene signatures of EMT-associated fibrosis and highlighted each analytical characteristic.

EMT is a fundamental process governing morphogenesis in multicellular organisms that is reactivated in a variety of diseases, including fibrosis and cancer metastasis. After discovery of the involvement of the Snail/Slug and ZEB-1/SIP1 families in EMT, the molecular mechanisms were rapidly elucidated by experiments designed around these families as hubs (1-3). Our understanding of EMT has grown profoundly, but no information is available to interpret a sequential molecular mechanism underlying it. To address this challenge, we employed an established EMT-associated fibrosis model induced by cytokine stimulation on human RPE cells and comprehensively analyzed the mRNA expression by DNA microarray (Takahashi E et. al., in press).

MATERIALS AND METHODS

Cell culture and cytokine treatment. Human RPE cell line ARPE-19 was obtained from American Type Culture Collection and maintained in Dulbecco's modified Eagle's medium–nutrient mixture F-12 (Sigma) supplemented with 10% fetal bovine serum. A subconfluent monolayer cell culture was treated with both TGF-β2 (5 ng/ml) and TNF-α (10 ng/ml), and the cells were harvested at each time point (0, 1, 6, 24, 42, 60 hours) after the stimuli.

RNA extraction, microarray analysis, and data manipulation. Total RNA was extracted with the use of Trizol reagent (Invitrogen, Carlsbad, CA, USA) according to the manufacturer's instructions. Biotin-labeled cRNA probes were hybridized to a GeneChip® Human Genome U133 Plus 2.0 Array (Affymetrix, Santa Clara, CA, USA) and then visualized by streptavidin-phycoerythrin. We followed the detailed experimental procedures provided by the manufacturer when using the GeneChip®. Raw intensity data for each experiment was first normalized by the MAS5.0 Statistical algorithm (Affymetrix). The data for each experiment was log2 transformed and then used for the calculation of Z scores. Z scores were calculated by subtracting the overall average gene intensity (within a single experiment) from the raw intensity data for each gene, and dividing that result by the SD of all of the measured intensities, according to the previously established formula (4). Further statistical analyses were performed by ANOVA (5). All the raw data described in this study were deposited into the Gene Expression Omnibus (http://www.ncbi.nlm.nih.gov/geo/) under the accession number GSE12548.

Result and Discussion

To comprehensively identify the gene expression signatures, we applied DNA microarrays to a human RPE cell line (ARPE-19). We subjected the ARPE-19 cell line to a combination treatment of TGFβ and TNFα, which induced phenotypic changes *in vitro* that mimic the EMT, and then analyzed the sequential changes of over 47,000 transcripts. The data from 8278 probes that met our requirements (ANOVA p-value<0.0001, RefDIC rank = A (6)) were extracted, and we drew the heat map and the dendrogram of a hierarchical clustering (Figure 1). Exact clustering was performed with the commercially available software DecisionSite^{TM} (Spotfire, Inc., Somerville, MA) using the default conditions (Similarity measurement, Euclidean distance; Ordering function, Average value). For the first analysis, we used the UPGMA method for the hierarchical clustering. Although we could have applied other methods such as Ward's method (7), we selected UPGMA for this study because this algorithm gives the most common hierarchical clustering for DNA microarray analysis (8). Using hierarchical clustering, it is easy to understand the relationship between the genes, and this is a suitable analysis for a study with an emphasis on speed and convenience. However, since this type of analysis leans heavily on degree of similarity (or dissimilarity) between the individual gene-expression intensities, the comparison between grouped genes may have different analytical results from those of other comparisons. Furthermore, the hierarchical clustering is an agglomerative (bottom-up) approach and the clustering process starts with each gene as an individual cluster. These clusters are then successively merged to form new, larger clusters until all of the genes are in one big cluster. Therefore, it is difficult to clarify any hidden correlations between initially differentially clustered genes. To address these difficulties, we applied a non-hierarchical clustering method to our sequentially accumulated DNA microarray data for the next analysis, in which we used the SOM algorithm. Although the *K*-means algorithm is a representative algorithm for non-hierarchical clustering, we used SOM in this study because our laboratory has a close relationship with Dr. T Ikemura's laboratory, which established and reinforced this algorithm for use in genome research (9-11). The same data from 8278 probes that met our requirements were used, and the clustering was performed on the free GEDI software using the default conditions (http://www.childrenshospital. org/research/ ingber/ GEDI/gedi-home.htm) (Figure 2A) (12, 13). The non-hierarchical clustering using SOM requires pattern recognition of thousands of gene expressions rather than relying on the strength of each gene expression, and SOM is thus suitable for dynamic analysis such as the comparison of expression profile time evolution in response to a certain stimulus such as a series of drugs. Furthermore, the SOM algorithm is more robust against noise, thereby enabling us to extract gene groups with similar expression patterns at a high level of precision. We show representative gene groups with similar expression patterns against cytokine TGFB1 and proto-oncogene MET (Figure 2B and Table 1).

Table 1. Genes belonging to the same gene group with TGFB1 and MET

Official Symbol	Name	Official Symbol	Name
TGFB1	transforming growth factor, beta 1	MET	met proto-oncogene
WIPI1	WD repeat domain, phosphoinositide interacting 1	PCM1	pericentriolar material 1
IFI6	interferon, alpha-inducible protein 6	RHOBTB3	Rho-related BTB domain containing 3
BNIP1	BCL2/adenovirus E1B 19kDa interacting protein 1	ZNF185	zinc finger protein 185
TULP3	tubby like protein 3	STAU2	staufen, RNA binding protein, homolog 2
YAF2	YY1 associated factor 2	F3	coagulation factor III
CD70	CD70 molecule	MITF	microphthalmia-associated transcription factor
PPP3CC	protein phosphatase 3 (formerly 2B), catalytic subunit	EZR	ezrin
TAPBP	TAP binding protein	NR2F2	nuclear receptor subfamily 2, group F, member 2
CYB5A	cytochrome b5 type A	DENND3	DENN/MADD domain containing 3
HTRA2	HtrA serine peptidase 2	FNBP1L	formin binding protein 1-like
MED8	mediator complex subunit 8	TOX3	TOX high mobility group box family member 3
CD47	CD47 molecule	OSBPL10	oxysterol binding protein-like 10
FXYD5	FXYD domain containing ion transport regulator 5	NPR3	natriuretic peptide receptor C/guanylate cyclase C
PDPN	podoplanin	PIGP	phosphatidylinositol glycan anchor biosynthesis, class P
IL27RA	interleukin 27 receptor, alpha	AGGF1	angiogenic factor with G patch and FHA domains 1
SRPRB	signal recognition particle receptor, B subunit	TMEFF2	transmembrane protein with EGF-like and two follistatin-like domains 2
TINF2	TERF1 (TRF1)-interacting nuclear factor 2	DIRC2	disrupted in renal carcinoma 2
RNASEK	RNASEK ribonuclease, RNase K	IER5L	immediate early response 5-like
PATL1	protein associated with topoisomerase II homolog 1	DTNA	dystrobrevin, alpha
FAM101A	family with sequence similarity 101, member A	C10orf104	chromosome 10 open reading frame 104
IL4I1	interleukin 4 induced 1	NRXN3	neurexin 3
TAF4B	TAF4b RNA polymerase II, TBP-associated factor, 105kDa	SERAC1	serine active site containing 1
		RGS7BP	regulator of G-protein signaling 7 binding protein
		PRR16	proline rich 16

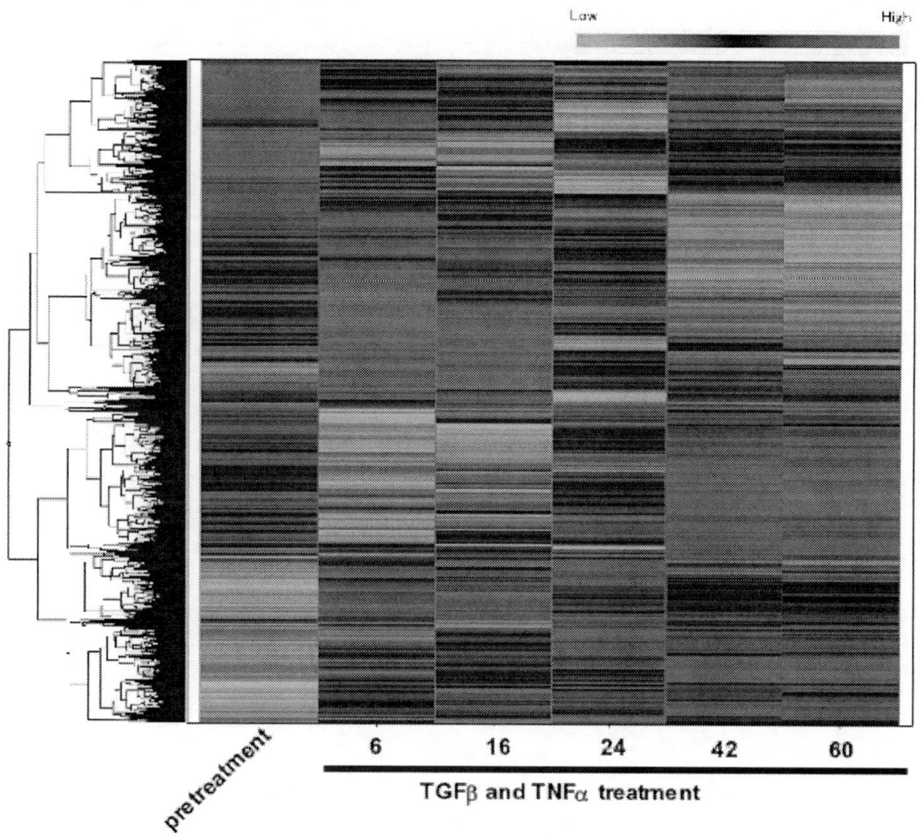

Figure 1. A hierarchical clustering of 8278 probes in response to combination treatment of TGFβ and TNFα was constructed from the log2-converted intensities by the UPGMA method. Color changes from green to red correspond to increases of expression. Each row indicates the average of three independent experiments at each time point.

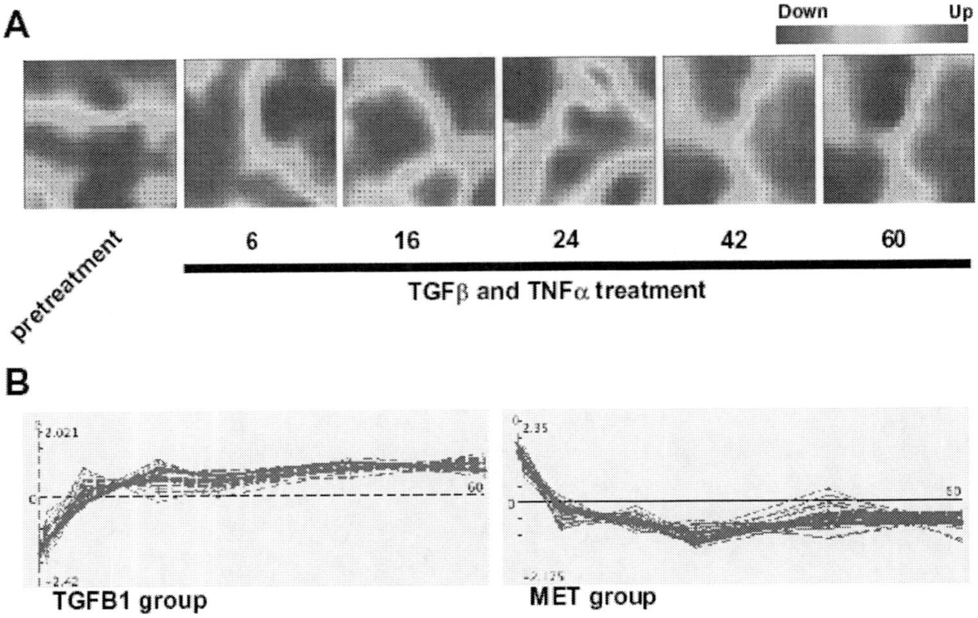

Figure 2. A, GEDI maps of 8278 probes in response to combination treatment of TGF□ and TNF□ were constructed using the SOM algorithm. Color changes from blue to red correspond to down- and up-regulation against the overall average. B, Average temporal profiles of 2 groups obtained using SOM clustering. The X-axis is a temporal axis, and the Y-axis indicates the Z scores calculated from log2-converted averages from three independent experiments at each time point.

For the last analysis, we applied the most recently generated method, GSEA (http://www.broad.mit.edu/gsea/) (14). GSEA makes a strong showing in the comparison between two independently derived gene expression data sets (possibly obtained with different platforms). Furthermore, GSEA is a powerful analytical method when comparing gene sets on the same biological pathway that can be simultaneously extracted. On the other hand, GSEA seems relatively weak for dynamic analysis using sequentially accumulated DNA microarray data. However, the advantage of GSEA for analyzing gene clustering based on an enrichment score (ES) is worthy of special mention. We applied GSEA to the comparison of data gathered at pretreatment and at each time point. Following the original instructions for using GSEA, we calculated the false discovery rate (FDR) (15, 16) and excluded data with FDR > 0.25 as false positives. We were able to extract several up- and down-regulated cellular pathways from the comparison between pretreatment and treatment (after 24 hours) (Figure 3 and Table 2). Interestingly, down-regulated genes were significantly (FDR = 0.23863675) enriched in the "MELANOMA" pathway, including the CDH1 gene (also known as E-cadherin; ECAD in Figure 4). Since the down-regulation of CDH1 is a hallmark of EMT, like the up-regulation of vimentin and α-SMA (17), we assume that the biological mechanisms behind EMT are faithfully reflected on the clustering by GSEA. The "CYTOKINE_CYTOKINE_RECEPTOR_ INTERACTION" pathway is listed on the top of the up-regulated gene set, which further increases confidence in this method. To facilitate the use of GSEA, the analyzed results are linked with the freely available MSigDB software (http://www.broad.mit.edu/gsea/msigdb/ index.jsp) in which the user can easily access the widely used pathway database KEGG (http://www.genome.jp/kegg/) (18) (Table 2, Figure 4). This kind of functional coordination of available knowledge may provide novel insight into the interpretation of DNA microarray results. Although we used gene sets derived from KEGG in this experiment, it is also possible to use any kind of gene set. One can generate customized gene sets according to other clustering methods and utilize GSEA for the validation.

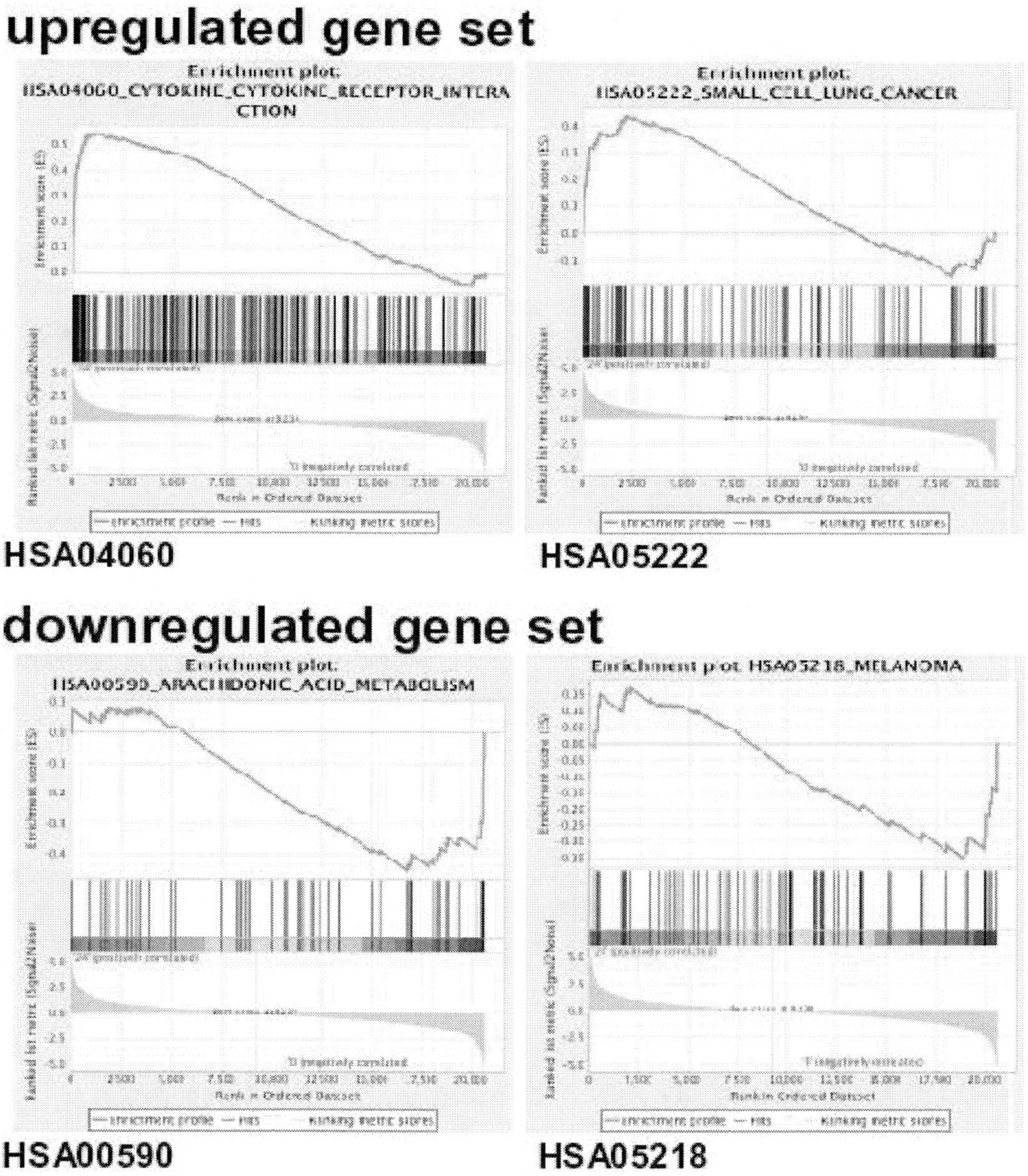

Figure 3. A, Representative results of GSEA. In the top of the panels, the X-axis indicates a ranked list of genes and the Y-axis indicates the enrichment score (ES). The ES is the maximum deviation from zero encountered in the random walk; it corresponds to a weighted Kolmogorov–Smirnov-like statistic. The middle panels show where the members of the gene set appear in the ranked list. Names of the gene set according to MSigDB are also indicated at the bottom of each panel (these are identical to that in Table 2).

Table 2. Gene sets under the influence of a combination treatment of TGFβ and TNFα.

Standard name in MSigDB	FDR q-val	core enrichment/ totall gene number	link to KEGG
upregulated gene set			
HSA04060_CYTOKINE_CYTOKINE_RECEPTOR_INTERACTION	0.23280188	51/244	http://www.kegg.jp/dbget-bin/show_pathway?HSA04060
HSA05222_SMALL_CELL_LUNG_CANCER	0.23685357	24/87	http://www.kegg.jp/dbget-bin/show_pathway?HSA05222
HSA05219_BLADDER_CANCER	0.2452125	8/41	http://www.kegg.jp/dbget-bin/show_pathway?HSA05219
HSA00790_FOLATE_BIOSYNTHESIS	0.24768238	7/39	http://www.kegg.jp/dbget-bin/show_pathway?HSA00790
HSA05212_PANCREATIC_CANCER	0.24875754	23/72	http://www.kegg.jp/dbget-bin/show_pathway?HSA05212
downregulated gene set			
HSA00590_ARACHIDONIC_ACID_METABOLISM	0.19128749	19/52	http://www.kegg.jp/dbget-bin/show_pathway?HSA00590
HSA05218_MELANOMA	0.23863675	10/71	http://www.kegg.jp/dbget-bin/show_pathway?HSA05218

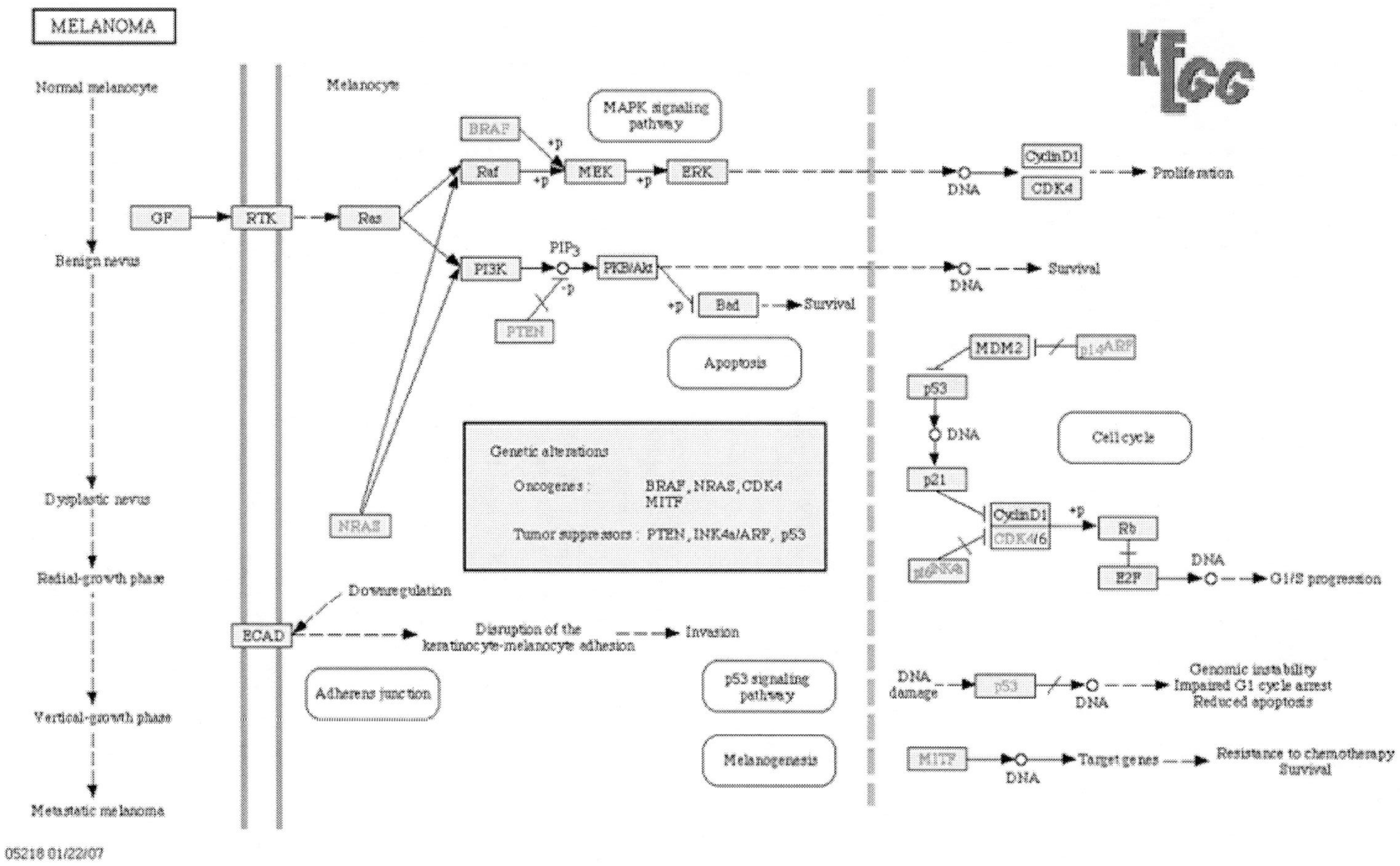

Figure 4. The KEGG pathway map corresponding to the HSA05218 gene set in Figure 3. The rectangles indicate the molecules in this melanoma pathway.

Concluding Remarks

In this short communication, we introduced three different clustering methods and applied these methods to the same DNA microarray data that we had sequentially accumulated. We cannot state which method is the most efficient, since we did not examine the performance of these methods. Nevertheless, we did observe that each method has characteristic analytical advantages, so one must select the most suitable method for one's own analytical purpose or combine multiple methods to extract genes with real biological meaning. Development and refinement of analytical methods are constantly advancing this field of study, and experimental biologists will be well served by paying attention to the trends. We should also keep firmly in mind that the collaboration of experimental biologists is essential for the validation and the reinforcement of newly established analytical methods.

Acknowledgments

We thank Dr. Hideyuki Saya (Keio University School of Medicine) for the material providing and for his helpful discussions of EMT. This study was supported by a grant from the Genome Network Project of the Ministry of Education, Culture, Sports, Science and Technology, Japan. This study was also supported by grants from the Kazusa DNA Research Institute.

References

[1] Barrallo-Gimeno, A; Nieto, MA. The Snail genes as inducers of cell movement and survival: implications in development and cancer. *Development*, 2005, 132 (14), 3151-61.

[2] Peinado, H; Olmeda, D; Cano, A. Snail, Zeb and bHLH factors in tumour progression: an alliance against the epithelial phenotype? *Nat Rev Cancer*, 2007, 7(6), 415-28.

[3] Zhou, BP; Hung, MC. Wnt, hedgehog and snail: sister pathways that control by GSK-3beta and beta-Trcp in the regulation of metastasis. *Cell Cycle*, 2005, 4(6), 772-6.

[4] Cheadle, C; Vawter, MP; Freed, WJ; Becker, KG. Analysis of microarray data using Z score transformation. *J. Mol. Diagn.*, 2003, 5(2), 73-81.

[5] Kerr, MK; Martin, M; Churchill, GA. Analysis of variance for gene expression microarray data. *J. Comput. Biol.*, 2000, 7(6), 819-37.

[6] Hijikata, A; Kitamura, H; Kimura, Y; et al. Construction of an open-access database that integrates cross-reference information from the transcriptome and proteome of immune cells. *Bioinformatics*, 2007, 23(21), 2934-41.

[7] On, SL; Dorrell, N; Petersen, L; et al. Numerical analysis of DNA microarray data of Campylobacter jejuni strains correlated with survival, cytolethal distending toxin and haemolysin analyses. Int *J Med Microbiol*, 2006, 296(6), 353-63.

[8] Datta, S. Comparisons and validation of statistical clustering techniques for microarray gene expression data. *Bioinformatics*, 2003, 19(4), 459-66.

[9] Abe, T; Kanaya, S; Kinouchi, M; Ichiba, Y; Kozuki, T; Ikemura, T. A novel bioinformatic strategy for unveiling hidden genome signatures of eukaryotes: self-organizing map of oligonucleotide frequency. *Genome Inform.*, 2002, 13, 12-20.

[10] Abe, T; Kanaya, S; Kinouchi, M; Ichiba, Y; Kozuki, T; Ikemura, T. Informatics for unveiling hidden genome signatures. *Genome Res.*, 2003, 13(4), 693-702.

[11] Kanaya, S; Kinouchi, M; Abe, T; et al. Analysis of codon usage diversity of bacterial genes with a self-organizing map (SOM): characterization of horizontally transferred genes with emphasis on the E. coli O157 genome. *Gene*, 2001, 276(1-2), 89-99.

[12] Guo, Y; Eichler, GS; Feng, Y; Ingber, DE; Huang, S. Towards a holistic, yet gene-centered analysis of gene expression profiles: a case study of human lung cancers. *J Biomed. Biotechnol.*, 2006, 2006(5), 69141.

[13] Eichler, GS; Huang, S; Ingber, DE. Gene Expression Dynamics Inspector (GEDI): for integrative analysis of expression profiles. *Bioinformatics*, 2003, 19(17), 2321-2.

[14] Subramanian, A; Tamayo, P; Mootha, VK; et al. Gene set enrichment analysis: a knowledge-based approach for interpreting genome-wide expression profiles. *Proc. Natl. Acad. Sci.*, U S A 2005, 102(43), 15545-50.

[15] Benjamini, Y; Drai, D; Elmer, G; Kafkafi, N; Golani, I. Controlling the false discovery rate in behavior genetics research. *Behav. Brain. Res.*, 2001, 125(1-2), 279-84.

[16] Reiner, A; Yekutieli, D; Benjamini, Y. Identifying differentially expressed genes using false discovery rate controlling procedures. *Bioinformatics*, 2003, 19(3), 368-75.

[17] Venkov, CD; Link, AJ; Jennings, JL; et al. A proximal activator of transcription in epithelial-mesenchymal transition. *J. Clin. Invest.*, 2007, 117(2), 482-91.

[18] Ogata H; Goto S; Sato K; Fujibuchi W; Bono H; Kanehisa M. KEGG: Kyoto Encyclopedia of Genes and Genomes. *Nucleic Acids Res.*, 1999, 27(1):29-34.

In: Encyclopedia of DNA Research (3 Volume Set)
Editors: S. J. Duncan and P. H. Wiley

ISBN 978-1-61324-305-3

Chapter 35

Application of Genomic Fingerprinting Methods for Molecular Characterization of Bioleaching Microorganisms

Bidyut R. Mohapatra**, *W. Douglas Gould,
Orlando Dinardo and David W. Koren
CANMET Mining and Mineral Sciences Laboratory, Natural Resources Canada, 555 Booth Street, Ottawa, Ontario K1A0G1, Canada

Abstract

The acidophilic microorganisms (defined as those which can grow at pH ≤3) play a major role in the bioleaching processing of low grade ores for extraction of base and precious metals, including Cu, Zn, Ni, Co, Ag and Au. The bioleaching microorganisms act as the biocatalyst in the oxidation and dissolution of metal sulfide ores by the production of sulfuric acid and ferric ions. The application of bioleaching microorganisms in ore processing has several distinct advantages over conventional mineral processing methods. Microbial extraction processes are believed to be eco-friendly methods owing to the requirement of lesser amounts of energy compared to roasting or smelting, and no production of sulfur dioxide or other gases detrimental to the environment. Bioleaching microorganisms have also attracted significant industrial interest for treatment of toxic heavy metals- and sulfide-containing industrial waste effluents. On the other hand, these microorganisms have a negative impact on public and environmental health with the generation of highly toxic acid mine drainage (AMD). AMD is generated by the subsurface and the surface (open pit) mining of metal ores and coal with the resulting exposure of sulfide minerals to air and water. AMD degrades the quality of surface and groundwater intended for drinking, recreation, crop irrigation and aquaculture with elevated levels of sulfate, and toxic heavy metals and metalloids. In order to improve the mineral oxidizing activity of bioleaching microorganisms and to inhibit the production of AMD, it is essential to develop robust genotypic tools for speciation as well as for determination of the biological activities of bioleaching

* Corresponding address: Bidyut Ranjan Mohapatra, #203-Fantable Ninomiya-I, Ninomiya 1-10-19, Tsukuba, Ibaraki 305-0051, JAPAN, Fax: +81-29-861-6733, E. mail: m_bidyut@yahoo.com.

microorganisms. In this chapter, we have provided an overview of the molecular techniques including different genomic fingerprinting methods that have been used for the characterization of bioleaching microorganisms.

1. Introduction

The global sulfur cycle responsible for the mineralization of sulfur compounds is predominated by the oxidative and reductive reactions of the sulfur-oxidizing and sulfate-reducing microorganisms (Postgate, 1984; Canfield and Raiswell, 1999). Dissimilatory sulfate reduction is carried out by a group of phylogenetically diverse anaerobic microorganisms (SRM) belong to the domain bacteria and archaea (Gibson, 1990; Stackebrandt et al., 1995). Dissimilatory sulfate reduction by SRM produces hydrogen sulfide. The resulting hydrogen sulfide reacts with different metals and precipitates as insoluble metal sulfides. The dissimilatory sulfate reduction reactions are considered as the principal biogeochemical agent for mineral deposition in the crust of earth. These metal sulfides have been converted to soluble metal sulfates by sulfur-oxidizing microorganisms. The reactions of dissolution and oxidation of insoluble metal sulfides are accelerated by bioleaching acidophilic iron- and sulfur-oxidizing microorganisms (reviewed in Gould and McCready, 1994; Mohapatra et al., 2008). These bioleaching microorganisms have been considered as important biocatalyst for the mining industry in the processing of sulfidic ores and concentrates to recover precious and base metals including Cu, Zn, Ni, Co, Ag and Au (McCready and Gould, 1990; Brierley and Brierley, 1999; Olson et al., 2003; Dinardo et al., 2004; Reith et al., 2007). The bioleaching processes for metal extraction have many advantages over the traditional pyrometallurgical methods including no harmful emission, simple procedures, cost-effectiveness, ability to operate at mild conditions and lower energy requirements. In addition, the mine tailings and wastes generated through bioleaching methods are less reactive due to the reduced sulfur content. The bioleaching methods are economically feasible methods for processing low-grade ores and mine tailings. These microorganisms have been emerging as efficient biocatalysts for the desulfurization of coal and in the environmentally friendly abatement of toxic and corrosive sulfide- and heavy metals-containing industrial sludge and waste effluents generated by petrochemicals refineries, liquefied petroleum gas plants, tanneries, textile mills, pulp and paper mills, mining industries and sewage treatment plants (Gould and McCready, 1994; Basu et al., 1996, Whiteley and Lee, 2006). On the other hand, bioleaching microorganisms are responsible for the generation of acid mine drainage (AMD), a potentially severe form of pollution associated with active and abandoned mining operations (Nordstrom and Alpers, 1999; Gould and Kapoor, 2003). As AMD is extremely acidic and contains toxic heavy metals and metalloids, it is detrimental to the public and environmental health. AMD is also a serious threat to aquatic life, and to the quality of surface water and groundwater utilized for drinking, recreation, aquaculture and crop irrigation. These microorganisms are also a risk to the safety, reliability and integrity of concrete and metallic infrastructures, including municipal sewer pipes and high pressure oil and natural gas distribution and transmission pipelines by inducing microbiologically influenced corrosion (MIC) (Droffelaar and Atkinson, 1995; Little et al., 2000).

Biomining has also a clear advantage in the extraction of metals from certain low-grade ores (Rawlings, 2002). For example, the extraction of copper can be achieved from low-grade ores and dumps of abandoned mines using the microbial activity generated through controlled irrigation of the dump (Watling, 2006). Many of these metals are not economically recoverable by conventional mineral processing methods. Where ore-type and geological features permit, metal recovery using leaching solutions produced and regenerated by microorganisms can be carried out *in situ* with obvious cost advantages and minimal disturbance to the surrounding environment (McCready and Gould, 1990). Although the ability of microorganisms in bioleaching processes for recovery of metals, and bioremediation of industrial sludge and waste effluents is remarkable, the inadequate knowledge about the diversity of microorganisms that take part in the bioleaching processes has limited the potential for their use in industrial processes.

In this chapter we have provided an overview about the mechanism of bioleaching processes of metal dissolution, phylogenetic type of bioleaching microorganisms and bioleaching processes performed by the mining industry for the extraction of base and precious metals. Additionally, the molecular techniques including different genomic fingerprinting methods used for the characterization of bioleaching microorganisms have been also reported. Molecular characterization of bioleaching microorganism will not only assess the diversity of microorganisms that participate in bioleaching processes but also assist in the development of ideal industrial processes for efficient metals extraction, inhibition of AMD and bioremediation of toxic and corrosive sulfide- and heavy metal(s)-containing sludge and waste effluents.

2. Bioleaching Mechanisms

The ferric ion and protons are required for the dissolution and oxidation of sulfide ores in bioleaching processes. Although the exact role of microorganisms in bioleaching processes has not been elucidated yet, it has been reported that bioleaching microorganisms are required for the production of chemicals and conditions to accelerate the bioleaching reactions (Sand et al., 1995; Olson et al., 2003). Crundwell (2003) has proposed three types of bioleaching mechanism: indirect leaching, indirect contact and direct contact for the dissolution and oxidation of metals.

In the indirect leaching mechanism, microorganisms oxidize ferrous ion to ferric ion in bulk solution and the resulting ferric ion oxidizes the sulfide phase. In the indirect contact mechanism, the contact between microorganisms and mineral surface is an important criterion for the bioleaching processes. Microorganisms usually form a biofilm by producing extracellular polymeric substances (EPS) to attach on the mineral surface. The efficacy of the metal dissolution in the EPS layer primarily depends on the composition of the sulfide ores and generation of reactive intermediates during bioleaching reactions (Schippers et al., 1999). Biofilms on pyrite surface contains high concentrations of proton and dissolved and chelated ferric species, which are derived from the biooxidation of ferrous ions. The high concentration of ferric ion and proton escalate the attack on the valence bonds of the pyrite which is then degraded via the thiosulfate pathway.

In the direct contact mechanism, microorganisms directly attack the mineral surface by breaking the metal-sulfide bond with enzymes rather than with ferric and ferrous ions. To date, there is no evidence for the direct contact mechanism.

The bioleaching reactions of both the indirect leaching mechanism and indirect contact mechanism can proceed in two pathways based on the solubility of sulfide ores. The thiosulfate pathway is responsible for the oxidation and dissolution of acid-insoluble metal sulfides, such as pyrite (FeS_2), tungstenite (WS_2) and molybdenite (MoS_2); and the polysulfide pathway is responsible for the oxidation and dissolution of acid-soluble metal sulfides: sphalerite (ZnS), chalcopyrite ($CuFeS_2$), galena (PbS), arsenopyrite (FeAsS) and hauerite (MnS_2) (Rohwerder et al., 2003).

In the thiosulfate pathway of pyrite, solubilization of acid-insoluble metal sulfide is catalysed by ferric ion with the production of thiosulfate as the main intermediate (see Eq. 1) and sulfate as the main end-product (see Eq. 2).

$$FeS_2 + 6Fe^{3+} + 3H_2O \rightarrow S_2O_3^{2-} + 7Fe^{2+} + 6H^+ \quad \text{(Eq. 1)}$$

$$S_2O_3^{2-} + 8Fe^{3+} + 5H_2O \rightarrow 2SO_4^{2-} + 8Fe^{2+} + 10H^+ \quad \text{(Eq. 2)}$$

In the case of polysulfide pathway, protons and ferric ion are responsible for the dissolution and oxidation of acid-soluble metal sulfide (MS) with elemental sulfur as the main intermediate (see Eq. 4). The resulting elemental sulfur is further oxidized to sulfate by acidophilic sulfur-oxidizing microorganisms *Acidithiobacillus thiooxidans* (formerly *Thiobacillus thiooxidans*) and/or *Acidithiobacillus caldus* (formerly *Thiobacillus caldus*) (see Eq. 5).

$$MS + Fe^{3+} + H^+ \rightarrow M^{2+} + 1/2\ H_2S_n + Fe^{2+}\ (n \geq 2) \quad \text{(Eq. 3)}$$

$$1/2\ H_2S_n + Fe^{3+} \rightarrow 1/8\ S_8 + Fe^{2+} + H^+ \quad \text{(Eq. 4)}$$

$$1/8\ S_8 + 3/2\ O_2 + H_2O \xrightarrow{\text{sulfur-oxidizing microorganisms}} SO_4^{2-} + 2H^+ \quad \text{(Eq. 5)}$$

The iron-oxidizing microorganisms of the genera *Acidithiobacillus*, *Leptospirillum* and *Sulfobacillus* play a significant role in the solubilization of metal sulfides by producing sulfuric acid for a proton attack (see Eq. 5) and by providing ferric ion for an oxidative attack (see Eq. 6).

$$2Fe^{2+} + 1/2\ O_2 + 2H^+ \xrightarrow{\text{Iron-oxidizing microorganisms}} 2Fe^{3+} + H_2O \quad \text{(Eq. 6)}$$

The general functions of bioleaching microorganisms in solubilization of metal sulfides are to provide sulfuric acid for a proton attack, to keep iron and other metals in oxidized state, and to create a reactive environment for efficient bioleaching reactions.

3. Phylogenetic Types of Bioleaching Microorganisms

The bioleaching microorganisms that participate in the stirred-tank and heap bioleaching industrial processes for dissolution and oxidation of metals can grow at extremely acidic pH ($pH \leq 3$). These acidophiles are mostly autotrophic and oxidize inorganic sulfur compounds and/or Fe (II) ions for their growth and proliferation. As the bioleaching microorganisms are difficult to cultivate and enumerate by traditional culture-based methods, the bioleaching microorganisms in mining environment were previously considered to be poor in diversity. With the development of new molecular microbiological methods for identification and quantification, it is possible not only to assess the microbial diversity but also to perform the spatio-temporal characterization of microorganisms in mining biotopes (Johnson, 1998; Hallberg and Johnson, 2001). Eleven putative prokaryotic divisions have already been characterized in AMD environments (Baker and Banfield, 2003). The mesophilic iron- and sulfur-oxidizing *Acidithiobacillus ferrooxidans* and sulfur-oxidizing *Acidithiobacillus thioooxidans* and the moderately thermophilic sulfur-oxidizing *Acidithiobacillus caldus* are the Gram-negative γ-proteobacteria responsible for the bioleaching. The other bacteria that participate in the bioleaching processes are iron- and sulfur-oxidizing *Acidiphilium acidophilum* and iron-oxidizing leptospirilli (*Leptospirillum ferrooxidans* and *Leptospirillum ferriphilum*) (Hippe, 2000; Coram and Rawlings, 2002).

Gram-positive bioleaching bacteria of the genus *Acidimicrobium*, *Ferromicrobium*, and *Sulfobacillus* are moderate thermophillic bacteria (Clark and Norris, 1996; Norris et al., 1996; Johnson and Roberto, 1997).

The bioleaching archaea of the order *Sulfolobales* are thermophilic. The genera of thermophilic sulfur- and iron-oxidizing archaea that participate in bioleaching processes are *Sulfolobus*, *Acidianus*, *Metallosphaera*, and *Sulfurisphaera* (Fuchs, et al., 1995; Kurosawa et al., 1998; Norris et al., 2000). The other archaea that contribute significantly in bioleaching processes are the sulfur- and the iron-oxidizing mesophilic archaea: *Thermoplasmales Ferroplasma acidiphilum* (Golyshina et al., 2000) and *F. acidarmanus* (Edwards et al., 2000).

Leptospirillum ferrooxidans and *Leptospirillum ferriphilum* can utilize Fe (II) as the nutrient source, whereas *Acidithiobacillus ferrooxidans* can grow by oxidizing Fe (II) and inorganic sulfur compounds. The growth of *Acidiphilium acidophilum* and *Acidimicrobium ferrooxidans* is autotrophic with sulfur and Fe (II) compounds, heterotrophic with glucose and/or yeast extract, and mixotrophic with sulfur, Fe (II) compounds, glucose and yeast extract (Clark and Norris, 1996; Hiraishi et al., 1998). Furthermore, *Acidiphilium* spp. and *Acidisphaera rubrifaciens* were reported as phototrophic (Hiraishi et al., 2000; Hiraishi and Shimada, 2001) and *Ferromicrobium acidophilus* was reported as an obligate mixotrophic bacterium (Johnson and Roberto, 1997).

In bioleaching processes, the enrichment of bioleaching microorganisms is achieved by providing aeration to an iron- and/or sulfur-containing mineral suspension or the irrigation water. The depletion of nutrients (nitrogen, phosphate, potassium and trace elements) is avoided by addition of inorganic fertilizer (Rawlings et al., 2003). In addition, the bioleaching processes are not vulnerable to environmental microbial contamination owing to the nutrient limitation and low pH. In stirred-tank bioleaching processes, the continuous washing of bioreactors helps to sustain the growth of metabolically active bioleaching microbial communities.

The summary of the information on the diversity of bioleaching-microbial communities published so far indicate that the most important microorganisms involved in the dissolution and biooxidation of minerals belong to the domains bacteria and archaea (Johnson, 1998). These microorganisms are responsible for generation of the Fe^{3+} ions (see Eq. 5) and sulfuric acid (see Eq. 6) for the bioleaching reactions. The bioleaching microbial communities are mostly comprised of iron- and sulfur-oxidizing chemolithotrophic bacteria and archaea (see Fig. 1), and have the following characteristics: a) they grow in extreme acidic conditions (pH ≤ 3); b) they obtain energy with Fe^{2+} and/or inorganic sulfur compounds as an electron donor, and oxygen as the electron acceptor; and c) they are resistant to a variety of metal ions (Dopson and Lindström, 2004).

4. Industrial Processes for Bioleaching of Ores

Two types of bioleaching processes, such as stirred-tank bioleaching and heap bioleaching have been used by the mining industry for the recovery of metals from sulfide and/or iron-containing ores and mineral concentrates.

4.1. Stirred-Tank Bioleaching

Stirred-tank bioleaching is a sophisticated and controlled process for the dissolution of minerals. In stirred-tank bioleaching process, a series of highly aerated continuous-flow reactors are used for the extraction of metals. Before incorporation into the first tank of the continuous-flow reactor, the ore or the metal concentrates samples are finely milled and mixed with ammonia- and phosphate-containing fertilizers. These fertilizers act as the nutrient source for the enrichment of bioleaching microbial communities. The minerals suspension continuously flows through a series of tanks which strongly influence the phylogenetic diversity of indigenous microbial species. The pH, temperature and aeration in each tank have been controlled to achieve the optimum bioleaching kinetics (van Aswegen et al., 1991, Rawlings et al., 2003). The dissolution and oxidation of minerals in continuous stirred-tank bioleaching processes can be achieved in weeks compared to months in heap reactors. This higher rate of mineral degradation has been achieved by developing efficient microbial species with the flush out of the undesirable microorganisms. The formation of an efficient microbial community facilitates and/or creates favorable condition for rapid mineral degradation. The stirred-tank bioleaching process has been used successfully for the preoxidation of gold-containing arsenopyrite concentrates at Sansu in the Ashanti goldfields of Ghana, and at Fairview mine, Barberton, South Africa. Tank bioreactors are also used to extract cobalt from cobalt-containing pyrite ores at Kasese, Uganda (Rawlings et al., 2003).

The bioleaching efficacy of the stirred-tank reactors depends on temperature and pulp density of the ore. The optimal efficacy of the bioreactor is recorded in the temperature range of 40°C-50°C and 20% ore pulp density (Rawlings, 2002). It has been reported that when the ore pulp density is greater than 20%, there has been a significant change in the microbial compositions. This problem arises due to the increase in viscosity of the mineral suspension (liquor) which hinders efficient gas transfer. In addition, the increase in stirring rates of the tank caused physical damage to the microbial cells due to abrasion. Although stirred-tank reactor is a robust process for extraction of metals, high capital and operational costs has restricted its use for the processing of high-value ores and mineral concentrates (Rawlings et al., 2003).

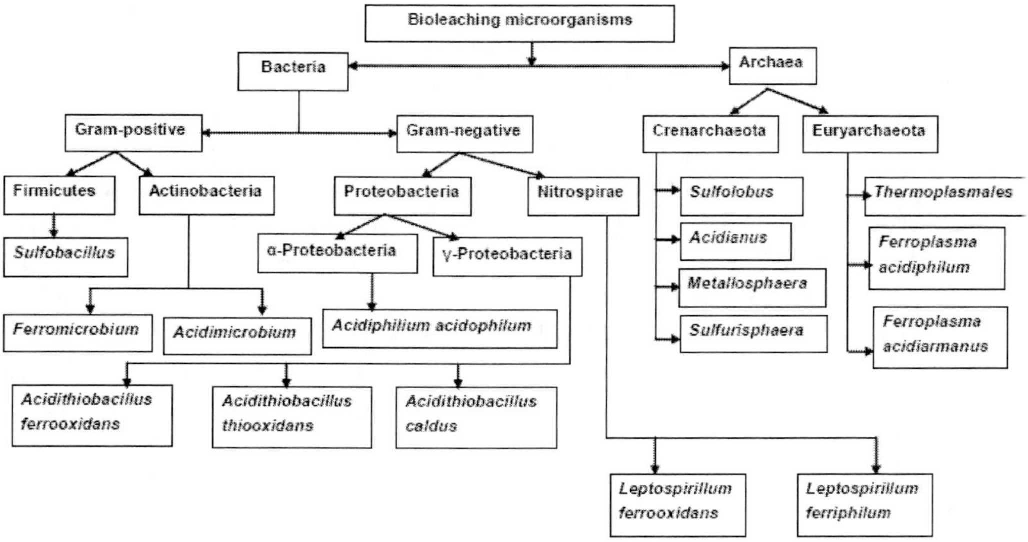

Figure 1. Phylogenetic diversity of bioleaching microorganisms.

4.2. Heap Bioleaching

Heap bioleaching processes have been used widely by the mining industry for the treatment of minerals because the initial investment for construction and operation is lower compared to stirred-tank bioleaching processes. Heap bioleaching has been performed with the low-grade ore, mine tailings and run-of-the mine (ROM) (Watling, 2006). The development of effective bioleaching microbial communities in heap bioleaching process depends on several factors, including type of minerals, aeration, moisture contents and temperature. Recently the efficiency of the under-performing heap has been improved by construction of specially design heap, and by providing irrigation and insulation to the heap (Rawlings and Johnson, 2007). The construction of the heap is usually performed by stacking carefully the crushed ore over an impermeable base to create a bed that exhibits good permeability for both solution and gases. The bed is irrigated by sprinklers or drippers arranged in a grid across the top of the heap. The base is equipped with a leach liquor collection and distribution system (Schnell, 1997; Rawlings et al., 2003). Acidic leaching solution is percolated through the crushed ore and microorganisms growing on the surface of the ore particles inside the heap generate ferric ion and acid. The acid and ferric ions assist in mineral dissolution and metal solubilization. Aeration in heap leaching bioreactors can be carried out by supplying air through pipes placed at the bottom of the heap. The metal-containing leach solution is collected regularly at the bottom of the heap and sent for the metal recovery through solvent extraction and electrowinning. A part of the leaching solution which is enriched with the bioleaching microorganisms is also recycled back to the top of the heap for ensuring the distribution of active bioleaching microbial population throughout the ore. Natural microorganisms present in the ore play a significant role in the heap bioleaching, however, the patented BIOPRO^{TM} recommends heap inoculation with bioleaching microorganisms during acid conditioning and agglomeration substantially improves the bioleaching rate (Brierley, 1997). The other methods that are proposed for heap inoculation include heap irrigation from a well-aerated pond enriched with P, K, N and Fe for the growth of bioleaching microorganism; inoculation of a biological contactor with iron-oxidizing microorganisms and feeding the ferric-rich effluents into the heap (Hunter, 2001); development of a viable culture of bioleaching microorganisms in a small portion of ore and subsequent scale up in bulk ore before heap piling (Hunter and Williams, 2002); inoculation of the heap with genetically tailored microorganisms so that they are functionally active in extreme conditions (increase in acidity, temperature, presence of toxic metal ions and organic compounds) generated inside the heap (Hunter, 2002); inoculation of the heap surface with temporarily non-attachable microorganisms (StickiBug^{TM}) for greater penetration into the heap (Gericke et al., 2005); and inoculation of gaseous suspension of ultra-small microorganisms in the heap aeration pipeline (du Plessis, 2003). The major drawbacks of heap bioleaching processes are the difficulty in providing air to the heap and controlling the pH gradients, and nutrient levels. Most of the heap bioleaching processes in the world are operated in Chile and Australia for extraction of copper. The Sociedad Contractual Minera El Abra and the Codelco Division Radimiro Tomic heap bioleaching operations in Chile produce 225,000 and 180,000 tons Cu per annum, respectively (Olson et al., 2003; Rawlings et al., 2003). Refractory gold ore and low grade gold ores are also pretreated by heap bioleaching at Newmont Mining, Nevada, USA.

5. Molecular Methods Used for Characterization of Bioleaching Microorganisms

As the availability of high-grade mineral resources is dwindling and the demand for base and precious metals has been growing, it is essential for mining and biotechnology industries to develop innovative, green and sustainable technology to process low-grade ores, wastes from current and past mining operations, and electronic scrap (Brandl et al., 2001; Brierley, 2008). Microbial-mediated mineral processing is an emerging technology that has attracted significant interest by the mining industry because the bioleaching process is easy-to-perform, has low operation costs, appropriate recovery of metals from low-grade ores and waste materials. In addition, the bioleaching microorganisms have been used for environmentally friendly treatment of toxic sulfide and heavy metals-containing industrial waste effluents and sludge.

The functional and numerical importance of bioleaching microorganisms in critical industrial processes of base and precious metals extraction and environmental remediation of toxic-sludge and wastewater require the development of robust tools to monitor the spatial and temporal changes in the bioleaching microbial community. Considerable efforts have been devoted towards the development of rapid and accurate genotypic methods for detection and enumeration of bioleaching microorganisms in mining environments to improve their biocatalytic efficacy (Johnson and Hallberg, 2007; Jerez, 2008). In general, the molecular methods which are applied for the characterization of bioleaching microorganisms are divided into thirteen types: DNA sequencing and cloning, ARDRA (amplified ribosomal DNA restriction analysis), PCR with bioleaching microbes-specific primers, DGGE (denaturing gradient gel electrophoresis), T-RFLP (terminal restriction fragment length polymorphism),

PFGE (pulsed-field gel electrophoresis), RAPD (random amplification of polymorphic DNA), rep-PCR (repetitive extragenic palindromic-polymerase chain reaction), RISA (ribosomal intergenic spacer analysis), SSCP (single-stranded confirmation polymorphism), DNA microarrays, FISH (fluorescent *in situ* hybridization) and real-time PCR targeting the bioleaching microorganisms-specific primers. FISH and real-time PCR have been used for the quantification of bioleaching microorganisms in mining environments.

5.1. Cloning and Sequencing of PCR-Amplified Genes

This method characterizes the bioleaching microorganisms by targeting the small subunit 16S rRNA genes. The 16S rRNA genes have several advantages compared to other candidate genes, e.g. they are essential component of ribosomes, they are highly conserved, the sequence of 16S rRNA genes are taxon specific, and existence of extensive rRNA sequence reference databases for phylogenetic studies. The use of 16S rRNA gene-based analysis to characterize bacteria and archaea has increased significantly owing to the realization that >99% of the microorganisms in the environment are uncultivable under standard techniques (Amann et al., 1995; Kaeberlein et al., 2002). This method consists of five steps: extraction of DNA from environmental sample, amplification of 16S rRNA genes by PCR, purification of PCR-amplicons, screening for unique clones (insertion of target genes) and sequencing of the clones by automated sequencer (ABI PRISM Sequencer) with commercially available methods (ABI PRISM BigDye Terminator Cycle Sequencing Kit) (Head, 1999). Taxonomic identification of the 16S rRNA gene sequence is performed by comparison with the available sequence database using common search algorithm BLAST (basic local alignment search tool). Harrison (1982) and Lane et al. (1992) used the 16S rRNA genes analysis to characterize the bioleaching microorganisms (sulfur- and iron-oxidizing bacteria). Goebel and Stackebrandt (1994) also evaluated the 16S rRNA gene to characterize the bioleaching microbial communities associated with acidic runoff of a chalcocite overburden heap and laboratory-scale (1 to 4-liter) batch and continuous bioreactors used for the commercial assessment of the bioleachability of zinc sulfide ore concentrates. This method was also applied for assessment of phylogenetic diversity of bioleaching microbial communities in acid mine drainages of Yunfu sulfide mine in China (He et al., 2007), in acid mineral bioleaching systems of Tong Shankou copper mine, China (Xie et al., 2007), in acid rock drainage from Japanese pyrite mine (Okabayashi et al., 2005), in mine tailings contaminated sediments of Sudbury in Canada (Dinardo and Mohapatra, 2008) and in copper heap bioleaching commercial plant of Santiago, Chile (Romero et al., 2003). Although PCR-clone-16S rRNA sequence analysis is a powerful culture-independent method for characterization of microbial population, identification of novel phylotypes and design of oligonucleotide primer for yet-to-cultured microorganisms, this method is labor intensive and time consuming to screen all the clones. To avoid sequencing of large numbers of clones present inside a library, ARDRA method (mentioned in sub-section 5.2) has been used to gather several clones into one group based on the restriction pattern(s), and subsequently perform the sequencing on representative clone of ARDRA-defined group.

5.2. ARDRA

ARDRA, also known as restriction fragment length polymorphism analysis, is a PCR-based genomic fingerprinting method that separates rDNA genes based on differences in both size and sequence. In ARDRA 16S or 23S rDNA gene is amplified by PCR, followed by restriction enzyme digestion. The digested DNA is size separated by electrophoresis either by using an agarose gel in case of single species/clones or by using a non-denaturing polyacrylamide gel in case of microbial community. The operating principle of this method is that divergences in the rDNA gene sequences of different species will create differences in restriction sites for various enzymes. Application of correct restriction enzymes will generate a unique fingerprint for each species or strain (Vaneechoutte et al., 1992). The major limitations of ARDRA are the selection of restriction enzymes and interpretation of the complex banding patterns generated during microbial community profiling. Additionally, the lack of sensitivity associated with gel staining may create difficulty in the identification of low abundance members in the target community. ARDRA have been used for rapid identification and differentiation of thiobacilli: *Thiobacillus thioparus*, *Acidithiobacillus cuprinus*, *Acidithiobacillus ferrooxidans* and *Acidithiobacillus thiooxidans* as well as the strains *Leptospirillum ferrooxidans* from leaching environments (Rawlings, 1995; Selenska-Pobell et al., 1998). Johnson et al. (2005) have developed an ARDRA protocol that can facilitate the differentiation and identification of all currently known phylogenetically distinct groups of mesophilic and moderately thermophilic iron-oxidizing acidophiles.

5.3. PCR with Bioleaching Microbes-Specific Primers

The design of oligonucleotide PCR primers for the diagnostic detection of microorganisms primarily depends on the presence of a variable region in the 16S rRNA primary structure. The oligonucleotide PCR primers designed for targeting the hypervariable regions of 16S rRNA sequence are found to be highly specific for identification of microorganisms at genus, species and subspecies level. Moreover, the presence of the conserved 16S rRNA sequence helps to develop PCR primers specific for genus, subdivision and phylum. The 16S rRNA sequence-based PCR method has simple procedural steps, such as extraction of genomic DNA from environmental sample, amplification of 16S rRNA genes with specific oligonucleotide primers, and gel electrophoresis to identify the size(s) of the amplicon(s). The bioleaching microbes-specific primers for specific amplification of 16S rRNA gene fragments are reported for six industrially important species: *Acidiphilium*, *Leptospirillum*, *Sulfobacillus*, *Acidithiobacillus caldus*, *Acidithiobacillus thiooxidans* and *Acidithiobacillus ferrooxidans* (De Wulf-Durand et al., 1997; Escobar et al., 2008). The major limitations of this method include the presence of multiple rRNA operons in environmental microorganisms, the absence of target sequence due to inhibition of expression of target gene, and the formation of chimeric PCR products. The authenticity of the results for identification of microorganisms in environmental samples using PCR-amplified 16S rRNA genes should be tested by cloning and sequencing (Head, 1999).

5.4. DGGE

DGGE is a technique that is used to resolve the PCR-amplified gene fragment on a denaturing gradient gel according to the differences in nucleotide sequences (Muyzer and Smalla, 1998). The experimental steps of this method include the extraction of genomic DNA, amplification of target gene using PCR with one of the specific primer set clamped with G-C, and electrophoresis of the amplicons on a special electrophoresis gel incorporated with a denaturing gradient (urea or formamide). When the amplicon (DNA) migrates from low to high concentration of denaturants, it starts to denature into single strands. Complete denaturation of double stranded DNA ceases the migration. The amplicon is unable to degrade completely due to the presence of G-C clamp in one of the primers. The migration of DNA amplicons primarily depends on the relative proportions of G-C and A-T contents. Since A-T bonds are more vulnerable to denaturation than G-C bonds, the differences in sequence between amplicons that result in differences in G-C content will cause DNA to migrate to different positions in the gel. DGGE is sensitive for identification a single nucleotide differences between the amplicons. The DNA present in the bands can be excised from the gel and sequenced for the phylogenetic studies. DGGE is considered as lesser labor intensive and lesser time consuming than the construction of a clone library. Furthermore, a broader spectrum of genetic diversity of microorganisms can be observed in DGGE gel owing to the absence of random process of picking up and sequencing of the clones. DGGE has been used extensively for studying the genetic diversity of bioleaching microorganisms by targeting the 5S rRNA gene and 16S rRNA gene fragment. DGGE analysis of 5S rRNA genes has been found to generate similar migration patterns for distinct species of bioleaching microorganisms (Stoner et al., 1996). Most of the DGGE studies for characterization of bioleaching microorganisms have used 16S rRNA genes. Kinnunen and Puhakka (2004) have applied the DGGE technique to characterize the bioleaching bacterial and archaeal populations in fluidized-bed reactors for the treatment of drainage water and sludge of the Pyhäsalmi FeS_2-$CuFeS_2$-α-(Zn,Fe)S mine. DGGE has also performed for molecular characterization of bioleaching microorganisms in stirred-tank processes for bioleaching of sulfidic mine tailings in Sweden (Rzhepishevska et al., 2005), in a low-grade copper sulfide ROM test heap of a project in Chile (Demergasso et al., 2005), in an industrial-scale chalcocite heap bioleaching operation in Myanmar (Hawkes et al., 2006), and in an engineered wetland system treating acid coal mine drainage in Ohio, United States of America (Nicomrat et al., 2008). Although DGGE is a successful tool for characterization of bioleaching microorganisms, the banding patterns mainly represent the major constituents of the complex environmental communities (Heuer et al., 1997). Small bioleaching microbial species that constitute <1% of the total population can not be readily detected by DGGE (Muyzer et al., 1993; Dar et al., 2005).

5.5. T-RFLP

In T-RFLP, the PCR is performed with genomic DNA extracted from environmental samples to amplify the 16S rDNA genes with one primer of the primer set which is labeled with a flurochrome (TET: 4,7,2',7'-tetrachloro-6-carboxyfluorescein or 6-FAM: phosphoramidite fluorochrome 5-carboxyfluorescein). As a result restriction digestion and gel or capillary electrophoresis will always produce a terminal fragment that is fluorescently labeled at the 5' end. The size and intensity of the terminally labeled restriction fragments (T-RF) is determined by an automated DNA sequencer. The plot of the T-RF size and intensity is considered as unique profile (fingerprint) for a microbial community. T-RF fingerprints are used to estimate the microbial species richness and diversity within a community (Zhou et al., 2002). Each T-RF corresponds to a single microbial species. The sequence of the T-RF is compared with the global 16S rDNA sequence database to identify the taxonomic affiliation of a particular T-RF obtained from a community (Ribosomal Database Project II: http://rdp8.cme.msu.edu/html/TAP-trflp.html; Microbial Community Analysis: http://-mica.ibest.uidaho.edu/). Although T-RFLP can provide information about species composition of a microbial community, experimental results have documented that T-RFLP can underestimate the species composition by identifying only 0.5% of the total 16S rDNA amplified (Liu et al., 1997; Dunbar et al., 2000). This problem arises due to complex fingerprint patterns generated by restriction enzyme digestion and the absence of a broader 16S rDNA databases for phylogenetic T-RFLP pattern analysis. T-RFLP has been applied by Escobar et al. (2008) to assess the diversity of bioleaching microorganisms in solutions from laboratory bioleaching columns and pregnant leaching solutions of a bioleaching heap processes. The T-RFLP profiles were obtained by PCR-amplification of 16S rDNA gene with the universal primers 341f (5'-CCTACGGGAGGCAGCAG-3') and 1100r (5'-GGGTTGCGCTCGTTG-3'), followed by digestion with *RsaI* restriction enzyme. The 341f primer was fluorescently labeled with 6-FAM. The predicted T-RF lengths using 341f and 1100r in combination with the *RsaI* restriction enzyme were 121bp, 551 bp, 247 bp, 309 bp, 146 bp, 146 bp, 142 bp for bioleaching bacterial species *Acidiphilium*, *Acidithiobacillus caldus*, *Acidithiobacillus ferrooxidans*, *Acidithiobacillus thiooxidans*, *Leptospirillum ferriphilium*, *Leptospirillum ferrooxidans* and *Sulfobacillus*, repectively.

5.6. PFGE

PFGE is a genomic fingerprinting method in which the microbial cells are mixed with molten agarose and poured into small molds. The intact DNA in the agarose plugs are obtained by detergent-enzyme treatment. The plugs are washed in buffers and subjected to digestion with rare-cutting restriction endonucleases. The restriction enzyme digested plugs are inserted into an agarose gel. The electrophoresis of agarose gel is performed in a platform in which the polarity of the current is changed at predetermined intervals. The contour-clamped homogeneous electric field (CHEF) type biopulsator of BioRad Inc. has been widely used for PFGE analysis of microorganisms. The pulsed-field assists the separation of large-molecular length DNA fragments ranging from 10 to 800 kb. The banding patterns of the ethidium bromide stained-gel can be digitally captured by an image acquisition and analysis system fitted with ultraviolet light. A library of PFGE patterns of selected microorganisms can be created with reference strains and the PFGE fingerprint of any new strain can be compared with the fingerprint of other strains present in the reference library for identification of its phylogenetic relationship (Olive and Bean, 1999). PFGE has been used to study the phylogenetic heterogeneity of the species of *Acidithiobacillus ferrooxidans* (Karavaiko et al., 2003), taxonomic identification of the genera *Acidithiobacillus* (formerly *Thiobacillus*), *Thiomonas* and *Leptospirillum* (Amils et al., 1998), the microbial diversity and distribution of leptospirilli along the iron gradient of Río Tinto river (a model of interest to biohydrometallurgy) in Spain (López-Archilla et al., 2001; García-Moyano et al., 2008), and for adaptive changes of *Acidithiobacillus thiooxidans* during growth in sewage sludge (Matlakowska and Sklodowska, 2006). The major limitations of this method include the selection of the restriction endonucleases and interpretation of complex fingerprint produces in certain species of bacteria.

5.7. RAPD

RAPD assay, also known as arbitrarily primed PCR, is a PCR-based genomic fingerprinting method in which the 10-mer oligonucleotide primers are used to amplify the arbitrarily chosen chromosomal DNA sequence. The basis of this method is that as the primers are short in length, it will hybridize at a low annealing temperature to complementary and/or partial complementary sequences of a single microbial genome. The number and location of these random 10-mer oligonucleotide sites vary for different strains of a bacterial species. The separation of the PCR amplicons of the RAPD assay by an agarose gel electrophoresis will generate a banding pattern which is a unique fingerprint for a particular bacterial strain (Welsh and McClelland, 1990). RAPD fingerprinting method is found to be promising for the differentiation of bacterial strains among diverse species (Olive and Bean, 1999). It has also been used for the microbial community analysis (Franklin et al., 1999). Although RAPD fingerprinting method is rapid, cost-effective and easy-to-perform, the disadvantages of this method include lack of reproducibility and standardization in genotyping of certain bacterial species. As the primers in this method are not specifically targeted to a particular site in the genomic DNA, slight variation in the annealing temperature and magnesium chloride concentration in the PCR reaction will result in imperfect hybridization between the primer and target sequence. RAPD method has been applied for the characterization of bioleaching microorganisms, especially to assess genetic variability, rapid identification and taxonomical categorization of *Acidithiobacillus ferrooxidans*, *Acidithiobacillus thiooxidans*, *Acidithiobacillus cuprinus*, *Thiobacillus thioparus* and *Acidithiobacillus caldus* (Novo et al., 1996; Selenska-Pobell et al., 1998; Selenska-Pobell et al., 2001; Akbar et al., 2005; Waltenbury et al., 2005; Ni et al., 2008). Recently, Ni et al. (2008) have also applied the RAPD fingerprinting method to study the spatio-temporal characterization of *Acidithiobacillus* species isolated from coal, copper and molybdenum mine wastes in different regions of China.

5.8. Rep-PCR

Rep-PCR is a genotypic method that differentiates microorganisms by using primers complementary to interspersed repetitive consensus sequences. These non-coding sequences are present in multiple copies in the genomes of most Gram-negative and Gram-positive bacteria (Versalovic et al., 1994). Examples of these repetitive elements are the repetitive extragenic palindromic (REP) sequences, the enterobacterial repetitive intergenic consensus (ERIC) sequences, the BOX sequences and the polytrinucleotide $(GTG)_5$ sequences (Versalovic et al., 1994). These repetitive units are considered to be highly conserved because repetitive sites are crucial protein-DNA interaction sites and/or these sequences may disseminate themselves as selfish DNA by gene conversion. Amplification of the distinct repetitive sites produce diverse-sized DNA fragments that can be separated by agarose gel electrophoresis and the resulting banding patterns unique for specific bacterial strain can be compared (Rademaker and de Bruijn, 1997). Rep-PCR has been widely used for bacterial taxonomy, spatio-temporal analysis, diagnostic and epidemiological applications owing to its reproducibility, cost-effectiveness and easy operational procedures (Versalovic et al., 1994; Mohapatra et al., 2007a). Five methods, namely REP-PCR (primer set Rep1R-I and Rep2-I), ERIC-PCR (primer set ERIC1R and ERIC2), ERIC2-PCR (primer ERIC2), BOX-PCR (primer BOX A1R), and (GTG)5-PCR [primer (GTG)5] are frequently used in rep-PCR fingerprinting analysis (Rademaker and de Bruijn, 1997; Mohapatra et al., 2007b). Selenska-Pobell et al., (1998) have reported fast identification and taxonomical categorization of thiobacilli: *Thiobacillus thioparus*, *Acidithiobacillus cuprinus*, *Acidithiobacillus ferrooxidans* and *Acidithiobacillus thiooxidans* using BOX-PCR and ERIC2-PCR. Rep-PCR has also been used for molecular characterization of bioleaching microbial communities in coal, copper, gold and uranium mine wastes in Brazil (Paulino et al., 2001), in soil samples from two uranium waste piles in East Germany (Selenska-Pobell et al., 2001), and in coal, copper and molybdenum mine wastes in China (Ni et al., 2008). The disadvantage of this method is the production of complex rep-PCR fingerprint in certain bacteria which is statistically difficult to interpret accurately.

5.9. RISA

RISA is based on the PCR amplification of the intergenic spacer region between the small (16S) and large (23S) rRNA genes of the total microbial community DNA with oligonucleotide primers targeted to conserved regions in the 16S and 23S. This 16S-23S intergenic spacer region is heterogeneous with respect to length and nucleotide sequence compared to flanking 16S and 23S rRNA genes. The size of the intergenic spacer region which ranges from 400 to 1200 bp is used as the tool for strain characterization, subtyping and community-profiling (Nagpal et al., 1998; Garcia-Martinez et al., 1999). The gel electrophoresis of PCR amplicons generates a community-specific banding pattern with each band corresponding to a single organism of the target community. The lack of sensitivity of the gel electrophoresis results requires the development of an automated RISA (ARISA) (Fisher and Triplett, 1999). The procedural steps for ARISA are same as RISA except that a fluorescently labeled primer is used in the PCR. The electrophoresis is carried out in an automated instrument equipped with a laser for the detection of fluorescent DNA fragments. The demerit of this method includes the non-specific amplification of shorter DNA fragments which results in the generation of a complex fingerprint pattern. Cardinale et al. (2004) examined different primer sets with respect to universality, sensitivity and reproducibility for ARISA profiling of environmental microbial communities. RISA has been used for the characterization of bacterial populations in copper bioleaching systems (Pizarro et al., 1996; Vásquez and Espejo, 1997), in soil samples from two uranium waste piles in East Germany (Selenska-Pobell et al., 2001), and in coal, copper and molybdenum mine wastes in China (Ni et al., 2008). RISA has also been found to be a useful tool for detection, rapid differentiation and identification of acidithiobacilli (Bergamo et al., 2004) and leptospirilli (Coram and Rawlings, 2002).

5.10. SSCP

SSCP analysis is a PCR-based method that detects sequence variations between different DNA fragments, which are usually PCR amplified from variable regions of the 16S rRNA gene. This technique separates PCR amplicons based on the sequence-dependent differential intramolecular folding of single-stranded DNA. After denaturation of the double-stranded PCR products by heat or chemicals or lambda exonulcelases, the resulting single-stranded DNA is loaded on a non-denaturing polyacrylamide gel. Under non-denaturing conditions, a stable secondary structure is formed which is dependent on intramolecular interaction of the nucleotide sequence. Based on the migration of secondary structure in the gel, the PCR product with similar size can be separated and visualized (Orita et al., 1989; Lee et al., 1996). The advantages of SSCP technique over DGGE are non-requirement of GC-clamp primers and denaturing gradient gel. The major limitation of this method is the formation of more than one stable conformations of one single-stranded DNA which result in multiple bands on the gel. To avoid this limitation, an automated SSCP has been implemented. In automated SSCP, PCR is performed with fluorescent primers, and the PCR products are resolved using an automated DNA sequencer fitted with a laser for detection of fluorescent DNA fragments. SSCP has been used for the monitoring of pyrite-oxidizing bacterial population in stirred-tank bioreactors, suspended-solids column reactors and bubble column reactors (Battaglia-Brunet et al., 2002; Foucher et al., 2003; d'Hugues et al., 2008).

5.11. DNA Microarrays

DNA microarrays were originally developed for the analysis of gene expression in biomedical research. Recently DNA microarrays have been emerging as a powerful diagnostic tool for high-throughput detection and characterization of complex community of environmental microorganisms (Bodrossy and Sessitsch, 2004). In this method, taxonomic specific oligonucleotides (oligoprobes) are immobilized as a spot on the surface of a microscopic slide. Usually 250,000 oligoprobes can be spotted on the slides by automated dispenser for hybridization. The taxonomic-specific oligoprobes can be designed by aligning the 16S rDNA sequence available for a species in the global database (NCBI, http://www.ncbi.nlm.nih.gov). The genomic DNA extracted from the environmental sample, and/or the PCR amplified 16S rRNA gene fragment sequence are added to each spot on the slide for hybridization with the oligoprobes. The fluorescence signal of the hybridizing spot can be recorded by an automated microarray scanner (Affymetrix Inc.) equipped with a laser light source. The signal of hybridization can be used to create a complex qualitative profile of various taxonomic groups present in an environmental sample. Although DNA microarrays are remarkable for speciation of complex environmental communities in a short time, the variation in hybridization intensities for individual spots due to cross-hybridization of probes with closely related species and false results owing to secondary structure of target oligoprobe, steric hindrance of spot density and other hybridization conditions restrict their reliable application for characterization of complex microbial communities present in the environments. A comprehensive 50-mer-based oligonucleotide DNA microarrays has been developed for effective monitoring of the composition, structure, dynamics and functional activities of bioleaching microbial communities. The microarrays contained most of the known genes of bioleaching microorganisms. A total of 1072 probes were selected in which 571 represents 16S rRNA and 501 represents functional genes: carbon metabolism (158), nitrogen metabolism (72), sulfur metabolism (39), iron metabolism (68), DNA replication and repair (97), metal-resistance (27), membrane-related gene (16), transposon (13) and IST sequence (11) (Yin et al., 2007). The developed microarrays have been successful for the qualitative detection of metabolically active bioleaching microorganisms in AMD and bioleaching system of Dexing copper mine, Jiangxi Province, China (Yin et al., 2007).

5.12. FISH

FISH is a technique in which fluorescently labeled oligonucleotide of known sequence permeate through the whole bacterial and archaeal cells and hybridized to complementary 16S rRNA sequence of a particular group of microorganisms presents in a heterogeneous population of microorganisms. FISH is a culture-independent approach for identification and enumeration of bacteria and archaea cells with an epifluorescence microscope without isolation and cultivation of microbial cells (Amann et al., 1995). The oligonucleotide probes used for FISH are designed to target the 16S rRNA sequence of a particular taxonomic group of microbes because 16S rRNA sequence are highly conserved, and present in high copy numbers (10^3–10^4 ribosomes/cell) in metabolically active cells. The important aspect of the FISH technique is the design and synthesis of oligonucleotide probes for the specific-taxonomic groups of microorganisms (domain-, kingdom- and phylum-specific probes). Probes are usually designed from a sequence database using *in silico* selection and alignment of rRNA sequence. The probes are usually labeled at 5'-position with the fluorochrome (fluorescein or rhodamine). The procedural steps for FISH include preparation of gelatin-coated slides, cell fixation and permeabilization with paraformaldehyde and/or ethanol, hybridization of the fluorescently labeled oligonucleotide probes with the fixed cells in a sealed moisture chamber to avoid evaporation of hybridization buffer, thorough washing of the microscopic slides after hybridization to decrease the background fluorescence, and visualization of the hybridized microbial cells using epifluorescence microscopy (O'Donnell and Whiteley, 1999). Oligonucleotide probes, such as LF655 (specific for *Leptosprillum* species groups I, II and III), ACM732 (specific for *Acidimicrobium* and related species), SUL228 (specific for *Sulfobacillus* species), Acdp821 (specific for *Acidiphilium* species), Thio820 (specific for *Acidithiobacillus ferrooxidans* and *Acidithiobacillus thiooxidans*) and TF539 (*specific for Acidithiobacillus ferrooxidans*) have been designed for identification and enumeration of different genera of industrially important bioleaching microorganisms (Schrenk et al., 1998; Peccia et al., 2000; Bond and Banfield, 2001). Although FISH is an easy-to-perform method for the characterization of environmental microorganisms without cultivation, problems associated with cell permeability to the oligonucleotide probe, low copy numbers of ribosomes, absence of complementary sequences in the cell and inhibition of hybridization signal restrict the application of FISH for efficient characterization of organisms that comprise <0.1% of the samples (Amann et al., 1995). Some improvements to the sensitivity of the FISH technique have been developed recently by the use of catalyzed reporter deposition (CARD-FISH). In CARD-FISH, the FISH assay is performed with horseradish peroxidase-labeled oligonucleotide probes and flurochrome-labeled tyramides. The tyramides are responsible for the enhanced fluoroscence signal. CARD-FISH has been used for quantitative analysis of bioleaching microbial community in three different sulfidic and acidic mine waste tailing dumps located in Botswana, Germany and Sweden (Kock and Schippers, 2008). The CARD-FISH technique was also performed to study the microbial diversity and distribution of leptospirilli along the iron gradient of Río Tinto River (a model of interest to biohydrometallurgy) in Spain (García-Moyano et al., 2008).

5.13. Real-Time PCR

Real-time PCR monitors in real-time the accumulation of amplified target DNA sequences throughout the amplification (denaturation-annealing-extension) cycles. In conventional PCR the amplified PCR product is characterized by agarose gel electrophoresis and staining of the gel with DNA-binding dye ethidium bromide. In real-time PCR, the concentration of the amplicon is detected in real-time by the addition of a DNA-binding dye, such as SYBR Green I, and/or fluorescent labeled oligonucleotide primers. SYBR Green I is mostly used for detection of amplicons in real-time PCR because of the cost and easy assay protocol. SYBR Green I binds with the amplicon in each cycle without causing damage to it. The fluorescent signal is the real-time indicator of the concentration of the amplicon in each cycle and increases proportionally with increases in the accumulation of the target amplicon. The main advantages of real-time PCR are the accurate measurement of initial template copy numbers and reduced chances of contamination because the reactions are performed in a sealed tube. Real-time PCR has been emerging as a robust tool for culture-independent enumeration of environmental microorganisms by plotting the fluorescent signal of the amplified target gene against the copy numbers of the target gene (Zhang and Fang, 2006). The deficiency of this method includes the lack of specificity of DNA-binding dye SYBR Green I that may cause accumulation of non-specific PCR products during real-time reaction. Real-time PCR with SYBR Green I fluorescent dye has been used for the enumeration of bioleaching microorganisms in pure culture (Zammit et al., 2008), mine waste tailings (Kock and Schippers, 2006; Kock et al., 2007; Kock and Schippers, 2008; Schippers et al., 2008), column leaching environments (Liu et al., 2006) and industrial bioleaching heaps (Remonsellez et al., 2007) by targeting the 16S rRNA genes.

6. Conclusions

The numerical and functional importance of bioleaching microorganisms in the cost-effective and eco-friendly extraction of base and valuable metals, bioremediation of toxic and corrosive sulfide- and heavy metals-containing industrial sludge and wastewater, generation of extremely acidic and toxic AMD, and induction of MIC in concrete and metallic infrastructures necessitate the development of robust molecular tools for accurate, reliable and rapid monitoring of bioleaching microorganisms. Several molecular techniques including different DNA fingerprinting methods have been developed for the identification, characterization and enumeration of bioleaching microorganisms. These methods have merits and demerits. The accurate and reliable surveillance of bioleaching microorganisms can only be achieved by integration of different molecular methods. The standardization of genomics, proteomics and high-throughput OMICS methods, such as metagenomics, metaproteomics, metatranscriptomics and metabolomics will certainly help to evaluate the comprehensive ecophysiological functions of these microorganisms in extreme acidic environments, to implement best management practice(s) to control AMD and MIC, and to develop ideal industrial bioprocesses for the recovery of metals from low-grade ores, mine tailings and wastes by identifying novel strains and/or by genetically tailoring microbial strains with desired biocatalytic properties.

REFERENCES

[19] Akbar, T., Akhtar, K., Ghauri, M. A., Anwar, M. A., Rehman, M., Rehman, M., Zafar, Y. and Khalid, A. M. (2005). Relationship among acidophilic bacteria from diverse environments as determined by randomly amplified polymorphic DNA analysis (RAPD). *World J. Microbiol. Biotechnol.*, *21*, 645–648.

[20] Amils, R., Irazabal, N., Moreira, D., Abad, J. P. and Marín, I. (1998). Genomic organization analysis of acidophilic chemolithotrophic bacteria using pulsed field gel electrophoresis techniques. *Biochimie*, *80*, 911–921.

[21] Amann, R. I., Ludwig, W. and Schleifer, K. H. (1995). Phylogenetic identification and *in situ* detection of individual microbial cells without cultivation. *Microbiol. Rev.*, *59*, 143–169.

[22] Baker, B. J. and Banfield, J. F. (2003). Microbial communities in acid mine drainage. *FEMS Microbiol. Ecol.*, *44*, 139–152.

[23] Basu, R., Clausen, E. C. and Gaddy, J. L. (1996) Biological conversion of hydrogen sulfide into elemental sulfur. *Environ. Prog.*, *15*, 234–238.

[24] Battaglia-Brunet, F., Clarens, M., d'Hugues, P., Godon, J. J., Foucher, S. and Morin, D. (2002). Monitoring of a pyrite-oxidising bacterial population using DNA single-strand conformation polymorphism and microscopic techniques. *Appl. Microbiol. Biotechnol.*, *60*, 206–211.

[25] Bergamo, R. F., Novo, M. T. M., Veríssimo, R. V., Paulino, L. C., Stoppe, N. C., Sato, M. I. Z., Manfio, G. P., Prado P. I., Garcia, O. Jr. and Ottoboni, L. M. M. (2004). Differentiation of *Acidithiobacillus ferrooxidans* and *A. thiooxidans* strains based on 16S-23S rDNA spacer polymorphism analysis. *Res. Microbiol.*, *155*, 559–567.

[26] Bodrossy, L and Sessitsch, A. (2004). Oligonucleotide microarrays in microbial diagnostics. *Curr. Opin. Microbiol.*, *7*, 245–254.

[27] Bond, P. L. and Banfield, J. F. (2001). Design and performance of rRNA targeted oligonucleotide probes for in situ detection and phylogenetic identification of microorganisms inhabiting acid mine drainage environments. *Microb. Ecol.*, *41*, 149–161.

[28] Brandl, H., Bosshard, R. and Wegmann, M. (2001). Computer-munching microbes: metal leaching from electronic scrap by bacteria and fungi. *Hydrometallurgy*, *59*, 319–326.

[29] Brierley, J. A. (1997). Heap bioleaching of gold-bearing deposits: theory and operational description. In D.E. Rawlings (Ed.), *Biomining: Theory, Microbes and Industrial Processes*, pp. 103–115, Berlin, Springer.

[30] Brierley, J. A. (2008). A perspective on developments in biohydrometallurgy. *Hydrometallurgy*, *94*, 2–7.

[31] Brierley, J. A. and Brierley, C. L. (1999). Present and future commercial applications of biohydrometallurgy. In R. Amils and A. Ballester (Eds.), *Biohydrometallurgy and The Environment Toward The Mining of The 21st Century, IBS99*, pp. 81–89, Amsterdam, Elsevier.

[32] Canfield, D. E. and Raiswell, R. (1999). The evolution of sulfur cycle. *American J. Sci*, *299*, 697–723.

[33] Cardinale, M., Brusetti, L., Quatrini, P., Borin, S., Puglia, A. M., Rizzi, A., Zanardini, E., Sorlini, C., Corselli, C. and Daffonchio, D. (2004). Comparison of different primer sets for use in automated ribosomal intergenic spacer analysis of complex bacterial communities. *Appl. Environ. Microbiol.*, *70*, 6147–6156.

[34] Coram, N. J. and Rawlings, D. E. (2002). Molecular relationship between two groups of the genus *Leptospirillum* and the finding that *Leptospirillum ferriphilum* sp. Nov. dominates South African commercial biooxidation tanks that operate at 40^0C. *Appl. Environ. Microbiol.*, *68*, 838–845.

[35] Clark, D. A. and Norris, P. R. (1996). *Acidimicrobium ferrooxidans* gen. nov., sp. nov.: mixed-culture ferrous iron oxidation with *Sulfobacillus* species. *Microbiology*, *142*, 785–790.

[36] Crundwell, F. K. (2003). How do bacteria interact with minerals? *Hydrometallurgy*, *71*, 75–81.

[37] Dar, S. A., Kuenen, J. G. and Muyzer, G. (2005). Nested PCR-denaturing gradient gel electrophoresis approach to determine the diversity of sulfate-reducing bacteria in complex microbial communities. *Appl. Environ. Microbiol.*, *71*, 2325–2330.

[38] Demergasso, C. S., Galleguillos, P. A. P., Escudero, L. V. G., Zepeda, V. J. A., Castillo, D. and Casamayor, E. O. (2005). Molecular characterization of microbial populations in a low-grade copper ore bioleaching test heap. *Hydrometallurgy*, *80*, 241–253.

[39] De Wulf-Durand, P., Bryant, L. J. and Sly, L. I. (1997). PCR-mediated detection of acidophilic bioleaching-associated bacteria. *Appl. Environ. Microbiol.*, *63*, 2944–2948.

[40] d'Hugues, P., Joulian, C., Spolaore, P., Michel, C., Garrido, F. and Morin, D. (2008). Continuous bioleaching of a pyrite concentrate in stirred reactors: population dynamics and exopolysaccharide production vs. bioleaching performance. *Hydrometallurgy*, *94* 34–41.

[41] Dinardo, O. and Mohapatra, B. R. (2008). Identification of four bacterial species cultured from sediment samples collected at two mines in the Sudbury area based on 16S rRNA gene sequence analysis. *CANMET Mining and Mineral Sciences Laboratory Internal Report*, pp. 16.

[42] Dinardo, O., Hogan, C., Kapoor, A., Koren, D., Kwong, J., MacLeod, A., Pawlak M. Lortie, L. and Gould, W. D. (2004). Bioleaching in the mining industry – a literature review. *CANMET Mining and Mineral Sciences Laboratory Internal Report*, pp. 160.

[43] Dopson, M. and Lindström E. B. (2004). Analysis of community composition during moderately thermophilic bioleaching of pyrite, arsenical pyrite, and chapcopyrite. *Microb. Ecol.,* 48, 19–28.

[44] Droffelaar, H. V. and Atkinson, J. T. N. (1995). Corrosion and Its Control: An Introduction to the Subject. NACE International, Houston, TX.

[45] Dunbar, J., Ticknor, L. and Kuske, C. (2000). Assessment of microbial diversity in four southwestern United States soils by 16S rRNA gene terminal restriction fragment analysis. *Appl. Environ.Microbiol., 66,* 2943–2950.

[46] du Plessis, C. (2003). Delivery system for heap bioleaching. *World Patent WO, 2, 003, 068, 999,* 21 August.

[47] Edwards, K. J., Bond, P. L., Gihring, T. M. and Banfield, J. F. (2000). An archaeal iron-oxidizing extreme acidophile important in acid mine drainage. *Science, 287,* 1796–1799.

[48] Escobar, B., Bustos, K., Morales, G. and Salazar, O. (2008). Rapid and specific detection of *Acidithiobacillus ferrooxidans* and *Leptospirillum ferrooxidans* by PCR. *Hydrometallurgy, 92,* 102–106.

[49] Fisher, M. M. and Triplett, E. W. (1999). Automated approach for ribosomal intergenic spacer analysis of microbial diversity and its application to freshwater bacterial communities. *Appl. Environ. Microbiol., 65,* 4630–4636.

[50] Foucher, S., Battaglia-Brunet, F., d'Hugues, P., Clarens, M., Godon, J. J. and Morin, D. (2003). Evolution of the bacterial population during the batch bioleaching of a cobaltiferous pyrite in a suspended-solids bubble column and comparison with a mechanically agitated reactor. *Hydrometallurgy, 71,* 5–12.

[51] Franklin, R. B., Taylor, D. R. and Mills, A. L. (1999). Characterization of microbial communities using randomly amplified polymorphic DNA (RAPD). *J. Microbiol. Methods, 35,* 225–235.

[52] Fuchs, T., Huber, H., Teiner, K., Burggraf, S. and Stetter, K. O. (1995). *Metallosphaera prunae,* sp. nov., a novel metal-mobilizing, thermoacidophilic archaeum, isolated from a uranium mine in Germany. *Syst. Appl. Microbiol., 18,* 560–566.

[53] Garcia-Martinez, J., Acinas, S. G., Anton, A. I. and Rodriguez-Valera, F. (1999). Use of the 16S-23S ribosomal genes spacer region in studies of prokaryotic diversity. *J. Microbiol. Methods, 36,* 55–64.

[54] García-Moyano, A., González-Toril, E., Moreno-Paz, M., Parro, V. and Amils, R. (2008). Evaluation of *Leptrospirillum* spp. in the Río Tinto, a model of interest to biohydrometallurgy. *Hydrometallurgy 94,* 155–161.

[55] Gericke, M., Muller, H. H., Neale, J. W., Norton, A. E. and Crundwell, F. K. (2005). Inoculation of heap leaching operations. In S. T. L. Harrison, D. E. Rawlings and J. Petersen (Eds.), *Proc. Intl. Biohydrometallurgy Symposium (Cape Town),* pp. 255–264, Cape Town, IBS.

[56] Gibson, G. R. (1990). Physiology and ecology of the sulfate-reducing bacteria. *J. Appl. Bacteriol., 69,* 769–797.

[57] Goebel, B. M. and Stackebrandt, E. (1994). Cultural and phylogenetic analysis of mixed microbial populations found in natural and commercial bioleaching environments. *Appl. Environ. Microbiol., 60,* 1614–1621.

[58] Golyshina, O. V., Pivovarova, T. A., Karavaiko, G. I., Kondrateva, T. F., Moore, E. R., Abraham, W. R., Lunsdorf, H., Timmis, K. N., Yakimov, M. M. and Golyshin, P. N.

(2000). *Ferroplasma acidiphilum* gen. nov., sp. nov., an acidophilic, autotrophic, ferrous-iron-oxidizing, cell-wall-lacking, mesophilic member of the *Ferroplasmaceae* fam. nov., comprising a distinct lineage of the *Archaea*. *Int. J. Syst. Evol. Microbiol.*, *50*, 997–1006.

[59] Gould, W. D. and Kapoor, A. (2003). The microbiology of acid mine drainage. In J.L. Jambor, D.W. Blowes and A.I.M. Ritchie (Eds.), *Environmental Aspects of Mine Wastes*, pp. 203–226, Vancouver, British Columbia: Mineral Association of Canada.

[60] Gould, W. D. and McCready, R. G. L. (1994). Biotechnological treatment of liquid and solid inorganic wastes. In D.L. Stoner (Ed.), *Biotechnology for the Treatment of Hazardous Waste*, pp. 71–96, Boca Raton: Florida, CRC Press.

[61] Hallberg, K. B. and Johnson, D. B. (2001). Biodiversity of acidophilic prokaryotes. *Adv. Appl. Microbiol.*, *49*, 37–84.

[62] Harrison, A. P. Jr (1982) Genomic and physiological diversity amongst strains of *Thiobacillus ferrooxidans* and genomic comparison with *Thiobacillus thiooxidans*. *Arch. Microbiol.*, *131*, 68–76.

[63] Hawkes, R. B., Franzmann, P. D. and Plumb, J. J. (2006). Moderate thermophiles including "*Ferroplasma cupricumulans*". *Hydrometallurgy*, *83*, 229–236.

[64] He, Z., Xiao, S., Xie, X., Zhong, H., Hu, Y., Li, Q., Gao, F., Li, G., Liu, J. and Qiu, G (2007). Molecular diversity of microbial community in acid mine drainages of Yunfu sulfide mine. *Extremophiles*, *11*, 305–314.

[65] Head, I. M. (1999) Recovery and analysis of ribosomal RNA sequences from the environment. In C. Edwards (Ed.), *Environmental Monitoring of Bacteria*, pp. 139–174, Totowa, New Jersey: Humana Press.

[66] Heuer, H., Krsek, M., Baker, P., Smalla, K. and Wellington, E. M. H. (1997). Analysis of actinomycete communities by specific amplification of genes encoding 16S rRNA and gel-electrophoretic separation in denaturing gradients. *Appl. Environ. Microbiol.*, *63*, 3233–3341.

[67] Hippe, H. (2000). *Leptospirillum* gen. nov. (ex Markosyan 1972), nom. rev., including *Leptospirillum ferrooxidans* sp. nov. (ex Markosyan 1972), nom. rev. and *Leptospirillum thermoferrooxidans* sp. nov. (Golovacheva et al. 1992). *Int. J. Syst. Evol. Microbiol.*, *50*, 501–503.

[68] Hiraishi, A., Nagashima, K. V., Matsuura, K., Shimada, K., Takaichi, S., Wakao, N. and Katayama, Y. (1998). Phylogeny and photosynthetic features of *Thiobacillus acidophilus* and related acidophilic bacteria: its transfer to the genus *Acidiphilium* as *Acidiphilium acidophilum* comb. nov. *Int. J. Syst. Bacteriol.*, *48*, 1389–1398.

[69] Hiraishi, A., Matsuzawa, Y., Kanbe, T. and Wakao, N. (2000). *Acidisphaera rubrifaciens* gen. nov., sp. nov., an aerobic bacteriochlorophyll-containing bacterium isolated from acidic environments. *Int. J. Syst. Evol. Microbiol.*, *50*, 1539–1546.

[70] Hiraishi, A. and Shimada, K. (2001). Aerobic anoxygenic photosynthetic bacteria with zinc-bacteriochlorophyll. *J. Gen. Appl. Microbiol.*, *47*, 161–180.

[71] Hunter, C. J. (2001). A bacterially assisted heap leach. *World Patent WO, 0144, 519, 21*, 21 June.

[72] Hunter, C. J. (2002). A method for the bacterially assisted heap leaching of chalcopyrite. *World Patent, 2002 70757 (A1) (US 2004 091984).*

[73] Hunter, C. J. and Williams, T.L. (2002). Adaptation of bacteria for leaching. *World Patent WO, 02, 066, 689*, 29 August.

[74] Jerez CA (2008). The use of genomics, proteomics and other OMICS technologies for the global understanding of biomining microorganisms. *Hydrometallurgy*, *94*, 162–169.

[75] Johnson, D. B. and Hallberg, K. B. (2007). Techniques for detecting and identifying acidophilic mineral-oxidizing microorganisms. In D. E. Rawlings and D. B. Johnson (Eds.), *Biomining*, pp. 237–261, New York, Springer Berlin Heidelberg.

[76] Johnson, D. B. (1998). Biodiversity and ecology of acidophilic microorganisms. *FEMS Microbiol. Ecol.*, *27*, 307–317.

[77] Johnson, D. B. and Roberto, F. F. (1997). Heterotrophic acidophiles and their roles in the bioleaching of sulfide minerals. In D. E. Rawlings (Ed.), *Biomining: Theory, Microbes and Industrial Processes*, pp. 259–279, Berlin Heidelberg New York, Springer.

[78] Johnson, D. B., Okibe, N. and Hallberg, K. B. (2005). Differentiation and identification of iron-oxidizing acidophilic bacteria using cultivation techniques and amplified ribosomal DNA restriction enzyme analysis. *J. Microbiol. Methods 60*, 299–313.

[79] Kaeberlein, T., Lewis, K. and Epstein, S. S. (2002). Isolating "uncultivable microorganisms" in pure culture in a simulated natural environment. *Science 296*, 1127–1129.

[80] Karavaiko, G. I., Turova, T. P., Kondrat'eva, T. F., Lysenko, A. M., Kolganova, T. V., Ageeva, S. N., Muntyan, L. N. and Pivovarova, T. A. (2003). Phylogenetic heterogeneity of the species *Acidithiobacillus ferrooxidans*. *Int. J. Syst. Evol. Microbiol. 53*, 113–119.

[81] Kinnunen, P. H. M. and Puhakka, J. A. (2004). High-rate ferric sulfate generation by a *Leptospirillum ferriphilum*-dominated biofilm and the role of jarosite in biomass retainment in a fluidized-bed reactor. *Biotechnol. Bioeng. 85*, 697–705.

[82] Kock, D., Graupner, T., Rammlmair, D. and Schippers, A. (2007). Quantification of microorganisms involved in cemented layer formation in sulfidic mine waste tailings. *Adv. Mat. Res. 20-21*, 481–484.

[83] Kock, D. and Schippers, A. (2006). Geomicrobiological investigation of two different mine waste tailings generating acid mine drainage. *Hydrometallurgy 83*, 167–175.

[84] Kock, D. and Schippers, A. (2008). Quantitative microbial community analysis of three different sulfidic mine tailing dumps generating acid mine drainage. *Appl. Environ. Microbiol. 74*, 5211–5219.

[85] Kurosawa, N., Itoh, Y. H., Iwai, T., Sugai, A., Uda, I., Kimura, N., Horiuchi, T. and Itoh, T. (1998). *Sulfurisphaera ohwakuensis* gen. nov., sp. nov., a novel extremely thermophilic acidophile of the order *Sulfolobales*. *Int. J. Syst. Bacteriol. 48*, 451–456.

[86] Lane, D. J., Harrison, A. P., Stahl, D. Pace, B., Giovannoni, S. J., Olsen, G. J. and Pace, N.R. (1992). Evolutionary relationship among sulfur- and iron-oxidizing eubacteria. *J. Bacteriol. 174*, 269–278.

[87] Lee, D. H., Zo, Y. G. and Kim, S. J. (1996). Nonradioactive method to study genetic profiles of natural bacterial communities by PCR-single-strand-conformation polymorphism. *Appl. Environ. Microbiol.*, *62*, 3112–3120.

[88] Little, B. J., Ray, R. I. and Pope, R. K. (2000). Relationship between corrosion and the biological sulfur cycle: a review. *Corrosion (Houston, TX, U. S.) 56*, 433 – 443.

[89] Liu, C. Q., Plumb, J. and Hendry, P. (2006). Rapid specific detection and quantification of bacteria and archaea involved in mineral sulfide bioleaching using real-time PCR. *Biotechnol. Bioeng. 94*, 330–336.

[90] Liu, W. T., Marsh, T. L., Cheng, H. and Forney, L. J. (1997). Characterization of microbial diversity by determining terminal restriction fragment length polymorphisms of genes encoding 16S rRNA. *Appl. Environ. Microbiol., 63*, 4516–4522.

[91] López-Archilla, A.I., Marin, I. and Amils, R. (2001). Microbial community composition and ecology of an acidic aquatic environment: the Tinto River, Spain. *Microb. Ecol., 41*, 20–35.

[92] Matlakowska, R. and Sklodowska, A. (2006). Adaptive changes of chemolithoautotrophic acidophilic sulfur-oxidizing bacteria during growth in sewage sludge. *Can. J. Microbiol., 52*, 1189–1198.

[93] McCready, R. G. L. and Gould, W. D. (1990). Bioleaching of uranium. In H.L. Ehrlich and C.L. Brierley (Eds.), *Microbial Mineral Recovery*, pp. 107–125, New York. McGraw-Hill.

[94] Mohapatra, B. R., Broersma, K., Nordin, R. and Mazumder, A. (2007a). Evaluation of repetitive extragenic palindromic-PCR for discrimination of fecal *Escherichia coli* from humans, and different domestic- and wild-animals. *Microbiol. Immunol., 51*, 733–740.

[95] Mohapatra, B. R., Broersma, K. and Mazumder, A. (2007b). Comparison of five rep-PCR genomic fingerprinting methods for differentiation of fecal *Escherichia coli* from humans, poultry and wild birds. *FEMS Microbiol. Lett. 277*, 98–106.

[96] Mohapatra, B. R., Gould, W. D., Dinardo, O. and Koren, D. W. (2008). An overview of the biochemical and molecular aspects of microbial oxidation of inorganic sulfur compounds. *CLEAN-Soil, Air, Water 36*, 823–829.

[97] Muyzer, G. and Smalla, K. (1998). Application of denaturing gradient gel electrophoresis (DGGE) and temperature gradient gel electrophoresis (TGGE) in microbial ecology. *Antonie van Leeuwenhoek 73*, 127–141.

[98] Muyzer, G., de Waal, E. C. and Uitterlinden, A. G. (1993). Profiling of complex microbial populations by denaturing gradient gel analysis of polymerase chain reaction-amplified genes coding for 16S rRNA. *Appl. Environ. Microbiol. 59*, 695–700.

[99] Nagpal, M. L., Fox, K. F. and Fox, A. (1998). Utility of 16S–23S rRNA spacer region methodology: how similar are interspace regions within a genome and between strains for closely related organisms? *J. Microbiol. Methods 33*, 211–219.

[100] Ni, Y. Q., He, K. Y., Bao, J. T. Yang, Y., Wan D. S. and Li, H. Y. (2008). Genomic and phenotypic heterogeneity of *Acidithiobacillus* spp. strains isolated from diverse habitats in China. *FEMS Microbiol. Ecol., 64*, 248–259.

[101] Nicomrat, D., Dick, W. A., Dopson, M. and Tuovinen, O. H. (2008) Bacterial phylogenetic diversity in a constructed wetland system treating acid coal mine drainage. *Soil. Biol. Biochem., 40*, 312–321.

[102] Nordstrom, D. K. and Alpers, C. N. (1999). Negative pH, efflorescent mineralogy, and consequences for environmental restoration at the Iron Mountain Superfund site, California. *Proc. Natl. Acad. Sci., 96*, 3455–3462.

[103] Norris, P. R., Clark, D. A., Owen, J. P. and Waterhouse, S. (1996). Characteristics of *Sulfobacillus acidophilus* sp. nov. and other moderately thermophilic mineral-sulphide-oxidizing bacteria. *Microbiology 142*, 775–783.

[104] Norris, P. R., Burton, N. P. and Foulis, N. A. M. (2000). Acidophiles in bioreactor mineral processing. *Extremophiles 4*, 71–76.

[105] Novo, M. T. M., DeSouza, A. P., Garcia, O. Jr. and Ottoboni, L. M. M. (1996). RAPD genomic fingerprinting differentiates *Thiobacillus ferrooxidans* strains. *Syst. Appl. Microbiol.*, *19*, 91–95.

[106] O'Donnell, A. G. and Whiteley, A. S. (1999). Recovery and analysis of ribosomal RNA sequences from the environment. In C. Edwards (Ed.), *Environmental Monitoring of Bacteria*, pp. 221–236, Totowa, New Jersey: Humana Press.

[107] Okabayashi, A., Wakai, S., Kanao, T., Sugio, T. and Kamimura, K. (2005). Diversity of 16S ribosomal DNA-defined bacterial population in acid rock drainage from Japanese pyrite mine. *J. Biosci. Bioeng.*, *100*, 644–652.

[108] Olive, D. M. and Bean, P. (1999). Principle and applications of methods for DNA-based typing of microbial organisms. *J. Clin. Microbiol. 37*, 1661–1669.

[109] Olson, G. J., Brierley, J. A. and Brierley, C. L. (2003). Bioleaching review part B: Progress in bioleaching: applications of microbial processes by the mineral industries. *Appl. Microbiol. Biotechnol.*, *63*, 249–257.

[110] Orita, M., Iwahana, H., Kanazawa, H., Hayashi, K. and Sekiya, T. (1989). Detection of polymorphisms of human DNA by gel electrophoresis as single-strand conformation polymorphisms. *Proc. Natl. Acad. Sci.*, *86*, 2766–2770.

[111] Paulino, L. C., Bergamo, R. F., de Mello, M. P., Garcia, O. Jr., Manifo, G. P. and Ottoboni, L. M. M. (2001). Molecular characterization of *Acidithiobacillus ferrooxidans* and *A. thiooxidans* strains isolated from mine wastes in Brazil. *Antonie van Leeuwenhoek 80*, 65–75.

[112] Postgate, J. R. (1984). *The sulfate-reducing Bacteria*, pp. 1–7, New York, Cambridge University Press.

[113] Peccia, J., Marchand, E. A., Silverstein, J. and Hernandez, M. (2000). Development and application of small-subunit rRNA probes for assessment of selected *Thiobacillus* species and members of the genus *Acidiphilium*. *Appl. Environ. Microbiol. 66*, 3065–3072.

[114] Pizarro, J., Jedlicki, E., Orellana, O., Romero, J. and Espejo, R. T. (1996). Bacterial populations in samples of bioleached copper ore as revealed by analysis of DNA obtained before and after cultivation. *Appl. Environ. Microbiol.*, *62*, 1323–1328.

[115] Rademaker, J. L. W. and de Bruijn, F. J. (1997). Characterization and classification of microbes by rep-PCR genomic fingerprinting and computer-assisted pattern analysis. In G. Caetano-Anoll´es and P. M. Gresshoff (Eds.), *DNA Markers: Protocols, Applications, and Overviews*, pp. 151–171, New York , John Wiley and Sons.

[116] Rawlings, D. E. (1995). Restriction enzyme analysis of 16S rDNA genes for the rapid identification of *Thiobacillus ferrooxidans*, *Thiobacillus thiooxidans* and *Leptospirillum ferrooxidans* strains in leaching environments. In T. Vargas, C.A. Jerez, J.V. Wiertz, and H. Toledo (Eds.), *Biohydrometallurgy Processing, Vol. II*, pp. 9–18, Santiago, University of Chile.

[117] Rawlings, D. E. (2002). Heavy metal mining using microbes. *Annu. Rev. Microbiol.*, *56*, 65–91.

[118] Rawlings, D. E. and Johnson, D. B. (2007). The microbiology of biomining: development and optimization of mineral–oxidizing microbial consortia. *Microbiology 153*, 315–324.

[119] Rawlings, D. E., Dew, D. and du Plessis, C. (2003). Biomineralization of metal-containing ores and concentrates. *TIBTECH 21*, 38–44.

[120] Reith, F., Rogers, S. L., McPhail, D. C. and Brugger, J. (2007). Potential for the utilization of microorganisms in gold processing. *World Gold Conference (Cairns, QLD, 22-24 October)*, pp. 67–74.

[121] Remonsellez, F., Galleguillos, F., van Rensburg, S. J., Rautenbach, G. F., Galleguillos, P., Castillo, D. and Demergasso, C. (2007). Monitoring the microbial community inhabiting a low-grade copper sulphide ore by quantitative real-time PCR analysis of 16S rRNA genes. *Adv. Mat. Res., 20-21*, 539–542.

[122] Rohwerder, T., Gehrke, T., Kinzler, K. and Sand, W. (2003). Bioleaching review, Part A: Progress in bioleaching fundamentals and mechanisms of bacterial metal sulfide oxidation. *Appl. Microbiol. Biotechnol., 63*, 239–248.

[123] Romero, J., Yañez, C., Vásquez, M., Moore, E. R. B. and Espejo, R. T. (2003). Characterization and identification of an iron-oxidizing, *Leptospirillum*-like bacterium, present in the high sulfate leaching solution of a commercial bioleaching plant. *Res. Microbiol., 154*, 353–359.

[124] Rzhepishevska, O. I., Lindström, E. B., Tuovinen, O. H. and Dopson, M. (2005). Bioleaching of sulfidic tailing samples with a novel, vacuum-positive pressure driven bioreactor. *Biotechnol. Bioeng., 92*, 559–567.

[125] Sand, W., Gerke, T., Hallmann, R. and Schippers, A. (1995). Sulfur chemistry, biofilm, and the (in)direct attack mechanism—a critical evaluation of bacterial leaching. *Appl. Microbiol. Biotechnol. 43*, 961–966.

[126] Schippers, A., Rohwerder, T. and Sand, W. (1999). Intermediary sulfur compounds in pyrite oxidation: implications for bioleaching and biodepyritization of coal. *Appl. Microbiol. Biotechnol., 52*, 104–110.

[127] Schippers, A., Naggy, A. A., Kock, D., Melcher, F. and Gock, E. D. (2008). The use of FISH and real-time PCR to monitor the biooxidation and cyanidation for gold and silver recovery from a mine tailings concentrate (Ticapampa, Peru). *Hydrometallurgy, 94*, 77–81.

[128] Schnell, H. (1997). Bioleaching of copper. In D.E. Rawlings (Ed.), *Biomining: Theory, Microbes and Industrial Processes*, pp. 21–43, Berlin, Springer.

[129] Schrenk, M. O., Edwards, K. J., Goodman, R. M., Hamers, R. J. and Banfield, J. F. (1998). Distribution of *Thiobacillus ferrooxidans* and *Leptospirillum ferrooxidans*: implications for generation of acid mine drainage. *Science, 279*, 1519–1522.

[130] Selenska-Pobell, S., Otto, A. and Kutschke, S. (1998). Identification and discrimination of thiobacilli using ARDREA, RAPD and rep-APD. *J. Appl. Microbiol., 84*, 1085–1091.

[131] Selenska-Pobell, S., Kampf, G., Flemming, K., Radeva, G. and Satchanska, G. (2001). Bacterial diversity in soil samples from two uranium waste piles as determined by rep-APD, RISA and 16S rDNA retrieval. *Antonie van Leeuwenhoek, 79*, 149–161.

[132] Stackebrandt, E., Stahl, D. A. and Devereux, R. (1995). Taxonomic relationships. In L.L. Barton (Ed.), *Sulfate-reducing Bacteria*, pp. 49–87, New York, Plenum Press.

[133] Stoner, D. L., Browning, C. K., Bulmer, D. K., Ward, T. E. and MacDonell, M. T. (1996). Direct 5S rRNA assay for monitoring mixed-culture bioprocesses. *Appl. Environ. Microbiol. 62*, 1969–1976.

[134] van Aswegen, P. C., Godfrey, M. W., Miller, D. M. and Haines, A. K. (1991). Developments and innovations in bacterial oxidation of refractory ores. *Miner. Metallurg. Processing,* 8, 188–192.

[135] Vaneechoutte, M., Rossau, R., Vos, P. D., Gillis, M., Janssens, D., Paepe, N., Rouck, A. D., Fiers, T., Claeys, G. and Kersters, K. (1992). Rapid identification of bacteria of the *Comamonadaceae* with amplified ribosomal DNA restriction analysis (ARDRA). *FEMS Microbiol. Lett.*, *93*, 227–233.

[136] Vásquez, M. and Espejo, R. T. (1997). Chemolithotrophic bacteria in copper ores leached at high sulfuric acid concentration. *Appl. Environ. Microbiol.*, *63*, 332–334.

[137] Versalovic, J., Schneider, M., de Bruijn, F. J. and Lupski, J. R. (1994). Genomic fingerprinting of bacteria using repetitive sequence-based polymerase chain reaction. *Methods Mol. Cell. Biol.*, *5*, 25–40.

[138] Waltenbury, D. R., Leduc, L. G. and Ferroni, G. D. (2005). The use of RAPD genomic fingerprinting to study relatedness in strains of *Acidithiobacillus ferrooxidans*. *J. Microbiol. Methods*, *62*, 103–112.

[139] Watling, H. R. (2006). The bioleaching of sulphide minerals with emphasis on copper sulphides – A review. *Hydrometallurgy*, *84*, 81–108.

[140] Welsh, J. and McClelland, M. (1990). Fingerprinting genomes using PCR with arbitrary primers. *Nucleic Acids Res.*, *18*, 7213–7218.

[141] Whiteley, C. G. and Lee, D. J. (2006). Enzyme technology and biological remediation. *Enzyme Microb. Technol.*, *38*, 291–316.

[142] Xie, X., Xiao, S., He, Z., Liu, J. and Qiu, G. (2007). Microbiological populations in acid mineral bioleaching systems of Tong Shankou copper mine, China. *J. Appl. Microbiol.*, *103*, 1227–1238.

[143] Yin, H., Cao, L., Qui, G., Wang, D., Kellogg, L., Zhou, J., Dai, Z and Liu, X. (2007). Development and evaluation of 50-mer oligonucleotide arrays for detecting microbial populations in acid mine drainages and bioleaching systems. *J. Microbiol. Methods*, *70*, 165–178.

[144] Zammit, C. M., Mutch, L. A., Watling, H. R. and Watkin, E. L. J. (2008). Evaluation of quantitative real-time polymerase chain reaction for enumeration of biomining microorganisms in culture. *Hydrometallurgy*, *94*, 185–189.

[145] Zhang, T. and Fang, H. H. P. (2006). Applications of real-time polymerase chain reaction for quantification of microorganisms in environmental samples. *Appl. Microbiol. Biotechnol.*, *70*, 281–289.

[146] Zhou, J., Xia, B., Treves, D. S., Wu, L. Y., Marsh, T. L., O'Neill, R. V., Palumbo, A. V. and Tiedje, J. M. (2002). Spatial and resource factors influencing high microbial diversity in soil. *Appl. Environ. Microbiol.*, *68*, 326–334.

In: Encyclopedia of DNA Research (3 Volume Set)
Editors: S. J. Duncan and P. H. Wiley
ISBN 978-1-61324-305-3

Chapter 36

THE USE OF REP-PCR ON MEMBERS OF THE CLASS *ACTINOBACTERIA* STACKEBRANDT *ET AL.* 1997

***L.A. Maldonado*[1] *and E.T. Quintana*[2]**

[1] Instituto de Ciencias del Mar y Limnologia (ICMyL) Universidad Nacional Autónoma de México (UNAM) Circuito Exterior s/n Ciudad Universitaria 04510, México

[2] Escuela Nacional de Ciencias Biológicas Instituto Politécnico Nacional (IPN) Prolongación de Carpio y Plan de Ayala 11340, México

ABSTRACT

Advances in molecular biology have provided highly specific methods for classifying, identifying and typing isolates based on nucleic acid sequence data. Taxonomic systems are becoming increasingly important for characterizing microorganisms since they can be applied in epidemiological investigations of disease outbreaks to identify a potential common source and/or to establish grouping strategies for isolates recovered from several environments. Typing systems based on phenotypic tests have limitations as key phenotypic traits of microorganisms may be incosistently expressed or may not provide enough discriminative power to separate closely related microorganisms. Therefore, shortcomings of phenotypically based methods have pushed towards the development and employment of methods based on microbial genotypes or DNA sequences minimising problems with respect to reproducibility and, in some cases, enabling the establishment of large databases (a biological-genotypic database) of characterized organisms. Several molecular typing systems have been used to study the relationships within high G+C bacteria (aka *Actinobacteria*) and the current chapter focuses on the genera *Micromonospora* and *Streptomyces*. Repetitive Extragenic Palindromic Polymerase Chain Reactions (Rep-PCR) fingerprints bacterial genomes based on strain-specific patterns derived from PCR amplification of repetitive DNA elements present within bacterial genomes. The palindromic nature of repetitive elements (ie. BOX, ERIC and REP primers) across the microbial genome and their ability to form stem-loop structures leads to the generation of unique fingerprint patterns. Members of the genus *Micromonospora* cannot be separated on basis of their phenotypic properties, therefore, the use of molecular fingerprinting methods when large numbers of isolates belonging to this genera are urgently needed. The chapter exemplifies its use on nearly 200 micromonosporae recovered from aquatic ecosystems. On the other hand, studies on members of the genus *Streptomyces* -which currently holds nearly 600 species- makes comparative studies difficult, hence also the need for a reliable fingerprinting method. The chapter then deals with studies on streptomycetes isolated from clinical material (11 strains) and their fingerprint relationships. The use of REP-PCR to generate fingerprint patterns and the construction of a biological-genotypic database are provided for these *Actinobacteria* genera.

The accurate circumscription of subtypes within a species is becoming increasingly important in all branches of microbiology. Microbial fingerprinting is extensively used in diagnostic bacteriology (Oyarzabal et al., 1997), in ecological and evolutionary genetical studies (van Belkum et al., 2001) and in search and discovery programmes designed to detect new microbial products (Goodfellow and O'Donnell, 1989; Bull et al., 1992, 2000). The various molecular fingerprinting methods have advantages and disadvantages when applied to specific situations and objectives. Fingerprinting methods used to discriminate between strains within target species are easy to perform and data interpretation is relatively straighforward. Several molecular fingerprinting systems have been shown to be effective in the delineation of *Actinobacteria*, that is Gram positive bacteria of a high Guanine + Citosine content (> 55%), at and below the species level (Welsh and McClelland, 1990; Vaneechoutte et al., 1992; Gürtler and Stanisich, 1996; Yoon et al., 1997; Hall et al., 2001).

The introduction and application of molecular taxonomic procedures such as 16S rRNA gene sequencing (Woese, 1987; Olsen and Woese, 1993; Ludwig and Klenk, 2001), DNA fingerprinting (Vaneechoutte et al., 1998; Rademaker et al., 2000; Gürtler and Mayall, 2001; van Belkum et al., 2001), DNA:DNA hybridisation (Grimont, 1981; Stackebrandt et al., 2002), multilocus sequence typing (Maiden et al., 1998; Sails et al., 2003; Tavanti et al., 2003) and sequence analyses of complete genomes (Õmura et al., 2001; Bentley et al., 2002; Ikeda et al., 2003) are providing new insights into prokaryotic systematics (Woese, 1987; Olsen et al., 1994; Ludwig and Schleifer, 1999; Palys et al., 2000; Gürtler and Mayall, 2001; Kim et al., 2001; Stackebrandt et al., 2002), including the classification and identification of the *Actinobacteria* (Stackebrandt et al., 1997; Kim et al., 1999; Salazar et al., 2000; Zhang et al., 2001; Stach et al., 2003; Stevens et al., 2007).

The development of molecular taxonomic methods based on DNA analyses has made it possible to undertake extensive, rapid and precise characterisation of representatives of bacterial taxa isolated from diverse habitats (Moyer et al., 1994; Bull et al., 2000; Stach et al., 2003; Maldonado et al., 2005b, 2008, 2009; Stevens et al., 2007). Such techniques tend to give results that are more robust than those from chemosystematic studies which are often sensitive to small changes in cultivation conditions (Goodfellow and Minnikin, 1985; Goodfellow and O'Donnell, 1994; Hugenholtz et al., 1998). Besides, such techniques provide an insight of the whole genomic information from each strain. Many molecular taxonomic procedures are based on the use of the polymerase chain reaction, which is used to amplify target genes from either culturable isolates or microbial community DNA (Lane, 1991; Stackebrandt et al., 1997; Muyzer, 1999; Bull et al., 2000; Stach et al., 2003; Stevens et al., 2007).

The application of different molecular fingerprinting techniques to determine the degree of sequence conservation between bacterial genomes is based on the detection of naturally occurring DNA polymorphisms which are the result of either point mutations or rearrangements in genomic DNA (i.e. insertions or deletions) or fragments of DNA. DNA polymorphisms can be detected by scoring band presence against band absence in banding patterns generated either by restriction enzyme digestion or DNA amplification procedures (Versalovic et al., 1991; Moyer et al., 1994; Vaneechoutte et al., 1998; van Belkum et al.,

2001). The underlying premise is that variations in banding patterns are a direct reflection of genetic relationships between the tested strains hence banding patterns can be considered as genomic fingerprints for characterisation and identification purposes (Versalovic et al., 1994; Gürtler and Mayall, 2001). A robust fingerprinting technique can then become a dereplication molecular procedure to indicate whether a specific clinical and/or environmental isolate has been studied (cultured) previously (Maldonado et al., 2008).

REPETITIVE-PCR

Versalovic and colleagues (1991) described a method for fingerprinting bacterial genomes following examination of strain-specific patterns derived from PCR amplification of repetitive DNA elements present within bacterial genomes. Two main sets of repetitive elements are used for typing purposes. Repetitive extragenic palindromic (REP) elements are 38 bp sequences that consist of six degenerate positions and a 5 bp variable loop between each side of a conserved palindromic stem (Gilson et al., 1984; Stern et al., 1984). REP sequences have been described for numerous enteric bacteria (Gilson et al., 1984; Stern et al., 1984; Yang and Ames, 1988; Sharples and Lloyd 1990; Hulton et al., 1991; Versalovic et al., 1991, 1993, 1994, 1998; Versalovic and Lupski, 2002). The palindromic nature of REP elements and their ability to form stem-loop structures has led to multiple proposed functions for these highly conserved, dispersed elements (Newbury et al., 1987; Yang and Ames, 1988; Gilson et al., 1990). The second set of DNA sequences that have been used successfully for DNA typing are ERIC sequences. These sequences are 126 bp elements that contain a highly conserved central inverted repeat; they are located in extragenic regions of bacterial genomes (Sharples and Lloyd, 1990; Hulton et al., 1991). ERIC sequences have also been defined primarily on data obtained from *Escherichia coli* and *Salmonella typhimurium* strains (Sharples and Lloyd, 1990; Hulton et al., 1991).

ERIC and REP sequences are the most commonly used targets for DNA typing though another repetitive element, the BOX sequence, has been used to differentiate between *Streptococcus pneumoniae* strains (Martin et al., 1992; Koeuth et al., 1995). BOX elements consist of differentially conserved sequences which are located within intergenic regions, they can also form stem-loop structures due to their dyad symmetry. These mosaic repetitive elements are composed of various combinations of three subunit sequences referred to as boxA, boxB and boxC (Martin et al., 1992); the three subunit sequences have molecular lengths of 57, 43, and 50 nucleotides, respectively (Martin et al., 1992). BOX elements, which do not have any sequence relationship to either ERIC or REP sequences (Martin et al., 1992), were originally thought to be unique to *Streptococcus pneumoniae* but have now been found in members of several bacterial taxa (Louws et al., 1994; Versalovic et al., 1994; Kwon et al., 1998).

Rep-PCR can be carried out with DNA extracted from bacterial colonies or by a modified method using unprocessed whole cells (Woods et al., 1993). ERIC and REP amplifications can be performed with a single primer, a single set of primers, or by using multiple sets of primers. ERIC patterns are generally less complex than REP patterns though both give good discrimination at the strain level. The use of ERIC and REP in tandem allows increased separation of strains compared with the use of either technique alone. The combined technique is easy to perform and can be applied to large numbers of isolates.

Table 1. Primers commonly employed to generate REP-PCR fingerprints

Repetitive sequence	Primer ID	Sequence (5' to 3')	
BOX	BoxA1R1	CTACGGCAAGGCGAC GCTGACG	
ERIC	E1	ATGTAAGCTCCTGGG GATTCAC	
E2	AAGTAAGTGAC TGGGGTGAGCG		
	REP	Rep 1R1	iiiCGiGCiCATCiGGC
		Rep 21	iCGiCGTATCiGGCCTAC

Table 2. Comparison of some features between commonly employed PCR-based fingerprint techniques

Features	AFLP	RAPD	Rep-PCR
Small amount of DNA	√	√	√
Multiple primers	X	Sometimes	√
Reproducible results	√	After keeping multiple	√
		parameters constant	
Stringent PCR conditions[a]	√	X	X

[a] The term 'stringency' refers to experimental conditions to keep the alignment of the oligonucleotides to their corresponding DNA sequence. The higher the temperature during the annealing PCR conditions, the better the stringency.

Rep-PCR shows broader species applicability and better discriminatory power than either plasmid profiling or genomic fingerprinting (Georghiou et al., 1995; Rademaker et al., 2000).

Banding patterns based on ERIC and REP sequences have been used extensively to differentiate between *Acinetobacter baumanii* (Reboli et al., 1994; Dijkshoorn et al., 1993; Snelling et al., 1996; Martín-Lozano et al., 2002), *Bartonella* (Rodriguez-Baradas et al., 1995), *Burkholderia cepacia* (Hamill et al., 1995), *Citrobacter diversus* (Harvey et al., 1995), *Helicobacter pylori* (Kwon et al., 1998; Kidd et al., 2001), *Lactobacillus* (Gevers et al., 2001; Antonio and Hiller, 2003), *Legionella pneumophila* (Georghiuo et al., 1994), *Neisseria gonorrhoeae* (Poh et al., 1996), *Neisseria meningitidis* (Woods et al., 1993), *Pasteurella multocida* (Amonsin et al., 2002), *Rhizobium meliloti* (De Bruijin, 1992), *Streptococcus pneumoniae* (Versalovic et al., 1995; Zoe-Jordans et al., 1995), *Vibrio parahaemolyticus* (Wong and Lin, 2001) and *Xanthomonas* strains (Louws et al., 1995; Rademaker et al., 1997, 2000).

Rep-PCR has been found to give better discrimination when compared against other typing methods, for instance against multilocus enzyme electrophoresis (Woods et al., 1992), biochemical characterisations (Claridge et al., 1995) or ribotyping (Snelling et al., 1996; Appuhamy et al., 1997). Several studies show good correlation between Rep-PCR and pulsed field gel electrophoresis data though, in general, the former is slightly less discriminatory (Barbier et al., 1996; Liu and Wu, 1997).

The major drawback of Rep-PCR is similar to those found and reported for other PCR-based techniques. The annealing temperature, DNA concentration, genome size, primer concentration and strain to strain variation, can affect fingerprint patterns thereby making comparative studies between laboratories difficult (Gillings and Holley, 1997; Tyler et al., 1997). The technique has even been adapted to an automated format in which fluorescently labelled primers are used to prepare the ERIC or REP profiles prior to separation of the amplified sequences using a fluorescence-based DNA sequencer (DelVecchio et al., 1995; Versalovic et al., 1995). This method allows consistent pattern formation and storage of data as digitized images thereby allowing the identification of unknown strains.

The Class *Actinobacteria* Stackebrandt et al. 1997

Stackebrandt et al. (1997) proposed a new hierarchic classification for actinomycetes, including the assignment of the latter to the class *Actinobacteria* (still nowadays the term *actinomycetes* is employed when referring to any member of this class). This new classification system was based solely on analyses of small subunit (16S) rRNA genes, including specific nucleotide signatures for orders, families and genera.

Members of the *Actinobacteria* are of high economical impact worldwide. Two genera have been extensively (but not exclusively) studied: *Mycobacterium* and *Streptomyces*. The feared *M. tuberculosis* is a long time human pathogen and the appearance of multiresistant strains rises every year dramatically (Cole et al., 2005). On the other hand, the genus *Streptomyces* is one of the most studied genera worldwide because of its innate ability to produce secondary metabolites, notably antibiotics (Hopwood, 2007). At present time, the class *Actinobacteria* encompasses around 55 families and 262 genera all of which have been reported from nearly any habitat on earth.

There is no doubt that 16S rRNA gene sequencing have contributed enourmously to provide not only a better classification criteria but also to understand the ecological distribution of *Actinobacteria*. However, there are still two major drawbacks when working with some taxa of this class as outlined below:

a) The number of species assigned to some genera is a main obstacle for comparative analyses (eg. *Streptomyces*; Labeda, personal communication)
b) The 16S rRNA gene divergence found in some genera is too small to discriminate between closely related species (eg. genera *Micromonospora*, *Salinispora*, and *Tsukamurella*; Goodfellow and Maldonado, 2006; Maldonado et al., 2005a)

It is because of these major limitations that the use of fingerprinting techniques which evaluate the whole genome are advisable in order to discriminate rapidly and reproducible between novel and previously cultured strains. The fast and inexpensive Rep-PCR technique can be employed to construct a genotypic database comprising fingerprints of micromonosporae and streptomycetes.

Streptomyces

Waksman and Henrici (1943) introduced the term *Streptomyces* (Strep.to.my'ces. Gr Adj. *streptos* pliant, bent; *myces* fungus; M. L. masc. n.: *Streptomyces* pliant or bent fungus) a taxon that has had a turbulent taxonomic history partially due to the application of several systematic techniques. The genus encompasses aerobic, Gram-positive, non-acid alcohol fast actinomycetes which produce well developed vegetative hyphae that rarely fragment. Members of the genus are widely distributed in natural habitats and are particular abundant in soil, including composts though marine streptomycetes have been recently reported (Bredholt et al., 2007; Gontang et al., 2007; Maldonado et al., 2009). Some strains are pathogenic to animals, including humans, and to plants (Doering-Saad et al., 1992; Korn-Wendisch and Kutzner, 1992; McNeil and Brown, 1994; Brown and McNeil, 2003).

An unusual feature of the genus *Streptomyces* is the number of species that it contains. The taxon currently encompasses 565 validly described species (Euzéby, 2009) which can be assigned to lumpy groups based on 16S rRNA sequence data. However, the phylogenetic position of the resulting groups has been determined upon comparison with -some-incomplete 16S rRNA gene sequences as complete sequences are not available for all the validly described species of the genus. Nevertheless, there are good grounds to believe that the genus is still underspeciated (Atalan et al., 2000; Sembiring et al., 2000; Manfio et al., 2003) and that metabolically marine-adapted species do exist (Jensen et al., 2005; Hughes et al., 2008; Maldonado et al., 2009). This has only been reported for the genus *Salinispora* which is morphologically and phylogenetically related to *Micromonospora* (Mincer et al., 2002; Maldonado et al., 2005a).

There are only two validly described streptomycetes that can cause serious lesions to man. Actinomycetoma is a common disease in certain tropical and subtropical regions where it has a devastating effect on patients (and communities) as it often leads to deformities and amputation of the affected site. *Streptomyces somaliensis* was the only authenticated sterptomycete considered to be pathogenic to humans but *S. sudanensis* has also been recently recognised as a novel pathogenic species to human (Quintana et al., 2008).

The works by Sadowsky et al. (1996) and Clark et al. (1998) studied the resulting relationships among phyto- and disease-supressive pathogenic *Streptomyces* strains. BOX primed PCR reactions were found to be reproducible and stable producing unique patterns (Sadowsky et al., 1996). Clark and colleagues (1998) reported a higher variation of patterns in *S. ipomaeae* strains using the ERIC and REP primers while BOX-primed fingerprints showed higher similarities. It can be concluded from that work that REP-PCR fingerprinting is taxonomically discriminating at the strain and between strain and species levels respectively.

Micromonospora

The genus *Micromonospora* was first proposed in 1923 by Ørskov as a novel subgroup of the *Actinomycetales*. The genus was suggested to accommodate the monosporic species *Streptothrix chalcea* (Foulerton, 1905). The first species was considered by Ørskov as totally different from any other so-far known actinomycetes. This difference was mainly based on the absence of aerial mycelium and the formation of single spores; this would later be a key to the genus name. Several chemo- and numerical taxonomic studies placed the genus in the family *Micromonosporaceae* Krassilnikov 1938 but several gaps were observed in terms of the resulting intrageneric relationships between members of this taxa. These gaps are also not helped by the morphological unique features of the genus. Most species of *Micromonospora* produce small colonies that are originally orange, will become progressively dark covered by a mucoid substance and production of spores (Kawamoto, 1989).

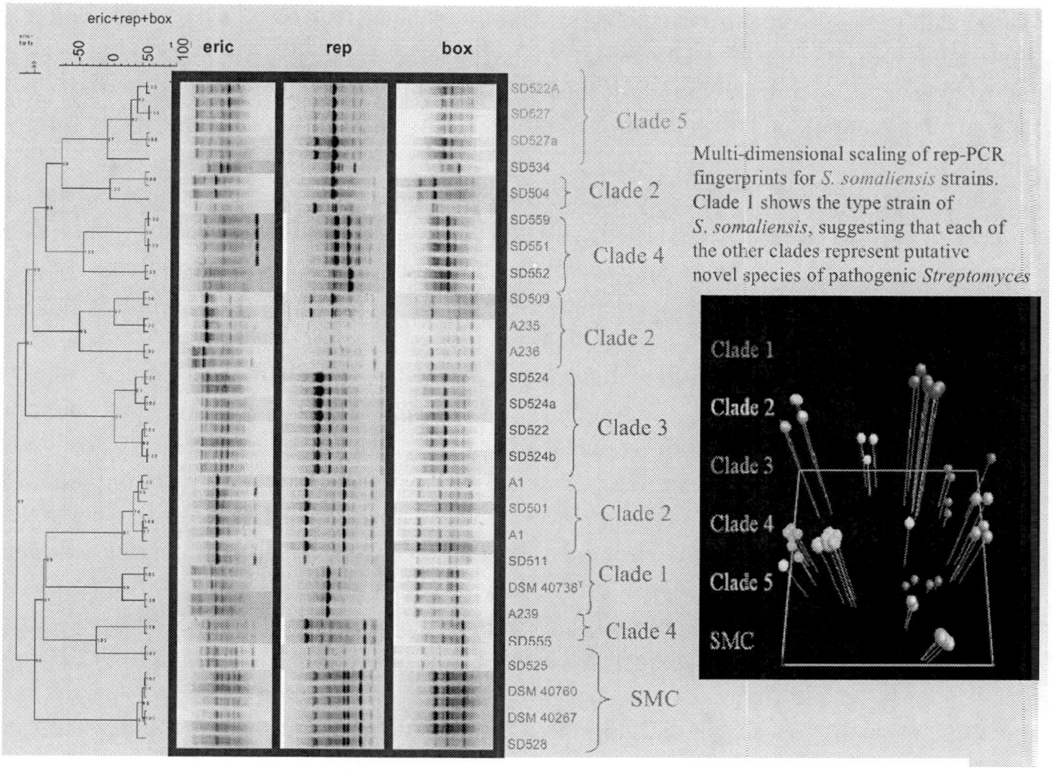

Figure 1. Fingerprint relationships using REP-PCR between streptomycetes isolated from clinical material of human origin.

Koch et al. (1996) studied the intrageneric relationship by determining the 16S rRNA gene sequences of type strains of micromonosporae but it was found that the genus showed sequence similarities of 98 to 98.5%. Such high 16S rRNA gene similarity values indicate a phylogenetically close grouping for which sequencing of other conserved genes may be recommended or suggesting the use of fingerprinting techniques that evaluate the entire genome (Figure 2).

Micromonospora have also proved to be an important source of secondary metabolites, notably antibiotics (Wagman and Weinstein, 1980; He et al., 2001; Ismet et al., 2004). Screening of micromonosporae goes back to 1964 after the discovery of gentamicin from strains of *Micromonospora purpurea* and *M. echinospora* (Weinstein et al., 1964). Since then, over 400 antibiotics from micromonosporae have been described (Goodfellow and O'Donnell, 1989). However, the search for new antibiotics is held back by the limits of selective isolation procedures in recovery of specific parts of the microbial community and the inadequate taxonomy of this genus (Vobis, 1989). Two thirds of the known aminoglycoside antibiotics are produced by non-*Streptomyces* species, with a significant number of those being isolated from the genus *Micromonospora* (Tilley et al., 1990). Indeed gentamicin is currently the only *Micromonospora* antibiotic commercially exploited, from nearly 150 aminoglycoside antibiotics isolated from the genus, highlighting the industrial potential of this genus. Additionally, representatives of practically every chemical family of antibiotics have been isolated from the genus *Micromonopora* (Wagman and Weinstein,

1980), with novel therapeutic compounds, such as calicheamicin and megalomicin receiving significant recent interest (Alarcon et al., 1988; Thorson et al., 2000).

Actinobacteria represent a significant part of the microflora of soils and their ability to form bioactive compounds has led to interest in their isolation from the natural environment (Goodfellow and O'Donnell, 1989) both of *Micromonospora* and *Streptomyces*. Jensen (1930) first reported the occurrence of *Micromonospora* in soil as their natural habitat and isolated 67 strains. Later, Kriss (1939) isolated further micromonosporae from soils in the Saratov region of Russia. *Micromonospora* have long been considered aquatic microorganisms because of the ocurrence of the single spores (hydrophilic) in sea and lake sediments which detached them from soil particles by water.

Umbreit and McCoy (1941) studied the microbial population of fifteen lakes in Wisconsin and reported that micromonosporae formed a large part of the microbial population (10-20%). These autors concluded that micromonosporae formed 40 to 50% of their recovered isolates and that they can act as mineralisers of lake bottom deposits, fullfilling a role similar to that of *Streptomyces* on terrestrial habitats. Micromonosporae have also been isolated from marine environments, such as beach sand, and deep marine sediments (Potter and Baker, 1956; Kawamoto, 1989; Maldonado et al., 2005b, 2008, 2009). It appears, however, that the spores of micromonosporae are easily washed into water reservoirs from soil where because of their innate resistance properties, they can remain dormant for extended periods (Johnston and Cross, 1976). The ability to tolerate reduced oxygen tensions (Watson and Williams, 1974) is in line with the fact that spores washed into aquatic ecosystems can survive in littoral and marine sediments though the possibility of indigenous micromonosporae and micromonosporae-like strains cannot be excluded from marine habitats (Maldonado et al., 2005a; Bredholt et al., 2007). It has been reported that micromonosporae in some environments are comparable to *Streptomyces* in terrestrial ecosystems (Zenova and Zvyagintsev, 1997; De Menezes et al., 2008; Maldonado et al., 2008). Studies on the properties of isolated strains suggest that they may have an effect in lacustrine ecology because of their ability to degrade cellulose and chitin (De Menezes, et al., 2008; Maldonado, unpublished data). Hence it is important to evaluate fingerprint methods to rapidly dereplicate isolates and construct a biological micromonosporae database based on their unique fingerprinting patterns. The absence of a practical subgeneric classification has made it complicated to select representatives for screening programmes or for measuring the extent of their microbial diversity, a prerequisite for studies on biogeography and community structure of these *Actinobacteria* (Maldonado et al., 2008).

FINGERPRINT ANALYSES

Analysis and evaluation of DNA fingerprinting data is not and should not be considered "straightforward". Simple visual examination of data may be sufficient if the objective is merely to compare a limited number of strains in order to determine whether a particular organism is different or similar to other strains. However, if the aim os to compare profiles of a large number of organisms in retrospective or prospective epidemiological studies, or to determine genetic relationships, considerable thought must be given to the design of the experiments, particularly to the acquisition and statistical analyses of data from autoradiograms, filters or gels. Run-to-run variation in the mobilities of DNA fragments need to be normalized against a standard(s) to allow comparison between gels.

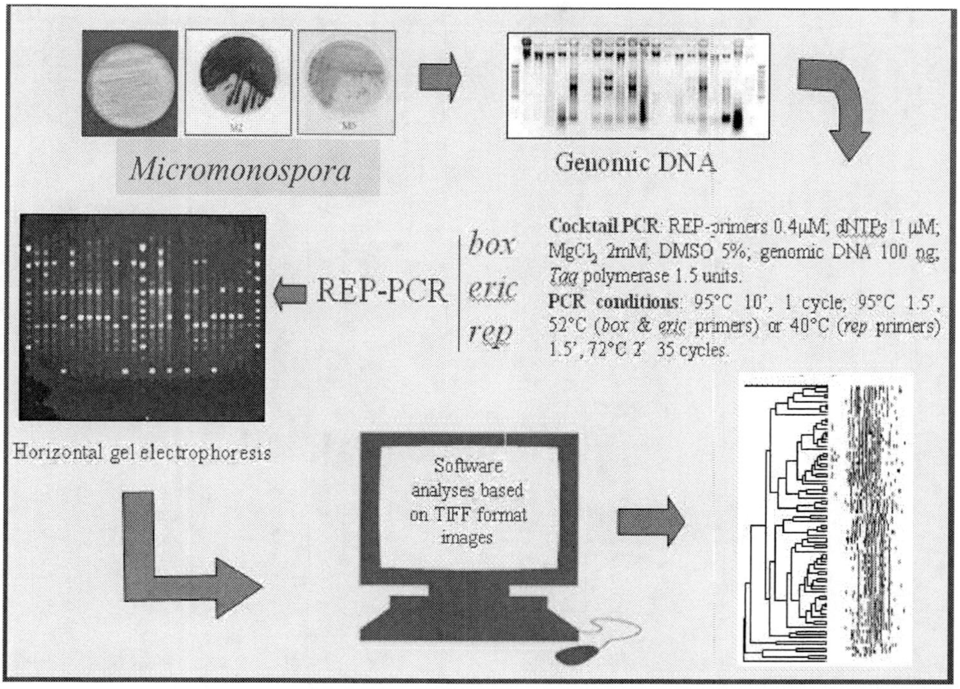

Figure 2. Scheme showing the steps employed to generate REP-PCR fingerprints from micromonosporae.

Genomic DNA fingerprinting data may be acquired in several ways. Agarose gels stained with ethidium bromide are usually photographed with negative or positive print. Autoradiograms and filters on which DNA fragments have been visualized by colorimetric methods can be used durectly for data acquisition. The first step in visual analyses is to determine the relative sizes of each fragment in a lane by comparing them against a molecular size standard run on the same gel. The molecular size standard should be chosen carefully to cover all of the fragments in the sample lanes. The relationship between the size and electrophoretic mobility of a DNA fragment is not linear. Several curve-fitting algorithms have been developed to determine fragment sizes, of these, the local reciprocal (Elder and Southern, 1983) and the cubic spline interpolation (Press et al., 1986) methods are the most accurate (Russel et al., 1991).

Genomic DNA fingerprinting data may also be acquired by using a digitizer or a flatbed scanner connected to a computer and by running appropriate software programs. Oner method involves the use of the Bio Rad Image System™, which consists of a charge coupled device camera connected to a PC computer (Bio rad, version 1.2, USA) equipped with Molecular Analyst/PC Software for data acquisition from autoradiograms, filters and gels. The Molecular Analyst package includes sophisticated automatic band detection and standard curve generation systems which allows the calculation of unknown concentrations, a molecular weight determination and lane-matching program which can be used to print profile graphs at any point in the band analyses.

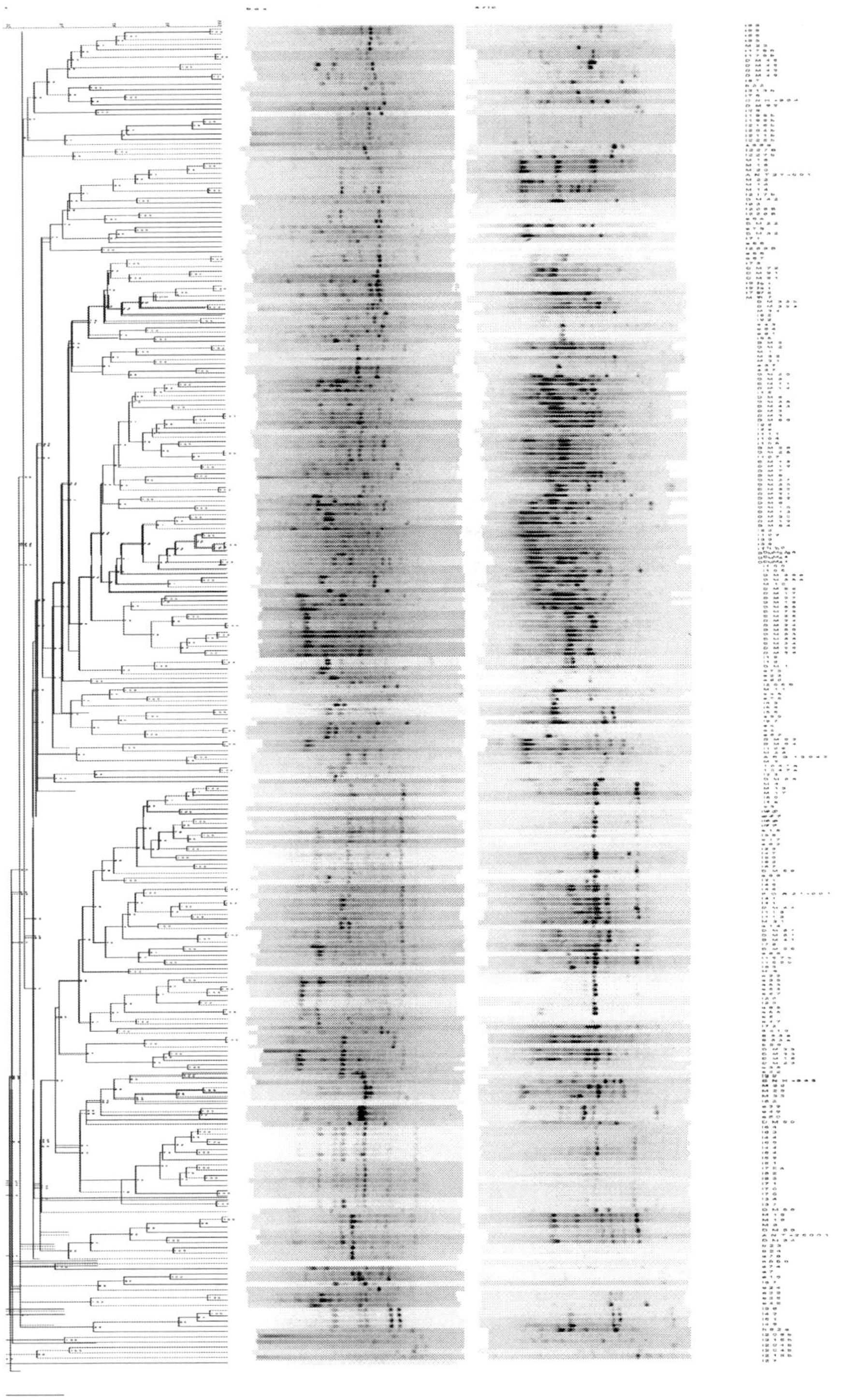

Figure 3. Fingerprint relationships using REP-PCR between micromonosporae recovered from aquatic environments.

The use of a PC-directed software package, for instance Gel Compar or BioNumerics (Applied Maths, Kortrijk Belgium), is required to analyse the resulting REP-PCR fingerprinting patterns. These computer programs use the Pearson correlation coefficient (r) to measure similarities on the basis of the degree of the linear relationship between two variables (Pearson, 1926). Where there is a perfect linear correlation between two variables, the observed values will lie on a straight line and (r) will equal 1.0. Conversely, the absence of a specific fragment is denoted by a score of 0. The formula for Pearson's product moment correlation coefficient (r) for variables x and y (Beaglehole et al., 1993) is:

$$r = \frac{\sum(Xi - X)(Yi - Y)}{\sqrt{\sum(Xi - X)^2 \sum(Yi - Y)^2}}$$

This coefficient is independent of the relative concentrations of patterns, is largely insensitive to differences in background and unlike most other coefficients it does not suffer from subjective band-detection and band-matching criteria as it is used to compare entire curves rather than band characteristics (Häne et al., 1993).

It is useful to generate dendrograms which shows relatedness between target strains in comparisons that involve many isolates or strains. The most common grouping method involves the calculation of a matrix of similarity coefficients between every pair of organisms, and the generation of the dendrogram from the similarity values by using the unweighted pair group method with arithmetic averages algorithm (UPGMA; Sneath and Sokal, 1973) though other methods commonly employed are also available.

Conclusions/Suggestions

It has become extremely important to assign systematically (taxonomically) related strains to groups to pre-select representative isolates for more comprehensive taxonomic studies and in order to prevent redundant screening efforts (Bull et al. 1992). Repetitive DNA fingerprinting has been employed successfully to assign large numbers of actinomycetes to taxonomically meaningful groups as exemplified herein to *Streptomyces* and *Micromonospora*. This is in line with works by other colleagues which include non-pathogenic species of *Streptomyces* (Lanoot et al., 2004) and *Nocardia* (Yamamura et al (2004). The latter is mostly composed by species of clinically significance that mostly affect immunocompromised patients.

Large numbers of isolates from several aquatic ecosystems can be assigned to rep-PCR groups based on BOX-AIR, ERIC primers and/or REP primers. The taxonomic integrity of the resulting groups can then be further analysed by detemining the corresponding 16S rRNA gene sequences of representatives of each group-clade. The resulting fingerprints shown here do suggest an unexpected degree of micromonosporae in the aquatic environments but also a need to re-evaluate our current view on clinically significant streptomycetes that can cause actinomycetoma.

The molecular fingerprint technique exemplified in this chapter is (a) rapid, (b) robust and (c) can provide reliable groups based on their fingerprint unique patterns. Besides, the technique can be coupled to a biological database that can be increased accordingly as more and more fingerprint patterns are obtained, analysed and added to the database. This provides great flexibility to the technique as reproducibility can be assessed at any time by including an 'internal' microorganism for which its fingerprint pattern has already been captured and analysed. The molecular fingerprint resulting groups can be stored and patterns obtained from recently described species can then be added and subsequently analysed. We believe this is indeed a unique intrinsic advantage of the technique in order to extend our current view on cultured microorganisms that can be added to a previous database. This, evidently, allows comparisons among several isolates to be performed easily -despite of their origin or souce of isolation- and with a reliable molecular fingerprint background.

ACKNOWLEDGMENTS

Some of the work presented here received financial support from (a) ICMyL-UNAM to LAMM, (b) IPN to ETQ and (c) CONACyT to ETQ and/or LAMM. The authors are indebted to unnamed colleagues for their comments and helpful discussions.

REFERENCES

Alarcon, B., Gonzalez, M. E. and Carrasco, L. (1988). Megalomycin-C, a macrolide antibiotic that blocks protein glycosylation and shows antiviral activity. *FEBS Letters*, *231*, 207-211.

Amonsin, A., Wellehan, J. F., Li, L. L., Laber, J. and Kapur, V. (2002). DNA fingerprinting of *Pasteurella multocida* recovered from avian sources. *Journal of Clinical Microbiology*, *240*, 3025-3031.

Antonio, M. A. and Hiller, S. L. (2003). DNA fingerprinting of *Lactobacillus crispatus* strain CTV-05 by repetitive elements sequence-based PCR analysis in a pilot study of vaginal colonization. *Journal of Clinical Microbiology*, *41*, 1881-1887.

Appuhamy, S., Parton, R., Coote, J. G. and Gibb, H. A. (1997). Genomic fingerprinting of *Haemophilus somnus* by a combination of PCR methods. *Journal of Clinical Microbiology*, *35*, 288-291.

Atalan, E., Manfio, G.P., Ward, A.C., Kroppenstedt, R.M. and Goodfellow, M. (2000). Biosystematic studies on novel streptomycetes from soil. *Antonie van Leeuwenhoek*, *77*, 337-353.

Barbier, N., Saulnier, P., Chachaty, E., Dumontier, S. and Andremont, A. (1996). Random amplified polymorphic DNA typing versus pulsed-field gel electrophoresis for epidemiological typing of vancomycin-resistant enterococci. *Journal of Clinical Microbiology*, *34*, 106-109.

Beaglehole, R., Bonita, R. and Kjellström. (1993). Basic Epidemiology. Macmillan / Eastern Press, England.

Bentley, S. D., Chater, K. F., Cerdeno-Tarraga, A. M., Challis, G. L., Thomson, N. R., James, K. D., Harris, D. E., Quail, M. A., Kieser, H., Harper, D., Bateman, A., Brown, S., Chandra, G., Chen, C. W., Collins, M., Cronin, A., Fraser, A., Goble, A., Hidalgo, J.,

Hornsby, T., Howarth, S., Huang, C. H., Kieser, T., Larke, L., Murphy, L., Oliver, K., O'Neil, S., Rabbinowitsch, E., Rajandream, M. A., Rutherford, K., Rutter, S., Seeger, K., Saunders, D., Sharp, S., Squares, R., Squares, S., Taylor, K., Warren, T., Wietzorrek, A., Woodward, J., Barrell, B. G., Parkhill, J. and Hopwood, D. A. (2002). Complete genome sequence of the model actinomycete *Streptomyces coelicolor* A3(2). *Nature, 417*, 141-147.

Bredholt, H., Galatenko, O.A., Engelhardt, K., Fjœrvik, E., Terekhova, L.P. and Zotchev, S.B. (2007). Rare actinomycete bacteria from the shallow water sediments of the Trondheim fjord, Norway: isolation, diversity and biological activity. *Environmental Microbiology, 9*, 2756-2764.

Brown, J.M. and McNeil, M. (2003). *Nocardia, Rhodococcus, Gordonia, Actinomadura, Streptomyces*, and Other Aerobic Actinomycetes., In Manual of Clinical Microbiology. Eight (Edt.), P. R. Murray, E. J. Baron, J. H. Jorgensen, M. A. Pfaller, and R. H.Yolken, (eds). Volume 1. pp. 502-531. ASM Press, Washington.

Bull, A. T., Goodfellow, M. and Slater, J. H. (1992). Biodiversity as a source of innovation in biotechnology. *Annual Review of Microbiology* 46, 219-252.

Bull, A. T., Ward, A. C. and Goodfellow, M. (2000). Search and discovery strategies for biotechnology: The paradigm shift. *Microbiology and Molecular Biology Reviews, 64*, 573-606.

Claridge, J. E. III, Raich, T. J., Pirwani, D., Simon, B., Tai, L., Rodriguez-Barradas, M. C., Regnery, R., Zollo, A., Jones, D. C. and Rambo, C. (1995). Strategy to detect and identify *Bartonella* species in routine clinical laboratory yields *Bartonella henselae* from human immunodeficiency virus-positive patient and unique *Bartonella* strain from his cat. *Journal of Clinical Microbiology, 33*, 2107-2113.

Clark, C.A., Chen, C., Ward-Rainey, N., Pettis, G.S. (1998). Diversity within *Streptomyces ipomoeae* based an inhibotory interactions, rep-PCR, and plasmid profiles *Phytopathology, 88*, 1179-1186.

Cole, S. T., Davis, K. E., McMurray, D. N. and Jacobs, W. R. (Eds). (2005). Tuberculosis and the tubercle bacillus. ASM Press. Washington

De Bruijn, F. J. (1992). Use of repetitive (repetitive extragenic palindromic and enterobacterial repetitive intergenic consensus) sequences and the polymerase chain reaction to fingerprint the genomes of *Rhizobium melitoli* isolates and other soil bacteria *Applied and Environmental Microbiology, 58*, 2180-2187.

De Menezes, A. B., Lockhart, R. J., Cox, M. J., Allison, H. E. and McCarthy, A. J. (2008) Cellulose degradation by Micromonosporas recovered from freshwater lakes and classification of these actinomycetes by DNA gyrase B gene sequencing. *Applied and Environmental Microbiology, 74*, 7080-7084.

DelVecchio, V. G., Petroziello, J. M., Gress, M. J., McClesky, F. K., Melcher, G. P., Crouch, H. K. and Lupski, J. R. (1995). Molecular genotyping of methicillin-resistant *Staphylococcus aureus* via fluorophore-enhanced repetitive-sequence PCR. *Journal of Clinical Microbiology, 33*, 2141-2144.

Dijkshoorn, L., Auken, H. M., Gerner-Smidt, P., Kaufmann, M. E., Ursing, J. and Pitt, T. L. (1993). Correlation of typing methods for *Acinetobacter* isolates from hospital outbreaks. *Journal of Clinical Microbiology, 31*, 702-705.

Doering-Saad, C., Kämpfer, P., Manulis, S., Kritzman, G., Schneider, J., Zakrzewska-Czerwinska, J., Schrempf, H. and Barash, I. (1992). Diversity among *Streptomyces* strains causing potato scab. *Applied and Environmental Microbiology*, *58*, 3932-3940.

Elder, J. K. and Southern, E. M. (1983). Measurement of DNA length by gel electrophoresis II: Comparison of methods for relating mobility to fragment length. *Analytical Biochemistry*, *128*, 227-231.

Euzéby, J. P. (2009). List of bacterial names with standing in nomenclature. http://www.bacterio.cict.fr/number.html

Foulerton, A. (1905). New species of *Streptothrix* isolated from the air. *Lancet* I, 1199-1200.

Georghiou, P. R., Doggett, A. M., Kielhofner, M. A., Stout, J. E., Watson, D. A., Lupski, J. R. and Hamill, R. J. (1994). Molecular fingerprinting of *Legionella* species by repetitive element PCR. *Journal of Clinical Microbiology*, *32*, 2989-2994.

Georghiou, P. R., Hamill, R. J., Wright, C. E., Versalovic, J., Koeuth, T., Watson, D. A. and Lupski, J. R. (1995). Molecular epidemiology of infections due to *Enterobacter aerogenes*: identification of hospital-associated strains by molecular techniques. *Clinical Infectious Diseases*, *20*, 84-94.

Gevers, D., Huys, G. and Swings, J. (2001). Applicability of *rep*-PCR fingerprinting for identification of *Lactobacillus* species. *FEMS Microbiology Letters*, *205*, 31-36.

Gillings, M. and Holley, M. (1997). Repetitive element PCR fingerprinting (rep-PCR) using enterobacterial repetitive intergenic consensus (ERIC) primers is not necessarily directed at ERIC elements. *Letters in Applied Microbiology*, *25*, 17-21.

Gilson, E., Clement, J. M., Brutlag, D. and Hofnung, M. (1984). A family of dispersed repetitive extragenic palindromic DNA sequences in *E. coli*. *EMBO Journal*, *3*, 1417-1421.

Gilson, E., Perrin, D. and Hofnung, M. (1990). DNA polymerase I and a protein complex bind specifically to *Escherichia coli* palindromic unit highly repetitive DNA: implications for bacterial chromosome organization. *Nucleic Acids Research*, *18*, 3941-3952.

Gontang, E. A., Fenical, W. and Jensen, P. R. (2007). Phylogenetic diversity of Gram positive bacteria cultured from marine sediments. *Applied and Environmental Microbiology*, *73*, 3272-3282.

Goodfellow, M. and Maldonado, L. A. (2006). The families *Dietziaceae*, *Gordoniaceae*, *Nocardiaceae* and *Tsukamurellaceae*. In Dworkin, M., Falkow, S., Rosenberg, E., Schleifer, K-H., Stackebrandt, E. (eds). The Prokaryotes. Archaea. Bacteria: Firmicutes, Actinomycetes. Springer New York. *3*, 843-888.

Goodfellow, M. and Minnikin, D. E. (Eds.) (1985). *Chemical Methods in Bacterial Systematics*. London: Academic Press.

Goodfellow, M. and O'Donnell, A. G. (1989). Search and discovery of industrially-significant actinomycetes. In: *Microbial Products: New Approaches*, pp. 343-383. Edited by S. Baumberg, I. Hunter and M. Rhodes. Cambridge: Cambridge University Press.

Goodfellow, M. and O'Donnell, A. G. (Eds.) (1994). *Chemical Methods in Prokaryotic Systematics*. Chichester: John Wiley and Sons.

Grimont, P. A. D. (1981). Use of DNA reassociation in bacterial classification. *Canadian Journal of Microbiology*, *34*, 541-546.

Gürtler, V. and Mayall, B. C. (2001). Genomic approaches to typing, taxonomy and evolution of bacterial isolates. *International Journal of Systematic and Evolutionary Microbiology*, *51*, 3-16.

Gürtler, V. and Stanisich, V. A. (1996). New approaches to typing and identification of bacteria using the 16S-23S rDNA spacer region. *Microbiology-UK*, *142*, 3-16.

Hall, V., Talbot, P. R., Stubbs, S. L. and Duerden, B. I. (2001). Identification of clinical isolates of *Actinomyces* species by amplified 16S ribosomal DNA restriction analysis. *Journal of Clinical Microbiology*, *39*, 3555-3562.

Hamill, R. J., Houston, E. D., Georghiou, P. R., Wright, C. E., Koza, M. A., Cadle, R. M., Goepfert, P. A., Lewis, D. A., Zenon, G. J. and Clarridge, J. E. (1995). An outbreak of *Burkholderia* (formerly *Pseudomonas*) *cepacia* respiratory tract colonization and infection associated with nebulized albuterol therapy. *Annals of Internal Medicine*, *122*, 762-766.

Häne, B. G., Jäger, K. and Drexler, H. G. (1993). The Pearson product-moment correlation coefficient is better suited for the identification of DNA fingerprint profiles than band matching algorithms. *Electrophoresis*, *14*, 967-972.

Harvey, B. S., Koeuh, T., Versalovic, J., Woods, C. R. and Lupski, J. R. (1995). Vertical transmission of *Citrobacter diversus* documented by DNA fingerprinting. *Infectious Control of Hospital Epidemiology*, *16*, 564-569.

He, H., Ding, W. D. and Bernan, T. (2001). Lomaiviticins A and B, potent antitumor antibiotics from Micromonospora lomaivitiensis. *JACS*, *123*, 5362-5363.

Hopwood, D. A. (2007). *Streptomyces* in nature and medicine. The antibiotic makers. Oxford University Press.

Hugenholtz, P., Goebel, B. M. and Pace, N. R. (1998). Impact of culture-independent studies on the emerging phylogenetic view of bacterial diversity. *Journal of Bacteriology*, *180*, 4765-4774.

Hughes, C. C., Prieto-Davo, A., Jensen, P. R. and Fenical, W. (2008). The marynopyrroles antibiotics of an unprecedented structure class from a marine *Streptomyces* sp. *Organic Letters*, *10*, 629-631.

Hulton, C. S., Higgins, C. F. and Sharp, P. M. (1991). ERIC sequences: a novel family of repetitive elements in the genomes of *Escherichia coli*, *Salmonella typhimurium* and other enterobacteria. *Molecular Microbiology*, *5*, 825-834.

Ikeda, H., Ishikawa, J., Hanamoto, A., Shinose, M., Kikuchi, H., Shiba, T., Sakaki, Y. Hattori, M. and Õmura, S. (2003). Complete genome sequence and comparative analysis of the industrial microorganism *Streptomyces avermitilis*. *Nature Biotechnology*, *21*, 526-531.

Ismet, A., Vikinesward, S., Paramaswari, S., Wang, W. H., Ward, A. C., Seki, T., Fiedler, H P. and Goodfellow, M. (2004). Production and chemical characterization of antifungal metabolites from *Micromonospora* sp. M39 isolated from mangrove rhizosphere soil *World Journal of Microbiological Biotechnology*, *20*, 523-528.

Jensen, H. L. (1930). The genus *Micromonospora* Orskov, a little known group of soil micro-organisms. *Proceedings of the Linnean Society of New South Wales.*, *55*, 231-248.

Jensen, P. R., Gontang, E. A., Mafnas, C., Mincer, T. C. and Fenical, W. (2005). Culturable marine actinomycete diversity from tropical Pacific Ocean sediments. *Environmental Microbiology*, *7*, 1039-1048.

Johnston, D. and Cross, T. (1976). The occurrence and distribution of actinomycetes in lakes of the English Lake District. *Freshwater Biology*, *6*, 457-463.

Kawamoto, I. (1989). *Micromonospora* In, Bergey's Manual of Systematic Bacteriology. Williams, S.T. (Ed). Williams and Wilkins.

Kidd, M., Atherton, J. C., Lastovica, A. J. and Louw, J. A. (2001). Clustering of South African *Helicobacter pylori* isolates from peptic ulcer disease patients is demonstrated by repetitive extragenic palindromic-PCR fingerprinting. *Journal of Clinical Microbiology*, *39*, 1833-1839.

Kim, W., Hong, Y. P., Yoo, J. H., Lee, W. B., Choi, C. S. and Chung, S. I. (2001). Genetic relationships of *Bacillus anthracis* and closely related species based on variable-number tandem repeat analysis and BOX-PCR genomic fingerprinting. *FEMS Microbiology Letters*, *207*, 21-27.

Kim, B. J., Lee, S. H., Lyu, M. A., Kim, S. J., Bai, G. H., Chae, G. T., Kim, E. C., Cha, C. Y. and Kook, Y. H. (1999). Identification of mycobacterial species by comparative sequence analysis of the RNA polymerase gene *(rpoB)*. *Journal of Clinical Microbiology*, *37*, 1714-1720.

Koch, C., Kroppenstedt, R. M., Rainey, F. A. and Stackebdrandt, E. (1996). 16S Ribosomal DNA analysis of teh genera *Micromonospora*, *Actinoplanes*, *Catellatospora*, *Catenuloplanes*, *Couchioplanes*, *Dactylosporangium*, and *Pilimelia* and emendation of the family *Micromonosporaceae*. *International Journal of Systematic Bacteriology*, *46*, 765-768.

Koeuth, T., Versalovic, J. and Lupski, J. R. (1995). Differential subsequence conservation of interspersed repetitive *Streptococcus pneumoniae* BOX elements in diverse bacteria. *Genome Research*, *5*, 408-418.

Korn-Wendisch, F. and Kutzner, H. J. (1992). The familiy Streptomycetaceae. In The Prokaryotes. A handbook on the Biology of Bacteria: Ecophysiology, isolation, identification, application. Second Edn., Vol. 1. pp. 921-995. Edited by Balows, H. G. Truper, M. Dworkin, W. Harder and K. H. Schleifer. Springer-verlag, New York.

Krassilnikov, N. A. (1938). Rayfungi and related organisms, Actinomycetales. Izdatel'stvo Akademii nauk, Moscow USSR.

Kriss, A. E. (1939). *Micromonospora* – an actinomycete like organism (*Micromonospora globosa* n. sp.). *Mikrobiologiya*, *8*, 178-185.

Kwon, D. H., El-Zaatari, F. A. K., Woo, J. S., Perng, C. L., Graham, D. Y., and Goo, M. F. (1998). Rep-PCR fragments as biomarkers for differentiating gastroduodenal disease specific-*Helicobacter pylori*. *Digest Diseases Sciences*, *43*, 980-987.

Lane, D. J. (1991). 16S/23S rRNA sequencing. In *Nucleic Acid Techniques in Bacterial Systematics*, pp. 115-148. Edited by E. Stackebrandt and M. Goodfellow. Chichester: John Wiley and Sons.

Lanoot B., Vancanneyt M., Dawyndt P., Cnockaert M., Zhang J., Huang Y., Liu Z. and Swings J. (2004). BOX-PCR fingerprinting as a powerful tool to reveal synonymous names in the genus *Streptomyces*. Emended descriptions are proposed for the species *Streptomyces cinereorectus*, *S. fradiae*, *S. tricolour*, *S. colombiensis*, *S. filamentosus*, *S. vinaceus* and *S. phaeopurpureus*. *Systematic and Applied Microbiology*, *27*, 84-92.

Liu, P. Y. F. and Wu, W. L. (1997). Use of different PCR-based DNA fingerprinting techniques and pulsed-field gel electrophoresis to investigate the epidemiology of

Acinetobacter calcoaceticus-Acinetobacter baumannii complex. *Diagnostic Microbiological Infectious Diseases*, *28*, 19-28.

Louws, F. J., Fulbright, D. W., Stephens, C. T. and de Bruijn, F. J. (1994). Specific genomic fingerprints of phytopathogenic *Xanthomonas* and *Pseudomonas* pathovars and strains generated with repetitive sequences and PCR. *Applied and Environmental Microbiology*, *60*, 2286-2295.

Louws, F. J., Fulbright, D. W., Stephens, C. T. and de Bruijn, F. J. (1995). Differentiation of genomic structure by rep-PCR fingerprinting to rapidly classify *Xanthomonas campestris* pv. *vesicatoria*. *Phytopathology*, *85*, 528-536.

Ludwig, W. and Klenk, H. P. (2001). Overview: A phylogenetic backbone and taxonomic framework for procaryotic systematics. In *Bergey's Manual of Systematic Bacteriology*, pp. 49-65. Edited by D. R. Boone, R. W. Castenholz and G. M. Garrity. New York: Springer-Verlag.

Ludwig, W. and Schleifer, K. H. (1999). Phylogeny of bacteria beyond the 16S rRNA standard. *ASM News*, *65*, 752-757.

Maiden, M. C. J., Bygraves, J. A., Feil, E., Morelli, G., Russell, J. E., Urwin, R., Zhang, Q. Zhou, J. J., Zurth, K., Caugant, D. A., Feavers, I. M., Achtman, M. and Spratt, B. G (1998). Multilocus sequence typing: A portable approach to the identification of clones within populations of pathogenic microorganisms. *Proceedings of the National Academy of Sciences USA*, *95*, 3140-3145.

Maldonado, L. A., Fenical, W., Jensen, P. R., Kauffman, C. A., Mincer, T. J., Ward, A. C., Bull, A. T. and Goodfellow, M. (2005a). *Salinispora arenicola* gen. nov., sp. nov. and *Salinispora tropica* sp. nov., obligate marine actinomycetes belonging to the family *Micromonosporaceae*. *International Journal of Systematic and Evolutionary Microbiology*, *55*, 1759-1766.

Maldonado, L. A., Fragoso-Yañez, D., Pérez-García, A., Rosellón-Druker, J. and Quintana, E. T. (2009). Actinobacterial diversity from marine sediments collected in Mexico. *Antonie van Leeuwenhoek.*, *95*, 111-120.

Maldonado, L. A., Stach, J. E. M., Pathom-aree, W., Ward, A. C., Bull, A. T. and Goodfellow, M. (2005b). Diversity of cultivable actinobacteria in geographically widespread marine sediments. *Antonie van Leeuwenhoek*, *87*, 11-18.

Maldonado, L. A., Stach, J. E. M., Ward, A. C., Bull, A. T. and Goodfellow, M. (2008). Characterisation of aquatic micromonosporae using molecular taxonomic methods. *Antonie van Leeuwenhoek.*, *94*, 289-298.

Manfio, G. P., Atalan, E., Zakrzewska-Czerwinska, J., Mordarski, M., Rodriguez, C., Collins, M.D. and Goodfellow, M. (2003). Classification of novel streptomycetes as *Streptomyces aureus* sp. nov., *Streptomyces laceyi* sp. nov. and *Streptomyces sanglieri* sp. nov. *Antonie van Leeuwenhoek*, *83*, 245-255.

Martin, B., Humbert, O., Camara, M., Guenzi, E., Walker, J., Mitchell, T., Andrew, P., Prudhomme, M., Alloing, G. and Hakenbeck, R. (1992). A highly conserved repeated DNA element located in the chromosome of *Streptococcus pneumoniae*. *Nucleic Acids Research*, *20*, 3479-3483.

Martin-Lozano, D., Cisneros, J. M., Becerril, B., Cuberos, L., Prados, T., Ortiz-Leyba, C., Canas, E. and Pachon, J. (2002). Comparison of a repetitive extragenic palindromic sequence-based PCR method and clinical and microbiological methods for determining strain sources in cases of nocosomial *Acinetobacter baumanii* bactercemia. *Journal of Clinical Microbiology*, *40*, 4571-4575.

McNeil, M. and Brown, J. (1994). The medically important actinomycetes: epidemiology and microbiology. Clinical Microbiological Reviews, *7*, 357-417.

Mincer, T. J., Jensen, P. R., Kauffman, C. A. and Fenical, W. (2002). Widespread and persistent populations of a major new marine actinomycete taxon in ocean sediments. *Applied and Environmental Microbiology*, *68*, 5005-5011.

Moyer, C. L., Dobbs, F. C. and Karl, D. M. (1994). Estimation of diversity and community structure through restriction fragment length polymorphism distribution analysis of bacterial 16S ribosomal RNA genes from a microbial mat at an active, hydrothermal vent system, Loihi Seamount, Hawaii. *Applied and Environmental Microbiology* 60, 871-879.

Muyzer, G. (1999). DGGE/TGGE a method for identifying genes from natural ecosystems. *Current Opinion in Microbiology*, *2*, 317-322.

Newbury, S. F., Smith, N. H., Robinson, E. C., Hiles, I. D. and Higgins, C. F. (1987). Stabilization of translationally active mRNA by prokaryotic REP sequences. *Cell*, *48*, 297-310.

Olsen, G. J. and Woese, C. R. (1993). Ribosomal RNA: a key to phylogeny. *FASEB Journal*, *7*, 113-123.

Olsen, G. J., Matsuda, H., Hagstrom, R. and Overbeek, R. (1994). fastDNAmL: a tool for construction of phylogenetic trees of DNA sequences using maximum likelihood. *Bioinformatics*, *10*, 41-48.

Õmura, S., Ikeda, H., Ishikawa, J., Hanamoto, A., Takahashi, C., Shinose, M., Takahashi, Y., Horikawa, H., Nakazawa, H., Osonoe, T., Kikuchi, H., Shiba, T., Sakaki, Y. and Hattori, M. (2001). Genome sequence of an industrial microorganism *Streptomyces avermitilis*: deducing the ability of producing secondary metabolites. *Proceedings of the National Academy of Sciences*, USA, *98*, 12215-12220.

Ørskov, J. (1923). Investigation into the morphology of the ray fungi. Levin and Munksgaard, 135. Copenhagen Denmark.

Oyarzabal, O. A., Wesley, I. V., Harmon, K. M., Schroeder-Tucker, L., Barbaree, J. M., Lauerman, L. H., Backert, S. and Conner, D. E. (1997). Specific identification of *Campylobacter fetus* by PCR targeting variable regions of the 16S rDNA. *Veterinary Microbiology*, *58*, 61-71.

Palys, T., Berger, E., Mitrica, I., Nakamura, L. K. and Cohan, F. M. (2000). Protein-coding genes as molecular markers for ecologically distinct populations: the case of two *Bacillus* species. *International Journal of Systematic and Evolutionary Microbiology*, *50*, 1021-1028.

Pearson, K. (1926). On the coefficient of racial likeness. *Biometrika*, *18*, 105-117.

Poh, C. L., Ramachandran, V. and Tasall, J. W. (1996). Genetic diversity of *Neisseria gonorrhoeae* IB-2 and IB-6 isolates revealed through whole-cell repetitive element sequence-based PCR. *Journal of Clinical Microbiology*, *34*, 292-295.

Potter, L. F. and Baker, G. E. (1956). The microbiology of Flathead and Rogers lakes. I. Preliminary survey of microbial populations. Ecology, *37*, 351-355.

Press, W. H., Flannery, B. P., Teukolsky, S. A. and Vetterling, W. T. (1986). Numerical Recipes: The Art of Scientific Computing. Cambridge University Press, Cambridge.

Quintana, E.T., Wierzbicka, K., Mackiewicz, P., Osman, A., Fahal, A. H., Hamid, M. E., Zakrzewska-Czerwinska, J., Maldonado, L. A. and Goodfellow, M. (2008). *Streptomyces sudanensis* sp. nov., a new pathogen isolated from patients with actinomycetoma. *Antonie van Leeuwenhoek, 93*, 305-313.

Rademaker, J. L. W., Louws, F. J. and Schultz, M. H., Rossbach, U., Vauterin, L., Swings, J. and de Bruijn, F. J. (1997). Molecular systematics of xanthomonads by rep-PCR genomic fingerprinting and computer-assisted pattern analysis. *Phytopathology, 87*, S81.

Rademaker, J. L. W., Hoste, B., Louws, F. J., Kersters, K., Swings, J., Vauterin, L., Vauterin, P. and de Bruijn, F. J. (2000). Comparison of AFLP and *rep*-PCR genomic fingerprinting with DNA- DNA homology studies: *Xanthomonas* as a model system. *International Journal of Systematic and Evolutionary Microbiology, 50*, 665-677.

Reboli, A. C., Houston, E. D., Monteforte, J. S., Wood, C. A. and Hamill, R. J. (1994). Discrimination of epidemic and sporadic isolates of *Acinetobacter baumannii* by repetitive element PCR-mediated DNA fingerprinting. *Journal of Clinical Microbiology, 22*, 2635-3640.

Rodriguez-Barradas, M. C., Hamill, R. J., Houston, E. D., Georghiou, P. R., Clarridge, J. E., Regnery, R. L. and Koehler, J. E. (1995). Genomic fingerprinting of *Bartonella* species by repetitive element PCR for distinguishing species and isolates. *Journal of Clinical Microbiology, 33*, 1089-1093.

Russel, P. J., Doenias, J. M. and Rusel, S. J. (1991). GELYMAC: A Macintosh application for calculating DNA fragment size from gel electrophoresis migration data. *CABIOS, 7* 265-266.

Sadowsky, M. J., Kinkel, L. L., Bowers, J. H. and Schottel, J. L. (1996). Use of repetitive intergenic DNA sequences to classify pathogenic and disease-suppressive *Streptomyces* strains. *Applied and Environmental Microbiology, 62*, 3489-3493.

Sails, A. D., Swaminathan, B. and Fields, P. I. (2003). Utility of multilocus sequence typing as an epidemiological tool for investigation of outbreaks of gastroenteritis caused by *Campylobacter jejuni. Journal of Clinical Microbiology, 41*, 4733-4739.

Salazar, O., Moron, R. and Genilloud, O. (2000). New genus-specific primers for the PCR identification of members of the genus *Saccharomonospora* and evaluation of the microbial diversity of wild-type isolates of *Saccharomonospora* detected from soil DNAs. *International Journal of Systematic and Evolutionary Microbiology, 50*, 2043-2055.

Sembiring, L., Ward, A. C. and Goodfellow, M. (2000). Selective isolation and characterisation of members of the *Streptomyces violaceusniger* clase associated with the roots of *Paraserianthes falcataria. Antonie van Leeuwenhoek, 78*, 353-366.

Sharples, G. J. and Lloyd, R. G. (1990). A novel repeated DNA sequence located in the intergenic regions of bacterial chromosomes. *Nucleic Acids Research, 18*, 6503-6508.

Sneath, P. and Sokal, R. R. (1973). Numerical Taxonomy: The Principles and Practice of Numerical Classification. W.H. Freeman, Baltimore.

Snelling, A. M., Gerner-Smidt, P., Hawkey, P. M., Heritage, J., Parnell, P., Porter, C., Bodenham, A. R. and Inglis, T. (1996). Validation of use of whole-cell repetitive extragenic palindromic sequence-based PCR (rep-PCR) for typing strains belonging to the *Acinetobacter calcoaceticus-Acinetobacter baumannii* complex and application of the method to the investigation of ahospital outbreak. *Journal of Clinical Microbiology, 34*, 1193-1202.

Stach, J. E. M., Maldonado, L. A., Masson, D. G., Ward, A. C., Goodfellow, M. and Bull, A. T. (2003). New primers for the class *Actinobacteria*: application to marine and terrestrial environments. *Environmental Microbiology*, *5*, 828-841.

Stackebrandt, E., Frederiksen, W., Garrity, G., Grimont, P. A. D., Kämpfer, P., Maiden, M. C. J., Nesme, X., Rosselló-Mora, R., Swings, J., Trüper, H. G., Vauterin, L., Ward, A. C. and Whitman, W. B. (2002). Report of the *ad hoc* committee for the reevaluation of the species definition in bacteriology. *International Journal of Systematic and Evolutionary Microbiology*, *52*, 1043-1047.

Stackebrandt, E., Rainey, F. A. and WardRainey, N. L. (1997). Proposal for a new hierarchic classification system, *Actinobacteria* classis nov. *International Journal of Systematic Bacteriology*, *47*, 479-491.

Stern, M. J., Ames, G. F., Smith, N. H., Robinson, E. C. and Higgins, C. F. (1984). Repetitive extragenic palindromic sequences: a major component of the bacterial genome. *Cell, 37*, 1015-1026.

Stevens, H., Brinkhoff, T., Rink, B., Vollmers, J. and Simon, M. (2007). Diversity and abundance of Gram positive bacteria in a tidal flat ecosystem. *Environmental Microbiology*, *9*, 1810-1822.

Tavanti, A., Gow, N. A. Senesi, S., Maiden, M. C. and Odds, F. C. (2003). Optimization and validation of multilocus sequence typing for *Candida albicans*. *Journal of Clinical Microbiology*, *41*, 3765-3776.

Thorson, J. S., Sievers, E. L., Ahlert, J., Shepard, E., Whitwam, R. E., Onwueme, K. C. and Ruppen, M. (2000). Understanding and exploiting nature's chemical arsenal: The past, present and future of calicheamicin research. *Current Pharmaceuticals Designs*, *6*, 1841-1879.

Tilley, B. C., Meyertons, J. L. and Lechevalier, M. P. (1990). MPPHIWR-2, a cryptic phage associated with *Micromonospora purpurea* ATCC-15835. *Journal of Industrial Microbiology*, *5*, 329-332.

Tyler, K. D., Wang, G., Tyler, S. D. and Johnson, W. M. (1997). Factors affecting reliability and reproducibility of amplification-based fingerprinting of representative bacterial pathogens. *Journal of Clinical Microbiology*, *35*, 339-346.

Umbreit, W. W. and McCoy, E. (1941). "Symposium on Hydrobiology". 106-114. Univ. Wisconsin, Madison, Wisconsin. Cited by Luedemann, 1970).

van Belkum, A., Struelens, A., de Visser, A., Verbrugh, H. and Tibayrenc, M. (2001). Role of genomic typing in taxonomy, evolutionary genetics and microbial epidemiology. *Clinical Microbiology Reviews*, *14*, 547-560.

Vaneechoutte, M., Boerlin, P., Tichy, H. V., Bannerman, E., Jager, B. and Bille, J. (1998). Comparison of PCR-based DNA fingerprinting techniques for the identification of *Listeria* species and their use for atypical *Listeria* isolates. *International Journal of Systematic Bacteriology*, *48*, 127-139.

Vaneechoutte, M., Rossau, R., De Vos, P., Gillis, M., Janssens, D., Paepe, N., De Rouck, A., Fiers, T., Claeys, G. and Kersters, K. (1992). Rapid identification of bacteria of the *Comamonadaceae* with amplified ribosomal DNA-restriction analysis (ARDRA). *FEMS Microbiology Letters*, *72*, 227-233.

Versalovic, J. and Lupski, J. R. (2002). Molecular detection and genotyping of pathogens: more accurate and rapid answers. *Trends in Microbiology*, *10*, S15-S21.

Versalovic, J., de Bruijn, F. J. and Lupski, J. R. (1998). Repetitive sequence-based PCR (*rep*-PCR) DNA fingerprinting of bacterial genomes. In *Bacterial Genomes: Physical Structure and Analysis*, pp 437-454. (Eds.), by. F. J. de Bruijn, J. R. Lupski and G. M. Weinstock. New York: Chapman and Hall.

Versalovic, J., Kapur, V., Koeuth, T., Mazurek, G. H., Whittam, T., Musser, J. and Lupski, J. R. (1995). DNA fingerprinting techniques of pathogenic bacteria by fluorophore-enhanced repetitive sequence-based polymerase chain reaction. *Archives in Pathological Laboratory Medicine*, *119*, 23-29.

Versalovic, J., Koeuth, T. and Lupski, J. R. (1991). Distribution of repetitive DNAsequences in *Eubacteria* and application to fingerprinting of bacterial genomes. *NucleicAcids Research*, *19*, 6823-6831.

Versalovic, J., Schneider, M., de Bruijn, F. J. and Lupski, J. R. (1994). Genomic fingerprinting of bacteria using repetitive sequence-based polymerase chain reaction. *Methods in Molecular and Cellular Biology*, *5*, 25-40.

Versalovic, J., Woods, C. R. Jr., Georghiou, P. R., Hamill, R. J. and Lupski, J. R. (1993). DNA-based identification and epidemiology typing of bacterial pathogens. *Archives Pathology Laboratory Medicine*, *117*, 1088-1098.

Vobis, G. (1989). The genus *Actinoplanes* and related genera. In, The Prokaryotes. Balows A., Truper, H. G., Dworkin, M., Harder, W. and Schleifer, K. H. (Eds). Springer-Verlag New York. Pp. 1029-1060.

Wagman, G. H. and Weinstein, M. J. (1980). Antibiotics from *Micromonospora*. *Annual Reviews in Microbiology*, *34*, 537-557.

Waksman, S. A. and Henrici, A. T. (1943). The nomenclature and classification of the actinomycetes. *Journal of Bacteriology*, *46*, 337-341.

Watson, E. T. and Williams, S. T. (1974). Studies on the ecology of actinomycetes in soil VII. *Soil Biology and Biochemistry*, *6*, 43-52.

Weinstein, M. J., Luedemann, G. M., Oden, E. M. and Wagman, G. H. (1964). Gentamicin, a new broad-spectrum antibiotic complex. *Antimicrobial Agents and Chemotherapy*, 1-7.

Welsh, J. and McClelland, M. (1990). Fingerprinting genomes using PCR with arbitrary primers. *Nucleic Acids Research*, *18*, 7213-7218.

Woese, C. R. (1987). Bacterial evolution. *Microbiological Reviews*, *51*, 221-271.

Wong, H. C. and Lin, C. H. (2001). Evaluation of typing of *Vibrio parahaemolyticus* by three PCR methods using specific primers. *Journal of Clinical Microbiology*, *39*, 4233-4240.

Woods, C. R., Versalovic, J., Koeuth, T. and Lupski, J. R. (1992). Analysis of relationships among isolates of *Citrobacter diversus* by using DNA fingerprints generated by repetitive sequence-based primers in polymerase chain reaction. *Journal of Clinical Microbiology*, *30*, 2921-2929.

Woods, C. R., Versalovic, J., Koeuth, T. and Lupski, J. R. (1993). Whole-cell repetitive element sequence-based polymerase chain reaction allows rapid assesment of clonal relationships of bacterial isolates. *Journal of Clinical Microbiology*, *31*, 1927-1931.

Yamamura H., Hayakawa M., Nakagawa Y. and Iimura Y. (2004). Characterization of *Nocardia asteroides* isolates from different ecological habitats on the basis of repetitive extragenic palindromic – PCR fingerprinting. *Applied and Environmental Microbiology*, *70*, 3149-3151.

Yang, Y. and Ames, G. F. (1988). DNA gyrase binds to the family of prokaryotic repetitive extragenic palindromic sequences. *Proceedings of the National Academy of Sciences, USA*, *85*, 8850-8854.

Yoon, J. H., Lee, S. T., Kim, S. B., Goodfellow, M. and Park, Y. H. (1997). Inter- and intraspecific genetic analysis of the genus *Saccharomonospora* with 16S to 23S ribosomal DNA (rDNA) and 23S to 5S rDNA internally transcribed spacer sequences. *International Journal of Systematic Bacteriology*, *47*, 661-669.

Zenova, G. M. and Zvyagintsev, D. G. (1997). The ecological status of actinomycetes of the genus *Micromonospora*. *Eurasia Soil Science*, *30*, 325-331.

Zhang, Z. S., Kudo, T., Nakajima, Y. and Wang, Y. (2001). Clarification of the relationship between the members of the family *Thermomonosporaceae* on the basis of 16S rDNA, 16S-23S rRNA internal transcribed spacer and 23S rDNA sequences and chemotaxonomic analyses. *International Journal of Systematic and Evolutionary Microbiology*, *51*, 373-383.

Zoe-Jordans, J., Paul, J., Bates, J., Beaumot, C., Kimari, J. and Gilks, C. (1995). Characterization of *Streptococcus pneumoniae* from human immunodeficiency virus-seropositive patients with acute recurrent pneumonia. *Journal of Infectious Diseases*, *172*, 983-987.

In: Encyclopedia of DNA Research (3 Volume Set)
Editors: S. J. Duncan and P. H. Wiley
ISBN 978-1-61324-305-3

Chapter 37

SNPs, Mini-STRs and Molecular Photofitting: Novel Directions for Forensic DNA Typing

Gerhard Mertens
Forensic DNA Laboratory, University of Antwerp, Antwerp, Belgium

Abstract

Since Alec Jeffreys in 1985 coined the term "individual-specific genetic fingerprints", DNA typing has become indispensable in forensic analysis, having as central dogma the evidential power of the "match" between stain and suspect. Technological advances in this field include the adoption of the polymerase chain reaction (PCR) to generate DNA profiles from minute biological samples, the use of mitochondrial DNA to obtain information from old bones and the application of Y chromosomal polymorphisms in sexual assault cases.

Present research focuses on genetic markers for external visible traits and on improving the utility of poor quality samples with degraded DNA.

The first valid marker for a physical trait was the sex marker amelogenin, which has been part of short tandem repeat (STR) multiplex PCRs for more than a decade. New markers are being investigated for human pigmentation, predicting hair, iris and skin colour. The relation between genes and pigmentation however is complex, involving numerous genes. Single nucleotide polymorphisms (SNPs) in these genes make promising candidates for predictive markers.

Also stature is a complex genetic trait where study subjects can be phenotyped easily. Linkage analysis has revealed several short stature-quantitative trait loci while polymorphisms in the fibrillin I gene are associated with tall stature.

An indirect approach of physical traits is inferring the geographic origin of an individual from DNA. Here, genome wide SNP panels have been used, correctly discriminating "European", "East Asian" and "Sub-Sahara African" ancestry, or even substructuring Europeans to the level of almost nationality, by typing 500 000 SNPs.

The ultimate goal is to arrive at "molecular photofitting", constructing a portrait of the person leaving a biological stain with molecular techniques.

The other direction of forensic DNA research addresses the difficulty of obtaining a DNA profile in a degraded crime stain.

One strategy to tackle this problem uses SNPs in stead of STRs. Due to the intrinsic small size of SNPs, they are well suited for degraded samples. There are two major drawbacks however. First, due to their bi-allelic nature, the resolution of a SNP is inferior to any STR. This can be overcome by using larger numbers of SNPs, yielding the same statistical power as 10 to15 STRs. Then, since currently DNA profiles of a standard set of only STRs are recorded in national forensic databases, a DNA profile consisting of only SNPs cannot be matched.

An alternative strategy still applies STRs, but with smaller amplicons. Primer pairs are chosen closer to the repeat region, yielding shorter PCR products than with classical primers. The STRs in question are partially "old", internationally used loci and for another part newly developed systems. Thus, comparison with DNA profiles in existing forensic databases remains possible.

INTRODUCTION

DNA profiling is a technique employed by forensic to assist in the identification of individuals on the basis of their respective DNA profiles. It all started with the seminal work of Alec Jeffreys who coined the term "individual specific genetic fingerprints" by combining restriction fragment length polymorphism (RFLP) analysis with multilocus probe technology [1]. Forensic application soon followed with as central dogma the evidential power of the "match" between stain and suspect.

The next leap forward came with the introduction of the polymerase chain reaction (PCR), permitting the analysis and generation of DNA profiles from minute biological samples [2]. Autosomal short tandem repeats (STR), co-amplified in multiplex-PCRs, became the cornerstone of forensic DNA testing [3], including national databases with DNA profiles of convicted felons.

Another major breakthrough was the application of the high copy number of mitochondrial DNA [4] so that valuable information could be obtained from (very) old specimens, such as bones and teeth.

A next advance came with the introduction of Y chromosomal STRs [5]. In forensics, this helps the resolution of sexual assault cases, discriminating between the (female) victim's own DNA profile and the one from the assailant. This technique complements the physico-chemical separation of the female and male fraction of a vaginal fluid sample.

We will review present research and future applications of forensic DNA typing, which focuses on two domains. The first relates to the use of markers to infer - directly or indirectly - external visible traits, the second aims to improve the usefulness of poor quality biological samples containing degraded DNA.

MOLECULAR PHOTOFITTING

Physical Traits and DNA

Current methods for forensic DNA analysis require an identified suspect for comparison purposes. However, in those instances when there is no suspect, the crime stain's DNA profile itself provides limited information to investigators, with gender being all that can currently be deduced. Indeed, the sex marker amelogenin [6] was the first valid marker for a physical trait and has been part of all standard forensic STR multiplex PCRs for over a decade.

The physical appearance of an individual is encoded in his genome as evidenced by the striking physical similarity of identical twins. The major physical descriptors of an individual, i.e. colouring, height and facial features are highly heritable. These traits are however not caused by a single variation in the genome or even a single gene.

From the genetic point of view, physical traits are complex traits, that is multifactorial traits where several genes interacting both between themselves and with the environment define the phenotype [7].

Genetic association studies are an important tool to assess correlations between genetic variants and differences in physical traits on a population scale [8]. A major leap forward in association studies came with the discovery of millions of SNPs in the human genome sequencing projects. Now more than 11 million SNPs have been gathered into the publicly accessible SNP database (www.ncbi.nlm.nih.gov/SNP/).

Skin Pigmentation

A great deal of work has focused on pigmentation, most notably skin pigmentation [9]. Here, the enzymes implicated in melanin synthesis are good contenders for genetic variation in causing diversity in pigmentation. The main candidate genes include genes for melanosome proteins (TYR, DCT, OCA2, MATP/SLC45A2), signal proteins (ASIP, MC1R) and proteins involved in melanosome uptake by keratinocytes (MYO5A, RAB27A).

Pigmentation results from the production and deposition of melanin which is synthesized from tyrosine as either black/brown eumelanin or yellow/red pheomelanin. The mechanism controlling the switch from eumelanin to pheomelanin is well understood. Stimulation of the MC1R receptor results in the production of melanin. Antagonism in this interaction by the ASIP signalling protein results in decreased production of pheomelanin.

MCR1 was the first human gene identified that showed an association with normal pigment variation. There are three common MCR1 polymorphisms - R151C, R160W and D294H - which account for 90% of red hair and fair skin phenotypes.

SNPs in the ASIP gene were studied by Voisey et al. [10]. The 8818A→G SNP was found at higher frequency in Africans as compared with other populations. Within the European population, a strong association was observed between the G allele and dark hair colour. In the MATP gene (recently renamed SLC45A2), two SNPs - Phe374Leu and Glu272Lys - are significantly associated with pigmentation. The 374Leu allele is more common in non-Caucasian populations, while - compared to Europeans - an increase frequency of the 272Lys allele is observed in Asian and African populations. Within the European populations, these polymorphisms showed highly significant associations with dark hair, skin and eye pigmentation.

The genetic basis of normal human pigmentation variation is decipherable and future forensic DNA tools will provide predictive investigation to crime scene officers [11].

Zaumsegel et al. [12] selected a set of 11 SNPs based on their relevance for human pigmentation and located in the OCA2 and MATP (SLC45A2) gene. 87 Germans and 71 Sub-Sahara Africans were genotyped for these SNPs by multiplex-PCR followed by SNaP shot. The two sample populations were chosen because of their clear difference in skin pigmentation. Five of the 11 SNPs proved informative, showing clearly different allele frequencies between both populations. Thus these five SNP can be considered pigmentation informative as well as ancestry informative, while the other six SNPs may have other effects which cannot be detected in the two groups studied.

Myles et al. [13] measured allele frequencies in Europeans, Chinese and Africans for 24 pigmentation genes from two large-scale SNP data sets. This revealed patterns of widely contrasting frequency differences between the populations. Notably the DCT gene was strongly associated with pigmentation control in Asians. The ASIP and OCA2 genes appeared to play important roles in pigmentation patterns of Europeans and Asians, while the MATP (SLC45A2) and TYR genes affect pigmentation only in Europeans [14].

Eye Colour

Iris pigmentation is related to skin pigmentation and is a major element in the facial features characterizing an individual. More than 70% of eye colour variation is due to quantitative trait loci (QTL) localised on chromosome 15q [15], where the OCA2 gene is localised. OCA2 is a transporter for tyrosine, an essential substrate in the production of melanin, determining the pigmentation phenotype. Branicki et al. [16] genotyped 390 Polish for 11 SNPs in the OCA2 gene, by multiplex PCR followed by SNaP shot. Based on conventional statistics and an evolutionary method known as "tree scanning", it was found that exonic position rs1800407 (Arg419Gln) is most significantly associated with iris colour.

Was He Tall or Short?

Adult height is a complex trait with high heritability and the possibility to phenotype study subjects by easy and accurate measurement. Still, studies of stature as a quantitative complex trait are less advanced than studies of pigmentation.

Several short stature-QTLs have been identified through genome-wide linkage analysis [17]: STQL1 on chromosome 6, STQL2 on chromosome 7, STQL3 on chromosome 12, STQL4 on chromosome 13, STQL5 on chromosome 3, STQL6 on chromosome X, STQL7 on chromosome 1 and STQL8 on chromosome 9.

Polymorphisms in the fibrillin I gene are associated with tall stature in normal individuals[18].

This research has not yet resulted in an assay that can be applied in the forensic setting.

Continuous Facial Morphology

A search of literature has not revealed any published work on the variation of continuous facial features in man. Some characteristics are anticipated to be monogenic, following Mendelian inheritance. These include chin dimple, facial dimples, hairy ears, earlobe attachment and freckles. The genes for these traits, nor associated SNPs, have yet been located.

Legal Implications

The concept of generating a physical description of an individual from the analysis of his or her DNA has been coined "Molecular Photofitting" [19]. While much is expected from this strategy, it will also require adaptation of the legislation on forensic DNA analysis in many countries. In most countries, only non-coding DNA can legally be applied in forensic examination and DNA based prediction of physical traits might be considered a violation of privacy. Actually, the Netherlands are the only country that has legally permitted the use of the molecular photofitting concept in criminal investigation [20].

Ancestry-Informative Markers

Predicting Geographic Origin from DNA

Recently, it has been demonstrated that an individual's geographic ancestries can be inferred from genetic markers, called ancestry-informative markers (AIM). In forensics, the prediction of geographic origin from body fluids recovered from a crime scene may provide meaningful information for investigators. It might narrow the pool of suspects, would be instrumental when eye witnesses are unavailable and could help for identification of corpses.

Y-chromosome loci and mitochondrial DNA variation [21] are haploid, uniparental inherited markers which are very interesting for population genetic studies, but can provide conflicting results within admixed populations. Autosomal markers reflect both paternal and maternal heritage and are therefore more suited for the prediction of geographic origin. Autosomal microsatellites do not exhibit large enough contrasts in allele frequencies between populations to be useful in numbers below 50 loci, mainly due to their mutational instability [22]. Autosomal SNPs have emerged as the best AIM due to their low mutation rate, density of distribution across the genome and full range of allele frequency patterns among populations [23].

Since the majority of human worldwide genetic diversity takes the form of geographic clines [24], it is essential to find the small number of SNPs that show the most pronounced allele frequency discontinuities between continental regions in order to create sets with population "diagnostic" genotypes.

In clinical genetics, AIM are an increasing part of genetic association studies but have not been fully developed into a practical forensic test. The following examples illustrate current advances in this area.

Resolution of SNPs

Phillips et al. [25] developed a single tube 34-plex SNP assay for assignment of ancestral origin by selecting AIM exhibiting highly contrasting allele frequency distribution between European, Sub-Sahara African and East Asian population groups. They developed a classification algorithm based on maximum likelihood using a database of 360 individuals sampled from two populations each from the major population groups (60 Spanish and 60

Danish for the European metapopulation, 60 Mozambican and 60 Somali for the Sub-Sahara African metapopulation, 60 Chinese and 60 Taiwanese for the East Asian metapopulation). The performance of the ancestry predictive algorithm was tested using Centre d'Étude du Polymorhisme Humain (CEPH)-Human Genome Diversity panel samples from 520 individuals with known geographic ancestry (160 Europeans, 123 Sub-Sahara Africans, 237 East Asians). Only 5 of these samples were erroneously assigned, comprising two Europeans as "African" and three Europeans as "East Asian", resulting in a practical error rate of 1%. The incorrectly classified individuals were not a random set but came from Sardinia (4 individuals) and from the Adygei population of the West-Caucasus (1 individual). Sardinia is a well-known genetic outlier in Europe and the Adygei live on the extreme eastern edge of the continent.

32 SNPs were selected by Chrystelle et al. [26] because of their allele frequency differences between the three major population groups: Sub-Sahara African, East Asian and European. The selected SNPs were genotyped using TaqMan technology on individuals from the three metapopulations (115 from each metapopulation) and also from two geographically intermediate groups (115 from North Africa and 155 from Sub-Continental Asia). The geographic affiliation of the samples was inferred using the clustering algorithm implemented in the STRUCTURE program. This analysis showed that individuals can be grouped into three genetic clusters (K = 3) which are strictly correlated with the three continental regions. The same accuracy of assignment was observed by using the nine loci with the highest genetic distance (measured as F_{ST}). The assumption of five genetic clusters (K = 5) allowed the distinction of North Africans and South Asians, but here the use of the whole panel of 32 SNPs is recommended to avoid false assignment.

16 SNPs were selected by Daniel et al. [27] to distinguish the major populations residing in present day Australia. They recorded geographical ancestry over three generations as well as eye and hair colour information for 242 Australians, comprising 51 East Asians, 33 Sub-Continental Asians, 25 Middle-Easterns, 79 Europeans, 20 North Africans and 34 Sub-Sahara Africans. The individuals were genotyped by PCR followed by SNaP Shot technology. Population assignment was assessed using likelihood-algorithms applied through the GenAlEx software package. This removes the individual sample from the dataset before it is assigned to a population. Using the 16 SNP set, a high degree of accuracy was achieved for the inference of Sub-Sahara African (100%), European (91%) and North African (90%) ancestry, while an intermediate degree of accuracy was observed for Sub-Continental Asian (79%) and Middle-Eastern (60%) origin. One of the16 SNPs, Leu374Phe, is located in the MATP (SLC45A2) gene which is a major determinant of pigmentation. As a crude classification of pigmentation, study subjects were classified into three groups: "dark", "medium" and "light". Using the GenAlEx assignment technique, a correct inference of the phenotype was obtained in 91% of the "dark" and 98% of the "light". Given the polygenic nature of pigmentation, Leu374Phe should be used in combination with other pigmentation-associated SNPs to make even more reliable phenotypic deductions.

While it thus has been proven that the major continental population groups can be differentiated using genetic markers, the question remains whether any reproducible substructure can be demonstrated within the European metapopulation. The question has been addressed in the setting of disease association studies where the association is tested between genetic markers and polygenic or complex diseases, such as rheumatoid arthritis or diabetes. Indeed, differences in population genetic structure between disease cases and controls can lead to false positive associations [28]. Since many studies focus on patients from European descent, the potential impact of European substructure on association testing has specifically engendered interest [29].

Array Technology for Increasing Detail

To assess the informativeness for within European ancestry, Price et al. [30] used the 500K Affymetrix GeneChip or the Illumina HumanHap bead array to genotype 583 SNPs on 667 samples from 7 European countries: 180 Swedish, 82 UK, 60 Polish, 60 Spanish, 124 Italian, 80 Greek and 81 US Ashkenazi Jewish. Principal component analysis clearly separated the samples into three clusters, corresponding to northwest European, southeast European and Ashkenazi Jewish ancestry. From these 583 SNPs, they then selected a subset of 300. Rerunning principal component analysis on the same samples using only these SNPs yielded the same results as before. So, these 300 SNPs can be considered an informative marker set distinguishing ancestries within Europe.

In a similar way, Seldin et al. [31] typed 928 individuals with European ancestry (Swedish, English, Irish, German, Ukrainian, Italian, Spanish, Portuguese and Greek) for 2657 SNPs using the Illumina bead array. Using the programme STRUCTURE, the samples were examined under different assumptions of the number of clusters, ranging from one to ten (K = 1, … K = 10).The data significantly favoured the assumption of K = 2, i.e. the distinction of two clusters. The analysis showed consistent assignment of the Italian, Spanish, Portuguese and Greek samples to the same cluster, while clustering the Swedish, English, Irish, German and Ukrainian samples in the other group. Thus, the 2657 SNPs consistently made the distinction between the "northern" and "southern" European population groups.

Lao et al. [32] used the 500K GeneChip array to type 500 568 SNPs in 2514 individuals from 23 sampling sites located in one of 20 different European countries. These comprehensive SNP genotype data from 23 European subpopulations provides a dense coverage at both the geographic and genomic level and represents the largest Europe-wide genetic study to date. It allowed for principal component analysis and to describe the genetic structure of the European metapopulation with the highest resolution. It could be concluded that although the amount of differentiation within the European autosomal gene pool is small, the existing genetic differences correlate well with geographic distances. The authors were the first to show that inferring geographic origin of an unknown person from autosomal DNA markers is feasible down to the level of Europeans sub-regions.

Mini-STRs

The Problem of DNA Degradration

It is common to encounter highly degraded DNA samples in a criminal investigation. The DNA may be so degraded that normal PCR amplification gives inconclusive results. The degradation processes affect the typing results in different ways: nucleases or oxidative damage affects the DNA structure itself, while indirectly co-extracted agents may inhibit the PCR reaction.

The extent of the degradation process depends on two factors: time and environmental conditions. Degradative processes accumulate with time while environmental conditions (temperature, humidity, pH, soil chemistry) modify the rate and aggressiveness of degradation [33]. Short amplicon approaches have recently been developed for the analysis of degraded DNA.

One way to improve the success rate for degraded DNA is to utilize redesigned STR primers, with the primers located closer to the repeat region, thus generating shorter amplicons. For these assays, the designation "mini-STR" was introduced in 2003 [34].

A Commercial Mini-STR Multiplex

Applied Biosystems was the first commercial supplier to market a mini-STR multiplex, the Minifiler [35]. This is a 9-plex mini-STR amplification kit, including the sex determining marker amelogenin and the most common problematic loci (above 200 bp) of the Identifiler kit (Applied Biosystems): FGA, D21S11, D18S51, D13S317, D7S820, D16S539, CSF1PO and D2S1338. Minifiler allowed to complete partial Identifiler profiles in 16 compromised samples (blood, saliva, tooth, bone and hair) in a study by Andrade et al. [36].

Developments by Research Institutes

Not only industry has developed mini-STR multiplexes. Camacho et al. [37] constructed a mini-SGM multiplex which encompasses TH01, FGA, D18S51, D6S539 and D2S1338, which are common STR markers in human identity testing. Using this method, a full DNA profile was obtained in two cases where the standard STR multiplex Powerplex 16 (Promega) showed drop-out of alleles greater than 300 bp (tissue preserved in a paraffin block) or gave no results at all (bone of corpse immersed in lake water for 15 months).

Another example of a home made mini-STR comes from Decorte et al. [38]. They developed a multiplex comprising amelogenin, four conventional STRs (TH01, D18S51, D21S11, FGA) with reduced amplicon size and four "new" mini-STRs (D1S1677, D2S441, D10S1248, D22S1045). The method was applied to a panel of 198 Belgians typed with Powerplex 16 (Promega). Full concordance was found for the common loci TH01, D18S51 and D21S11. A sensitivity study on a serial dilution of DNA samples under the standard conditions of 30 PCR cycles showed the occurrence of allele or locus dropout when the DNA input was less than 100 pg of DNA. However, by increasing the number of cycles to 34, a reproducible DNA profile was obtained when the amount of DNA was decreased to 30 pg. The sensitivity was two- to fourfold higher than the commercial standard STR multiplex PCRs SGM Plus (Applied Biosystems) and Powerplex 16 (Promega).

SNPs

International Consortia and the Advantages of SNPs

Research projects such as the SNP Consortium [39] and the International HapMap Consortium [40] identified millions of SNPs throughout the human genome and set up public databases allowing easy access to a huge amount of data. In a forensic context there is great interest in SNPs because of their abundance in the genome, low mutation rates and possibility of analysis in short amplicons improving successful amplification of degraded DNA.

The advantages of SNPs for forensic analysis include: small amplicon size, working better on degraded samples, lower mutation rate compared with STRs (10^{-8} vs. 10^{-3}), amenable to high throughput, abundant in the human genome, can provide specific information (individual identification, ancestry, lineage or phenotype).

There are also a number of limitations or challenges to be summed up: no commercial kits available at present, no widely established core loci, mixture resolution issues, large multiplex assays required, not likely to replace STRs in national DNA databases.

In their conclusion, a panel of experts agreed while SNPs would not replace STRs for most forensic applications anytime soon, SNP markers and assays should continue to be explored. SNPs may serve as an adjunct to STRs for solving special problems in forensic genetics [41].

Forensic Application

The SNP*for*ID consortium (www.snpforid.org) was established with support from the European Union with the aim of developing a SNP assay for forensic use [42]. A major achievement of this enterprise was the selection of 52 SNPs and their incorporation into a single multiplex PCR with amplicons ranging in size from 59 to 115 bp [43]. The analysis method of choice within the consortium was SNaPshot (Applied Biosystems). An example of a population genetic study using the SNP*for*ID 52-plex assay can be found in the work of Pereira et al. [44]. In the Portuguese population, they obtained a random match probability of 2.6×10^{-20}, demonstrating the forensic informativeness of the 52 SNPs.

The SNP*for*ID 52-plex also proved successful at analyzing degraded DNA, but there were problems with high background noise that could lead to incorrect allele calls. In order to tackle this problem, the SNP*for*ID Consortium developed the Genplex technology [45]. This was done through modification of the SNPlex system (Applied Biosystems) and utilising oligo-ligation of PCR products followed by probe hybridisation to generate dye-labeled allele-specific oligonucleotides which are detected by capillary electrophoresis. Genplex technology was used to type 48 SNPs plus amelogenin. The chosen SNPs provide match probabilities of 9.6×10^{-18} for Europeans and 6.9×10^{-16} for Africans. The success rate of Genplex was tested by typing 55 UV and naturally degraded tissue, bone and teeth samples in three laboratories. The samples were also examined with the standard PCR STR multiplexes Powerplex 16 (Promega) and SGMPlus (Applied Biosystems). It was shown that the Genplex technique provides a higher success rate than STR-based methods when typing degraded DNA.

Fondevilla et al. [46] compared the performance of established and novel genotyping methods in 15 skeletal samples found in the moist and acid soil of north west Spain. DNA extracts were quantified by real-time PCR, using the Quantifiler kit (Applied Biosystems), which includes an internal positive control to detect the presence of PCR inhibitors. Standard PCR typing was represented by the Identifiler (Applied Biosystems) and Powerplex 16 (Promega) kits. Two mini-STRs were tested: MiniNC01 (developed at NIST) and Minifiler (Applied Biosystems). SNP typing comprised two assays: the SNP*for*ID 52-plex human identification set and the SNP*for*ID 34-plex ancestry informative set.

This comparative study showed that simple dilution of the DNA extract enhanced the genotyping success in 13 of the 15 samples, suggesting that inhibitors play a critical role in reducing PCR efficiency in severely degraded samples. Still, the long amplicon systems Powerplex 16 and Identifiler, appeared to be more affected by inhibitors. In contrast, SNP typing showed relative immunity from PCR inhibition. If amplicon length is an important factor with or without inhibition control, then the success rate should be expected to rise as amplicon size diminishes. This was indeed observed with the standard amplicon STRs, Powerplex 16 showing the highest overall failure rate. This trend continued as the degree of degradation rose, and longer Indentifiler loci failed to amplify (amplicons up to 380 bp), followed by Minifiler (amplicons up to 300 bp). The performance of mini-STR loci with amplicons of 100 to 300 bp did not differ markedly, however PCR products below 100 bp, notably those of MiniNC01 (all amplicons < 120 bp) were resistant to even the most aggressive degradation in the examined samples. The authors concluded that for most samples, standard STR typing methods will suffice if inhibition is properly assessed and controlled prior to PCR. A further level of degradation, characterised by extremely aggressive environmental conditions over long periods of time can present the most challenging analyses. This appeared in the analysis of two femurs, one from a 35-year internment in a tomb, the other being buried in forest soil for 10 years. High molecular weight DNA was absent from these samples and all standard STRs failed, together with most mini-STR loci. In contrast, successful amplification of the shortest amplicon mini-STRs and all SNP multiplexes indicated that short, fragmented DNA was present was present in sufficient quantity for efficient PCR of these systems.

Thus, an optimal analysis method for degraded DNA can be chosen between standard STRs and short-amplicon STRs combined with SNP typing.

"NEW" STRs

European Strategies for the Future

In 2005 the ENFSI and EDNAP groups came forward with a number of recommendations for new European STR loci [47]:

1) Adopt mini-STRS to increase robustness and sensitivity of forensic DNA analysis.
2) The prime utilisation of SNPs will be in relation to mass disasters dealing with highly compromised samples, but can be carried out independent of national DNA databases.

3) Retain existing core-loci used in national DNA databases and convert into mini-STRs by re-engineering with primers built close to the repeat region.
4) Install the mini-STRs D10S1248, D14S1434 and D22S1045 as new core-loci.
5) A secondary list of loci (D12S391, D1S656, TPOX) can be considered by manufacturers to convert into useful mini-STRs.
6) A balance should be struck between the size of a PCR multiplex and its sensitivity.

Following up on these recommendations, two differing but parallel strategies have emerged [48]. The first employs a 13 STR multiplex incorporating 3 mini-STRs (D10S1248, D22S1045, D2S441) into the current multiplex test. The second strategy employs a multiplex of 6 high molecular weight STRs (in current use) modified to provide smaller amplicons, combined with two additional loci of high discriminating power (D1S1656, D12S391). Eventually, the two strategies will converge to provide a single multiplex of 15 STRs.

United States Initiatives

Also in the United States, the National Institute of Standards and Technology (NIST) is characterising additional STR loci with a number of potential uses, such as forensic DNA casework, but also missing persons/mass disaster sample testing and complex paternity testing. NIST have set out to find loci with narrow allele ranges, moderate to high heterozygosities and clean flanking sequences that can be used in mini-STR assays.

A first set of new mini-STRs developed by NIST are the loci D10S1248, DD14S1434, D22S1045 which can be co-amplified in a multiplex PCR named NC01 [49]. Population genetic studies, such as the one of Onofri et al. [50] have been performed, yielding population specific allele frequencies and showing the power of discrimination and other forensic parameters to be similar to that of the nine loci of Profiler Plus (Applied Biosystems), a standard STR multiplex.

D1S1677, D2S441 and D4S2364 constitute a second set of new mini-STRs developed at NIST, to be amplified in a multiplex PCR named NC02. A population genetic study in North-Western Italians by Peloso et al. [51] showed that these mini-STRs, combined with the NC01 triplex, have a power of discrimination of 0.9999.

Finally, a comprehensive multiplex assay has been developed, capable of amplifying 22 new autosomal STRs (D6S474, D12ATA63, D22S1045, D10S1248, D1S1677, D11S4463, D4S2364, D9S1122, D2S1776, D10S1435, D3S3053, D5S2500, D1S1627, D3S4529, D2S441, D6S1017, D4S2408, D171301, D1GATA113, D18S853, D20S482, D14S1434) and the amelogenin locus [52]. This 23-plex dubbed "Autoplex" uses five-dye chemistry to keep all PCR products below 400 bp in size. Three of these STRs - D10S1248, D2S441 and D22S1045 - have been recommended for extending the core European loci.

Conclusion

Over the past 20 years, DNA technology has truly revolutionised forensic science. Future developments will focus on the physical description of an individual from biological traces at a crime scene by the use of not only minute, but also degraded stain material. It seems only budgetary restraints and the inertia of international legal change may limit the tempo at which these developments come into daily practice.

REFERENCES

[1] Jeffreys, AJ; Wilson, V; Thein, SL. Individual-specific 'fingerprints *Nature*, 1985, 316, 76-9.

[2] Kimpton, CP; Gill, P; Walton, A; Urquhart, A; Millican, ES; Adams, M. Automated DNA profiling employing multiplex amplification of short tandem repeat loci. *PCR Methods Appl*, 1993, 3, 13-22.

[3] Gill, P; Brinkmann, B; d'Aloja, E; Andersen, J; Bar, W; Carracedo, A; Dupuy, B; Eriksen, B; Jangblad, M; Johnsson, V; Kloosterman, AD; Lincoln, P; Morling, N; Rand, S; Sabatier, M; Scheithauer, R; Schneider, P; Vide, MC. Considerations from the European DNA *Forensic Sci. Int.*, 1997, 87, 185-92.

[4] Butler, JM; Levin, BC. Forensic applications of mitochondrial *Trends Biotechnol.*, 1998, 6, 158-62.

[5] Jobling, MA; Pandya, A; Tyler-Smith, C. The Y chromosome Int *J. Legal Med.*, 1997, 110, 118-24.

[6] Pouchkarev, VP; Shved, EF; Novikov, PI. Sex determination of forensic *Electrophoresis*, 1998, 19, 76-9.

[7] Glazier, AM; Nadeau, JH; Aitman, TJ. Finding genes that underlie complex traits. *Science,* 2002, 298, 2345-9.

[8] Risch, N; Merikangas, K. The future of genetic studies of complex human diseases. *Science,* 1996, 273, 1516-7.

[9] McEvoy, B; Beleza, S; Shriver, MD. The genetic architecture of normal *Hum. Mol. Genet.*, 2006, 15, R176-81.

[10] Voisey, J; Box; van Daal, A. A polymorphism study of the human Agouti gene and its association with MC1R. *Pigment Cell Res.*, 2001, 14, 264-7.

[11] van Daal, A. The genetic basis of human pigmentation. *FSI Genetics Supplement Series 1*, 2008, 541-3.

[12] Zaumsegel, D; Rotschild, M; Schneider, P. SNPs for the analysis of pigmentation genes – a comparative study. *FSI Genetics Supplement Series 1*, 2008, 544-6.

[13] Myles, S; Somel, M; Tang, K; Kelso, J; Stoneking, M. Identifying genes underlying skin pigmentation differences among human populations. *Hum. Genet.*, 2007, 120, 613-21.

[14] Norton, HL; Kittles, RA; Parra, E; McKeigue, P; Mao, X; Cheng, K; Canfield, VA; Bradley, DG; McEvoy, B; Shriver, MD. Genetic evidence for the convergent evolution of light skin in Europeans and East Asians. *Mol. Biol. Evol.*, 2007, 24, 710-22.

[15] Zhu, G; Evans, DM; Duffy, DL; Montgomery, GW; Medland, SE; Gillespie, NA; Ewen, KR; Jewell, M; Liew, YW; Hayward, NK; Sturm, RA; Trent; Martin, NG. A genome scan for eye color in 502 twin families: most variation is due to a QTL on chromosome 15q. *Twin Res.*, 2004, 7, 197-210.

[16] Branicki, W; Szcerbinska, A; Brudnik, U; Wolanska-Nowak, P; Kupiec, T. The OCA2 gene as a marker for eye colour prediction. *FSI Genetics Supplement Series, 1*, 2008, 536-7.

[17] Hirschhorn, JN; Lindgren, CM; Daly, MJ; Kirby, A; Schaffner, SF; Burtt, NP; Altshuler, D; Parker, A; Rioux, JD; Platko, J; Gaudet, D; Hudson, TJ; Groop, LC; Lander, ES. Genomewide linkage *Am. J. Hum. Genet.*, 2001, 69, 106-16.

[18] Mamada, M; Yorifuji, T; Yorifuji, J; Kurokawa, K; Kawai, M; Momoi, T; Nakahata, T. Fibrillin I gene polymorphism is associated with tall stature of normal individuals. *Hum. Genet.*, 2007, 120, 733-5.

[19] Frudakis, T. *Molecular Photofitting: Predicting Ancestry and Phenotype Using DNA*. Burlington MA: Academic Press, 2008.

[20] Ossorio, PN. About face: forensic genetic testing for race and visible traits. *J. Law Med Ethics*, 2006, 34, 277-92.

[21] Jorde, LB; Watkins, WS; Bamshad, MJ; Dixon, ME; Ricker, CE; Seielstad, MT; Batzer, MA. The distribution of human genetic diversity: a comparison of mitochondrial; autosomal; and Y-chromosome data. *Am. J. Hum. Genet.*, 2000, 66, 979-88.

[22] Lowe, AL; Urquhart, A; Foreman, LA; Evett, IW. Inferring ethnic origin by means of an STR profile. *Forensic. Sci. Int.*, 2001, 119, 17-22.

[23] Rosenberg, NA; Li, LM; Ward, R; Pritchard, JK. Informativeness of genetic markers for inference of ancestry. *Am J Hum Genet.,* 2003, 73, 1402-22.

[24] Rosenberg, NA; Mahajan, S; Ramachandran, S; Zhao, C; Pritchard, JK; Feldman, MW Clines, clusters, and the effect of study design on the inference of human population structure. *PLoS Genet.*, 2005, 1, e70.

[25] Phillips, C; Sanchez, JJ; Fondevilla, M; Gomez-Tato, A; Alvarez-Dios, J; Calaza, M; Caseres de Cal, M; Ballard, D; Lereu, MV; Carracedo, A. Inferring ancestral origin using a single multiplex assay of ancestry-informative marker SNPs. *FSI Genetics*, 2007, 1, 273-80.

[26] Chystelle, R; Christelie, B; Poisan, JP. New set of markers for individual geographic origin. *FSI Genetics Supplement Series 1*, 2008, 482-3.

[27] Daniel, R; Sanchez, JJ; Nassif, NT; Hernandez, A; Walsh, SJ. SNPs associated with physical traits : a valuable tool for the inference of biogeographical ancestry. *FSI Genetics Supplement Series 1*, 2008, 538-40.

[28] Marchini, J; Cardon, LR; Phillips, MS; Donnelly, P. The effects of human population structure on large genetic association studies. *Nat Genet*, 2004, 36, 512-7.

[29] Bauchet, M; McEvoy, B; Pearson, LN; Quillen, EE; Sarkisian, T; Hovhannesyan, K; Deka, R; Bradley, DG; Shriver, MD. Measuring European population stratification with microarray genotype data. *Am. J. Hum. Genet* ., 2007, 80, 948-56.

[30] Price; Butler, J; Patterson, N; Capelli, C; Pascali, VL; Scarnicci, F; Ruiz-Linares, A; Groop, L; Saetta, AA; Korkolopoulou, P; Seligsohn, U; Waliszewska, A; Schirmer, C; Ardlie, K; Ramos, A; Nemesh, J; Arbeitman, L; Goldstein, DB; Reich, D; Hirschhorn, JN. Discerning the ancestry of European Americans in genetic association studies. *PLoS Genet.*, 2008, 4, e236.

[31] Seldin, MF; Shigeta, R; Villoslada, P; Selmi, C; Tuomilehto, J; Silva, G; Belmont, JW; Klareskog, L; Gregersen, PK. European population substructure: clustering of northern and southern populations. *PLoS Genet.*, 2006, 2, e143.

[32] Lao, O; Lu, TT; Nothnagel, M; Junge, O; Freitag-Wolf, S; Caliebe, A; Balascakova, M; Bertranpetit, J; Bindoff, LA; Comas, D; Holmlund, G; Kouvatsi, A; Macek, M; Mollet, I; Parson; Palo, J; Ploski, R; Sajantila, A; Tagliabracci, A; Gether, U; Werge, T; Rivadeneira, F; Hofman, A; Uitterlinden, AG; Gieger, C; Wichmann, HE; Rüther, A; Schreiber, S; Becker, C; Nürnberg, P; Nelson, MR; Krawczak, M; Kayser, M. Correlation between genetic and geographic structure in Europe. *Curr. Biol.*, 2008, 18, 1241-8.

[33] Burger, J; Hummel, S; Hermann, B; Henke, W. DNA preservation: a microsatellite-DNA study on ancient skeletal remains. *Electrophoresis,* 1999, 20, 1722-8.

[34] Butler, JM; Shen, Y; McCord, BR. The development of reduced size STR amplicons as tools for analysis of degraded DNA. *J. Forensic. Sci.*, 2003, 48, 1054-64.

[35] Mulero, JJ; Chang, CW; Lagacé, RE; Wang, DY; Bas, JL; McMahon, TP; Hennessy, LK. Development and validation *J. Forensic. Sci.*, 2008, 53, 838-52.

[36] Andrade, L; Bento, AM; Serra, A; Carvalho, M; Gamero, JJ; Oliveira, C; Batista, L; Lopes, V; Balsa, F; Corte-Real, F; Anjos, MJ. AmplFlSTR Minifiler PCR amplification kit: the new miniSTR multiplex kit. *FSI Genetics Supplement Series 1*, 2008, 89-91.

[37] Camacho, S; Vieira-Silva, C; Dario, P; Ribeiro, T; Espinheira, R; Geada, H. Mini-SGM multiplex in degraded samples. *FSI Genetics Supplement Series 1*, 2008, 100-1.

[38] Decorte, R; Liu, CF; Vanderheyden, N; Cassiman, JJ. Development of a novel miniSTR multiplex assay for typing degraded DNA samples. *FSI Genetics Supplement Series 1*, 2008, 112-4.

[39] Thorisson, GA; Stein, LD. The SNP Consortium website: past; present and future. *Nucleic Acids Res*., 2003, 31, 124-7.

[40] International HapMap Consortium, Frazer, KA; Ballinger, DG; Cox, DR; Hinds, DA; Stuve, LL; Gibbs, RA; Belmont, JW; Boudreau, A; Hardenbol, P; Leal, SM; Pasternak, S; Wheeler, DA; Willis, TD; Yu, F; Yang, H; Zeng, C; Gao, Y; Hu, H; Hu, W; Li, C; Lin, W; Liu, S; Pan, H; Tang, X; Wang, J; Wang, W; Yu, J; Zhang, B; Zhang, Q; Zhao, H; Zhao, H; Zhou, J; Gabriel, SB; Barry, R; Blumenstiel, B; Camargo, A; Defelice, M; Faggart, M; Goyette, M; Gupta, S; Moore, J; Nguyen, H; Onofrio, RC; Parkin, M; Roy, J; Stahl, E; Winchester, E; Ziaugra, L; Altshuler, D; Shen, Y; Yao, Z; Huang, W; Chu, X; He, Y; Jin, L; Liu, Y; Shen, Y; Sun, W; Wang, H; Wang, Y; Wang, Y; Xiong, X; Xu, L; Waye, MM; Tsui, SK; Xue, H; Wong, JT; Galver, LM; Fan, JB; Gunderson, K; Murray, SS; Oliphant, AR; Chee, MS; Montpetit, A; Chagnon, F; Ferretti, V; Leboeuf, M; Olivier, JF; Phillips, MS; Roumy, S; Sallée, C; Verner, A; Hudson, TJ; Kwok, PY; Cai, D; Koboldt, DC; Miller, RD; Pawlikowska, L; Taillon-Miller, P; Xiao, M; Tsui, LC; Mak, W; Song, YQ; Tam, PK; Nakamura, Y; Kawaguchi, T; Kitamoto, T; Morizono, T; Nagashima, A; Ohnishi, Y; Sekine, A; Tanaka, T; Tsunoda, T; Deloukas, P; Bird, CP; Delgado, M; Dermitzakis, ET; Gwilliam, R; Hunt, S; Morrison, J; Powell, D; Stranger, BE; Whittaker, P; Bentley, DR; Daly, MJ; de Bakker, PI; Barrett, J; Chretien, YR; Maller, J; McCarroll, S; Patterson, N; Pe'er, I; Price, A; Purcell, S; Richter, DJ; Sabeti, P; Saxena, R; Schaffner, SF; Sham, PC; Varilly, P; Altshuler, D; Stein, LD; Krishnan, L; Smith, AV; Tello-Ruiz, MK; Thorisson, GA; Chakravarti, A; Chen, PE; Cutler, DJ; Kashuk, CS; Lin, S; Abecasis, GR; Guan. W; Li. Y; Munro, HM; Qin, ZS; Thomas, DJ; McVean, G; Auton, A; Bottolo, L; Cardin, N; Eyheramendy, S; Freeman, C; Marchini, J; Myers, S; Spencer, C; Stephens, M; Donnelly, P; Cardon, LR; Clarke, G; Evans, DM; Morris, AP; Weir, BS; Tsunoda, T; Mullikin, JC; Sherry, ST;

Feolo, M; Skol, A; Zhang, H; Zeng, C; Zhao, H; Matsuda, I; Fukushima, Y; Macer, DR; Suda, E; Rotimi, CN; Adebamowo, CA; Ajayi, I; Aniagwu, T; Marshall, PA; Nkwodimmah, C; Royal, CD; Leppert, MF; Dixon, M; Peiffer, A; Qiu, R; Kent, A; Kato, K; Niikawa, N; Adewole, IF; Knoppers, BM; Foster, MW; Clayton, EW; Watkin, J; Gibbs, RA; Belmont, JW; Muzny, D; Nazareth, L; Sodergren, E; Weinstock, GM; Wheeler, DA; Yakub, I; Gabriel, SB; Onofrio, RC; Richter, DJ; Ziaugra, L; Birren, BW; Daly, MJ; Altshuler, D; Wilson, RK; Fulton, LL; Rogers, J; Burton, J; Carter, NP; Clee, CM; Griffiths, M; Jones, MC; McLay, K; Plumb, RW; Ross, MT; Sims, SK; Willey, DL; Chen, Z; Han, H; Kang, L; Godbout, M; Wallenburg, JC; L'Archevêque, P; Bellemare, G; Saeki, K; Wang, H; An, D; Fu, H; Li, Q; Wang, Z; Wang, R; Holden, AL; Brooks, LD; McEwen, JE; Guyer, MS; Wang, VO; Peterson, JL; Shi, M; Spiegel, J; Sung, LM; Zacharia, LF; Collins, FS; Kennedy, K; Jamieson, R; Stewart, J. A second generation human haplotype map of over 3.1 million SNPs. *Nature*, 2007, 449, 851-61.

[41] Butler, JM; Budowle, B; Gill, P; Kidd, KK; Phillips, C; Schneider, P; Vallone, PM; Morling, N. Report on ISFG SNP panel discussion. *FSI Genetics Supplement Series 1*, 2008, 471-2.

[42] Amigo, J; Phillips, C; Lareu, M; Carracedo, A. The SNPforID browser: an online tool for query and display of frequency data from the SNPforID project. *Int. J. Legal Med.* 2008, 122, 435-40.

[43] Sanchez, JJ; Phillips, C; Børsting, C; Balogh, K; Bogus, M; Fondevila, M; Harrison CD; Musgrave-Brown, E; Salas, A; Syndercombe-Court, D; Schneider, PM; Carracedo A; Morling, N. Sanchez, JJ; Phillips, C; Børsting, C; Balogh, K; Bogus, M; Fondevila M; Harrison, CD; Musgrave-Brown, E; Salas, A; Syndercombe-Court, D; Schneider, PM; Carracedo, A; Morling, N. A multiplex assay with 52 single nucleotide polymorphisms for human identification. *Electrophoresis,* 2006, 27, 1713-24.

[44] Pereira, R; Fondevila, M; Phillips, C; Amorim, A; Carracedo, A; Gusmao, L. Genetic characterisation of 52 autosomal SNPs in the Portuguese population. *FSI Genetics Supplement Series 1*, 2008, 358-60.

[45] Musgrave-Brown, E; Ballard, D; Fondevila, M; Fang, R; Harrison, C; Phillips, C; Prasad, Y; Sobrino, B; Thacker, C; Wiluhn, J; Carracedo, A; Schneider, P; Syndercombe Court, D. Forensic validation of the Genplex SNP typing system – results of an inter-laboratory study. *FSI Genetics Supplement Series 1*, 2008, 389-93.

[46] Fondevila, M; Phillips, C; Naveran, N; Cerrezo, M; Rodriguez, A; Calvo, R; Fernandez, LM; Carracedo, A; Lareu, MV. Chammenging DNA : assessment of a range of genotyping approaches for highly degraded forensic samples. *FSI Genetics Supplement Series 1*, 2008, 26-8.

[47] Gill, P; Fereday, L; Morling, N; Schneider, PM. The evolution of DNA databases--recommendations for new European STR loci. *Forensic Sci Int.*, 2006, 156, 242-4.

[48] Gill, P; Fereday, L; Morling, N; Schneider, PM. New multiplexes for Europe-amendments and clarification of strategic development. *Forensic. Sci. Int.*, 2006, 163, 155-7.

[49] Coble, MD; Butler, JM. Characterization of new miniSTR loci to aid analysis of degraded DNA. *J. Forensic. Sci.*, 2005, 50, 43-53.

[50] Onofri, V; Giocoma, I; Alessandrini, F; Chiara, Turchi; Buscemi, L; Tagliabracci, A. Population data for D10S1248; D14S1434; and D22S1045 miniSTRs loci from the Marches region (Central Italy). *FSI Genetics Supplement Series 1*, 2008, 353-5.

[51] Peloso, G; Grignani, P; Previdere, C. Genetic characterisation of six miniSTR loci in an Italian population sample. *FSI Genetics Supplement Series 1*, 2008, 356-7.
[52] Butler, JM; Hill, CR; Decker, AE; Kline, MC; Vallone, PM. New autosomal STR loci. *FSI Genetics Supplement Series 1*, 2008, 95-6.

In: Encyclopedia of DNA Research (3 Volume Set)
Editors: S. J. Duncan and P. H. Wiley
ISBN 978-1-61324-305-3

Chapter 38

APPLICATION OF PCR-DGGE FINGERPRINTING IN MOLECULAR ECOLOGY OF MARINE MICROBIAL SYMBIONTS

Zhiyong Li *

Marine Biotechnology Laboratory, School of Life Sciences and Biotechnology and Key Laboratory of Microbial Metabolism, Ministry of Education, Shanghai Jiao Tong University, 800 Dongchuan Road, Shanghai 200240, P.R. China

ABSTRACT

It is now well recognised that only a small fraction of microorganisms can be isolated and characterized, especially marine microbial symbionts because of the difficulty in cultivation. The application of molecular biological methods to study the diversity and ecology of microorganisms in natural environments has been practiced since the mid-1980s. Molecular biological techniques such as genetic fingerprinting techniques provide a pattern or profile of the genetic diversity in a microbial community, and are now frequently used in molecular ecology to explore the microbial diversity and to analyse the structure of microbial communities. Denaturing gradient gel electrophoresis (DGGE) of PCR amplified ribosomal DNA fragments has been introduced into microbial ecology in 1993. Combined with PCR amplification of marker genes or their transcripts, DGGE can give a direct display of the predominant constituents in microbial assemblages with the advantage of easiness, reproducibility, reliability, and speed. PCR-DGGE fingerprinting is suitable for revealing the diversity of uncultured marine microbial symbionts, comparing and monitoring the dynamic change of the community structure of marine microbial symbionts. In this chapter, the theoretical aspects of PCR-DGGE fingerprinting and its application in the molecular ecology of microbial symbionts of marine organisms such as sponge, coral, ascidians, bryozoans and algae are introduced.

* Corresponding author: E-mail: zyli@sjtu.edu.cn; Tel.: (+86)21-34204036.

1. Introduction

It is now well recognised that only a small fraction of microorganisms can be isolated and characterized in the laboratory. Marine microbial symbionts are much more difficult to cultivate because of their special environmental conditions. So, a cultivation-dependent approach is difficult for revealing the microbial diversity of marine microbial symbionts. Meanwhile, media used for the isolation and cultivation of microorganisms are selective and hence give a biased view of microbial community structure. Therefore, culture-independent molecular strategies are required to supplement the classical culture-based techniques.

The application of molecular biological methods to study the diversity and ecology of microorganisms in natural environments has been practiced since the mid-1980s. Since that time, many new insights into the composition of uncultivated microbial communities have been gained. Culture-independent molecular biological techniques such as genetic fingerprinting techniques provide a pattern or profile of the genetic diversity in a microbial community, and are now frequently used in molecular ecology to explore the microbial diversity and to analyse the structure of microbial communities. In 1993, genetic fingerprinting technique, denaturing gradient gel electrophoresis (DGGE) of PCR amplified ribosomal DNA fragments was introduced into microbial ecology by Muyzer et al.(1993) to profile community complexity of a microbial mat and bacterial biofilms. At present, PCR-DGGE fingerprinting has been widely used to study microbial communities in various habitats. In the case of marine microbial ecology, in 1995, Muyzer et al. (1995) used DGGE analysis of PCR amplified rDNA fragments to provide information on the genetic diversity of microbial communities found around hydrothermal vents. Sequencing of excised DGGE bands revealed sequences similar to those of members of the genus *Thiomicrospira*, sulfuroxidizing bacteria. In this chapter, we will mainly introduce the application of PCR-DGGE fingerprinting in the diversity revelation, dynamic monitor and detection of special microbes of marine microbial symbionts.

2. Theoretical Aspects of PCR-DGGE Fingerprinting

Denaturing gradient gel electrophoresis (DGGE) is a method by which fragments of DNA of the same length but different sequence can be resolved electrophoretically, allowing the separation of a heterogeneous mixture of PCR amplified genes on a polyacrylamide gel. The separation of DGGE DNA fragments of the same length but with different sequences is based on the decreased electrophoretic mobility of a partially melted doublestranded DNA molecule in polyacrylamide gels containing a linear gradient of DNA denaturants (a mixture of urea and formamide). The melting of DNA fragments proceeds in discrete so-called melting domains: stretches of base pairs with an identical melting temperature. Once a domain with the lowest melting temperature reaches its melting temperature (*Tm*) at a particular position in the denaturing gradient gel, a transition from a double stranded to a partially melted molecule occurs. The protruding single strands practically cause a halt of the molecules at that position. Sequence variation within such domains causes the melting temperatures to differ, and molecules with different sequences will stop migrating at different positions in the gel. 100% of the sequence variants can be detected in DNA fragments up to

500 bp by the attachment of a GC rich sequence (GC clamp) that acts as a high melting domain preventing the two DNA strands from complete dissociation into single strands to one side of the DNA fragment. A sequence of guanines (G) and cytosines (C) varying between 30 and 50 nucleotides is added to the 50 end of one of the PCR primers, coamplified and thus introduced into the amplified DNA fragments (Muyzer et al.1998).

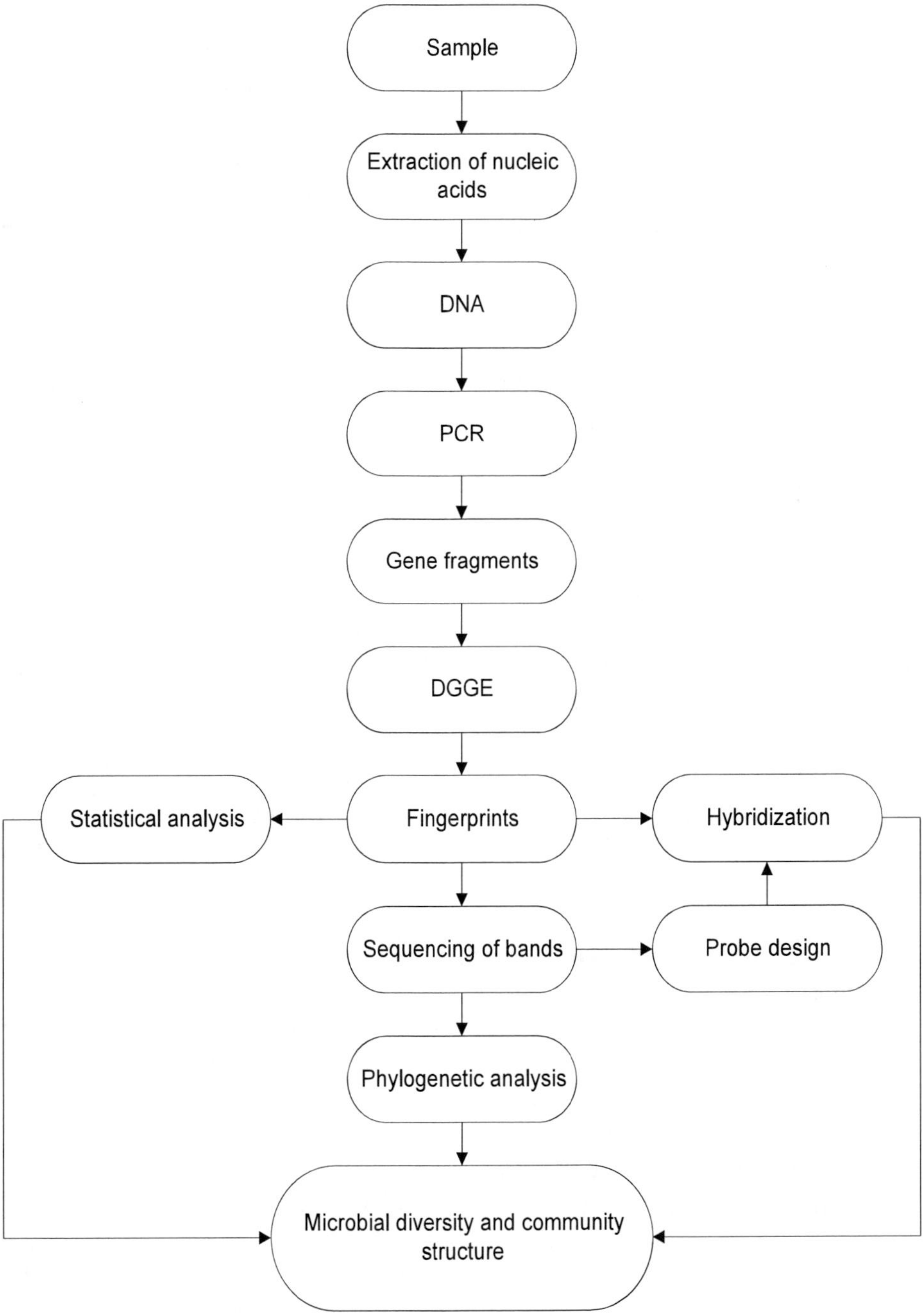

Figure 1. Flow diagram of PCR-DGGE fingerprinting analysis of microbial community.

Amplification of DNA extracted from mixed microbial communities with primers specific for 16Sr RNA gene fragments, 18Sr RNA gene fragments or ITS fragments result in mixtures of PCR products. The obtained different sequences with the same size by PCR can be separated in polyacrylamide gels containing a gradient of DNA denaturants. As the PCR products proceed through the gel, the denaturing conditions gradually become stronger, therefore PCR products with different sequences start melting at different positions (i.e. at different denaturant concentrations) in the gel. DNA bands in DGGE profiles can be visualised using ethidium bromide or silver staining. DGGE profiles can be blotted onto nylon membranes and hybridised with a radioactively-labelled oligonucleotide probe. Individual bands may be excised, reamplified and sequenced to give an indication of the composition and diversity of the microbial community. A general overview of PCR-DGGE fingerprinting of microbial community is shown in Figure 1. For a more comprehensive overview of the detailed technique of PCR-DGGE fingerprinting the reader is referred to Schäfer et al.(2001).

3. The Application of PCR-DGGE Fingerprinting in Molecular Ecology of Marine Microbial Symbionts

3.1. The Revelation of Microbial Community Structure and Diversity

Many marine macroorganisms such as sponges are known to acquire symbiotic microorganisms from the external environment (horizontal acquisition) or inherit their symbionts from the parent colony (maternal or vertical acquisition). Most of the marine microbial symbionts are uncultured, so, culture-independent PCR-DGGE fingerprint-based molecular methods have been widely used in the revelation of marine symbiotic microbial community structure and diversity. For instance, a stable and specific bacterial community was shown to be associated with the Mediterranean sponge *Chondrilla nucula* regardless of sampling time and geographical region (Thiel et al.2007a). As for four South China Sea sponges, *Stelletta tenuis*, *Halichondria rugosa*, *Dysidea avara*, and *Craniella australiensis*, sponge host-specific bacteria including Proteobacteria, Bacteroidetes, Firmicutes, and Actinobacteria were detected (Li et al.2006).

Besides sponge, the structure and composition of microbial communities inhabiting the soft coral *Alcyonium antarcticum* were investigated using DGGE by Webster et al.(2007). Diverse microbial communities including Alpha- and Betaproteobacteria, Bacteroidetes, Firmicutes, Actinomycetales, Planctomycetes, and Chlorobi and bacteria from the functional group of sulfate-reducing bacteria were revealed. The photobiology and distribution of dinoflagellates in the genus *Symbiodinium* was investigated for eight common reef coral species over a depth range of 1–25 m on a coral reef in Belize (Warner et al. 2006). PCR-DGGE of the internal transcribed spacer (ITS) 2 region revealed marked differences in host specificity and depth zonation for certain symbiont types.

Diversity of intratunical bacteria in the tunic matrix of the colonial ascidian *Diplosoma migrans* was also revealed using PCR-DGGE (Wichels, 2005). DGGE profiles based sequence data generated five different subgroups of intratunical proteobacteria such as

Ruminococcus flavefaciens, *R. flavefaciens*, *Pseudomonas saccherophilia*, *Flavobacterium* symbiont.

Species- and site-specific bacterial communities associated with four encrusting bryozoans from the North Sea, Germany, were investigated by Kittelmann et al.(2005). DGGE of PCR-amplified 16S rRNA gene fragments of associated bacteria displayed specific bacterial community profiles in the examined species *Aspidelectra melolontha*, *Conopeum reticulum*, *Electra monostachys* and *Electra pilosa*.

Culturable and nonculturable bacterial symbionts in the toxic benthic dinoflagellate *Ostreopsis lenticularis* was revealed by Ashton et al.(2003). DGGE analyses of extracted *Ostreopsis* associated bacterial DNAs indicated that there were three culturable and four non-culturable associated bacteria including *Pseudomonas*/*Alteromonas*, *Pseudomonas*/*Alteromonas* and *Acinetobacter*.

In the case of eukaryotic symbionts, phylogenetic analysis of sequence data from eukaryotic DGGE analysis revealed that the communities present in Antarctic sponges fell into diatom and dinoflagellate clusters with many sequences having no known close relatives. In addition, seven eukaryotic sequences that were not detected in seawater samples or other sponge species were observed in Antarctic sponge *Kirkpatrickia varialosa* (Webster et al.2004). DGGE analysis indicated that fungal communities differ among different sponge species (Suberites zeteki and Mycale armata) and also vary between sponges and seawater (Gao et al. 2008).

The diversity of dinoflagellate endosymbionts living with cnidarians was examined using denaturing gradient gel electrophoresis (PCR-DGGE) fingerprint analysis of the rDNA internal transcribed spacers (ITS) 1 and 2 (La Jeunesse et al.2008). Detection and identification of fungi intimately associated with the brown seaweed *Fucus serratus* were carried out by Zuccaro et al.(2008). The predominant DGGE bands obtained from healthy algal thalli belonged to the *Lindra*, *Lulworthia*, *Engyodontium*, *Sigmoidea/Corollospora* complex, and *Emericellopsis/Acremonium*-like ribotypes.

3.2. The Monitor of Spatial and Temporal Variability of Marine Microbial Symbionts

Distinct bacterial communities were found to inhabit the endosome and cortex of sponge *Tethya aurantium Pallas* 1766 by Thiel et al.(2007b), where a new phylotype of *Flexibacteria* was recovered only from the sponge cortex, while *Synechococcus* species were present mainly in the sponge endosome.

Using a combination of the internal transcribed spacer region 2 (ITS2) and denaturing gradient gel electrophoresis (DGGE), Macdonald et al.(2008) assessed the cladal and subcladal variability of *Symbiodinium* in the widely distributed species *Stylophora pistillata* along a latitudinal transect in southeast African waters which extended into high latitude locations, it appears that there is a shift in symbiont distributions with increasing latitude.

Frade et al.(2008) identified three symbiont genotypes with distributions that reveal patterns of host specificity and depth-based zonation using the same technique as above. Contrasting with variation over depth, strong functional within-colony uniformity in symbiont diversity was found. Relating symbiont distributions to measured physical factors (irradiance light spectral distribution, temperature), suggests depth-based ecological function and host specificity for *Symbiodinium* ITS2 types, even among closely related coral species.

PCR-DGGE fingerprinting has been used to monitor changes in bacterial communities when examining the feasibility of growing sponges in aquaculture systems (Mohamed et al.2008a; 2008b). DGGE analysis revealed that the diversity of the bacterial community of *M. laxissima* increased when sponges were maintained in aquaculture and that bacterial communities associated with wild and aquacultured *M. laxissima* were markedly different than those of the corresponding surrounding water (Mohamed et al.2008a). As for marine sponge *Ircinia strobilina*, according to DGGE analysis, populations affiliated with Beta- and Deltaproteobacteria, *Clostridia*, and *Planctomycetes* emerged in sponges maintained in aquaculture (Mohamed et al.2008b).

Highly stable symbioses in response to seasonal environmental situations among western Atlantic brooding corals were revealed by Thornhin et al.(2006a) using DGGE of the internal transcribed spacer 2 (ITS2) region. Multi-year, seasonal genotypic surveys of coral-algal symbioses reveal prevalent stability or post-bleaching reversion (Thornhill et al.2006b).

PCR-DGGE fingerprint-based molecular technique is also used to investigate how the microbial symbionts are acquired by their hosts. For example, the bacterial community profiles of adults, larvae, and juvenile Caribbean demosponge *Ircinia felix* were compared by Schmitt et al.(2007), as a result, it was shown that in *I. felix*, vertical transmission of microorganisms through the larvae is an important mechanism for the establishment of the sponge-microbe association.

3.3. The Detection of Special Marine Microbial Symbionts

A highly integrated, morphologically diverse bacterial community is associated with the dorsal surface of *Alvinella pompejana*, a polychaetous annelid that inhabits active high-temperature deep-sea hydrothermal vent sites along the East Pacific Rise (EPR). PCR amplification of the community with spirochete-specific primers used in conjunction with DGGE analysis identified two spirochete phylotypes (Campbell et al.2001). *Candidatus Endoecteinascidia frumentensis*, was found specifically associated to *E. turbinata* from the Caribbean and has also been found to be associated with *E. turbinata* from the Mediterranean (Govind 2007).

In the study of Webster et al. (2008), DGGE detected multiple sequences that were exclusively present in diseased sponges. A Deltaproteobacteria sequence with high homology to a coral black band disease strain was detected in all sponge lesions and was absent from all healthy and unaffected regions of diseased sponges. Based on 16S rRNA gene sequence analysis, only the diseased sponges were found to contain sequences belonging to the Epsilonproteobacteria and Firmicutes, and there was a much greater number of Bacteroidetes sequences within the diseased sponges. In contrast, only the healthy sponges contained sequences corresponding to the cyanobacteria and the healthy sponges were dominated by Chloroflexi and Gammaproteobacteria sequences. PCR-denaturing gradient gel electrophoresis was used to detect the epiphyte community on marine macroalgae by Ohkubo et al. (2006). Two phylotypes of Acaryochloris, a chlorophyll d-containing cyanobacterium, were found not only on red macroalgae but also on green and brown macroalgae.

4. Concluding Remarks

Combined with PCR amplification of marker genes or their transcripts, PCR-DGGE fingerprinting can give a direct display of the predominant constituents in microbial assemblages with the advantage of easiness, reproducibility, reliability, and speed and is suitable to reveal the microbial community, especially to compare and monitor the temporal dynamics or spatial variation of the complex microbial community structure. But, there are some limitations of PCR-DGGE fingerprinting. For instance, the PCR-DGGE fingerprinting is based on nucleic acid extraction, so a major limitation is quantitative recovery of nucleic acids from environmental samples. Selectivity in PCR amplification of rRNA genes is another source of bias that can affect the results of PCR-DGGE fingerprinting. Small differences in the sequence of universally conserved regions may result in selective amplification of some sequences, particularly when primer annealing is at high stringency (Head et al.1998). Even so, at present, PCR-DGGE fingerprinting technique is still an efficient technique in microbial ecology.

References

Ashton, M., Rosado, W., Govind, N. S. and Tosteson, T. R. (2003). Culturable and nonculturable bacterial symbionts in the toxic benthic dinoflagellate *Ostreopsis lenticularis*. *Toxicon*, *42*, 419-424.

Campbell, B. J. and Cary, S. C. (2001). Characterization of a novel spirochete associated with the hydrothermal vent polychaete annelid, *Alvinella pompejana*. *Applied and Environmental Microbiology*, *67*, 110-117.

Frade, P. R., De Jongh, F., Vermeulen, F., Van Bleijswijk, J. and Bak, R. P. M. (2008). Variation in symbiont distribution between closely related coral species over large depth ranges. *Molecular Ecology*, *17*, 691-703.

Gao, Z., Li, B., Zheng, C. and Wang, G. 2008). Molecular detection of fungal communities in the Hawaiian marine sponges. *Suberites zeteki* and *Mycale armata*. *Applied and Environmental Microbiology*, *74*, 6091-6101.

Govind, N. S. (2007). Bacterial diversity associated with the Caribbean tunicate *Ecteinascidia turbinate*. *Antonie van Leeuwenhoek*, *92*, 155-164.

Head, I. M., Saunders, J. R. and Pickup, R. W. (1998). Microbial evolution, diversity, and ecology: a decade of ribosomal RNA analysis of uncultivated microorganisms. *Microb Ecol.*, *35*, 1-21.

Kittelmann, S. and Harder, T. (2005). Species- and site-specific bacterial communities associated with four encrusting bryozoans from the North Sea, Germany. *Journal of Experimental Marine Biology and Ecology*, *327*, 201-209.

La Jeunesse, T. C., Bonilla, H. R., Warner, M. E., Wills, M., Schmidt, G. M. and Fitt, W. K (2008). Specificity and stability in high latitude eastern Pacific coral–algal symbioses. *Limnol. Oceanogr.*, *53*, 719-727.

Li, Z., He, L., Wu, J. and Jiang, Q. (2006). Bacterial community diversity associated with four marine sponges from the South China Sea based on 16S rDNA-DGGE fingerprinting. *Journal of Experimental Marine Biology and Ecology*, *329*, 75-85.

Macdonald, A. H. H., Sampayo, E. M., Ridgway, T. and Schleyer, M. H. (2008). Latitudinal symbiont zonation in *Stylophora pistillata* from southeast Africa. *Mar Biol., 154*, 209-217.

Mohamed, N. M., Enticknap, J. J., Lohr, J. E., McIntosh, S. M. and Hill1, R. T. (2008a). Changes in bacterial communities of the marine sponge *Mycale laxissima* on transfer into aquaculture. *Applied and Environmental Microbiology, 74*, 1209-1222.

Mohamed, N. M., Rao, V., Hamann, M. T., Kelly, M. and Hill, R. T. (2008b). Monitoring bacterial diversity of the marine sponge *Ircinia strobilina* upon transfer into aquaculture. *Applied and Environmental Microbiology, 74*, 4133-4143.

Muyzer, G., de Waal, E. C. and Uitterlinden, A. G. (1993). Profiling of complex microbial populations by denaturing gradient gel electrophoresis analysis of polymerase chain reaction-amplified genes coding for 16S rRNA. *Applied and Environmental Microbiology, 59*, 695-700.

Muyzer, G., Teske, A., Wirsen, C. O. and Jannasch, H. W. (1995). Phylogenetic relationships of *Thiomicrospira* species and their identification in deep-sea hydrothermal vent samples by denaturing gradient gel electrophoresis of 16S rDNA fragments. *Archives of Microbiology, 164*, 165-172.

Muyzer, G. and Smalla, K. (1998). Application of denaturing gradient gel electrophoresis (DGGE) and temperature gradient gel electrophoresis (TGGE) in microbial ecology. *Antonie van Leeuwenhoek, 73*, 127-141.

Ohkubo, S., Miyashita, H., Murakami, A., Takeyama, H., Tsuchiya, T. and Mimuro1, M. (2006). Molecular detection of epiphytic *Acaryochloris* spp.on marine macroalgae. *Applied and Environmental Microbiology, 72*, 7912-7915.

Schäfer, H. and Muyzer. G. (2001). Denaturing gradient gel electrophoresis in marine microbial ecology. In: J. Paul, (Eds.), *Methods in microbiology, Academic Press, London, 30*, 425-468.

Schmitt, S., Weisz, J. B., Lindquist, N. and Hentschel, U. (2007). Vertical transmission of a phylogenetically complex microbial consortium in the viviparous sponge *Ircinia felix*. *Applied and Environmental Microbiology, 73*, 2067-2078.

Thiel, V., Leininger, S., Schmaljohann, R., Brummer, F. and Imhoff, J. F. (2007a). Sponge-specific bacterial associations of the Mediterranean sponge *Chondrilla nucula* (Demospongiae, Tetractinomorpha). *Microbial Ecology, 54*, 101-111.

Thiel, V., Neulinger, S. C., Staufenberger, T., Schmaljohann, R. and Imhoff, J. F. (2007b). Spatial distribution of sponge-associated bacteria in the Mediterranean sponge *Tethya aurantium*. *FEMS Microbiol Ecol., 59* , 47-63.

Thornhill, D. J., Fitt, W. K. and Schmidt, G. W. (2006a). Highly stable symbioses among western Atlantic brooding corals. *Coral Reefs, 25*, 515-519.

Thornhill, D. J. and LaJeunesse, T. C. (2006b). Multi-year, seasonal genotypic surveys of coral-algal symbioses reveal prevalent stability or post-bleaching reversion. *Marine Biology, 148*, 711-722.

Warner, M. E., LaJeunesse, T. C., Robison, J. D. and Thur, R. M. (2006). The ecological distribution and comparative photobiology of symbiotic dinoflagellates from reef corals in Belize: Potential implications for coral bleaching. *Limnol. Oceanogr., 51*, 1887-1897.

Webster, N. S., Xavier, J. R., Freckelton, M., Motti, C. A. and Cobb, R. (2008). Shifts in microbial and chemical patterns within the marine sponge *Aplysina aerophoba* during a disease outbreak. *Environmental Microbiology, 10*, 3366-3376.

Webster, N. S. and Bourne, D. (2007). Bacterial communitystructure associatedwith the Antarctic soft coral, *Alcyonium antarcticum*. *FEMS Microbiol Ecol.*, *59*, 81-94.

Webster, N. S., Negri, A. P., Munro, M. M. H. G. and Battershill, C. N. (2004). Diverse microbial communities inhabit Antarctic sponges. *Environmental Microbiology*, *6*, 288-300.

Wichels, A. (2005). Diversity of intratunical bacteria in the tunic matrix of the colonial ascidian *Diplosoma migrans*. *Helgol Mar Res.*, *59*, 136-140.

Zuccaro, A., Schoch, C. L., Spatafora, J. W., Kohlmeyer, J., Draeger, S. and Mitchell, J. I. (2008). Detection and identification of fungi intimately associated with the brown seaweed *Fucus serratus*. *Applied and Environmental Microbiology*, *74*, 931-941.

INDEX

A

B

C

D

E

F

H

I

J

K

L

M

N

Q

R

S

T

U

V

W

X

Y

Z